Biology at the Single Molecule Level

Biology at the Single Molecule Level

Edited by

S.H. Leuba
National Cancer Institute, NIH,
Bethesda, MD, USA

and

J. Zlatanova
Department of Chemistry and Chemical Engineering
Polytechnic University Brooklyn,
New York, NY, USA

2001
PERGAMON
AMSTERDAM – LONDON - NEW YORK – OXFORD – PARIS – SHANNON - TOKYO

ELSEVIER SCIENCE Ltd
The Boulevard, Langford Lane
Kidlington, Oxford OX5 1GB, UK

First edition 2001.

Reprinted from *Progress in Biophysics & Molecular Biology* issue 74/1-2 and 77/1

Library of Congress Cataloging in Publication Data
A catalog record from the Library of Congress has been applied for.

British Library Cataloguing in Publication Data
A catalogue record from the British Library has been applied for.

ISBN: 0-08-044031-2

∞ The paper used in this publication meets the requirements of ANSI/NISO Z39.48-1992 (Permanence of Paper).
Printed in The Netherlands.

CONTENTS

Editorial

Single-molecule biochemistry coming of age

Nowadays, we are witnessing a true revolution in the way we can study the structure and function of biological macromolecules. The advent of instrumentation and techniques that allow us to investigate the dynamic behavior of single macromolecules or their complexes in real time has given birth to a whole new field: single-molecule biochemistry.

We have attempted to compile a group of reviews that would cover most of the significant developments in the new field, both with respect to the technical advances in instrumentation, and the kinds of results that may be obtained by studying and/or manipulating single molecules or macromolecular complexes. The range of topics is by no means exhaustive, but it may serve as an introduction to this new and quickly developing field. The large number of papers could not be accommodated in one issue, hence the two issues.

It was our intent to give the authors the full freedom to not only review the field, but also to express their own views and speculations in a way that is difficult, sometimes impossible, in original experimental papers. The authors willingly made use of this freedom; the stringent reviews of the manuscripts that followed was only meant to make sure that the presentation of the facts was correct, with little impingement on this freedom. We hope that the readers will be the beneficiaries of this editorial approach.

Finally, we would like to mention that a similar compendium of single-molecule biochemistry reviews appeared as a minireview series in the Journal of Biological Chemistry in 1999; the prologue to that series by the series editor Kensal van Holde coveys all the excitement of the new field (K. van Holde, J. Biol. Chem. 274, 1999, 14515).

S. Leuba, J. Zlatanova
Department of Chemical Engineering, Chemistry and Materials Science
Polytechnic University, Six Metro Tech Center
Brooklyn, NY 11201, USA
E-mail address: zlatanoj@everest.bim.anl.gov

Part I

Biology at the Single Molecule Level
S.H. Leuba and J. Zlatanova (Eds)

Review

Protein structural dynamics by single-molecule fluorescence polarization

Joseph N. Forkey*, Margot E. Quinlan, Yale E. Goldman

School of Medicine, University of Pennsylvania, Physiology Department, Pennsylvania Muscle Institute, D700 Richards Building, 3700 Hamilton Walk, Philadelphia, PA 19104-6083, USA

Contents

---

*Corresponding author.

Keywords: Single molecule; Fluorescence; Polarization; Evanescent wave; Microscopy; Bifunctional; Dynamics; Total internal reflection

1. Introduction

The astonishing progress made during the last two decades toward understanding the machinery of cells has been fueled in large part by developments in structural biology, molecular biology, and biophysics. Thousands of protein structures have been solved at atomic resolution and the amino acid sequences and roles of many specific motifs are now known. While application of these techniques has greatly enhanced our understanding of the structure and biochemistry of proteins, elucidation of the quantitative relationships between structural changes in the proteins, enzymatic activity, signaling, energy conversion and macromolecular interactions will require further development of novel methods.

In this chapter we describe a new technique, single-molecule fluorescence polarization (SMFP), that has the potential to fill a void in conventional methods for determining the structural basis of protein activity. Using state-of-the-art optical microscopes and novel fluorescent probes attached to specific protein domains, SMFP makes it possible to quantify angular rotations that mediate function in individual protein molecules. To demonstrate the importance of protein domain rotational motions, this chapter begins with several examples of biophysical systems in which orientational changes are essential to their mechanism. This is followed by a discussion of the advantages of detecting such rotations with individual molecules and a brief synopsis of the development of SMFP and early experiments. The next two sections are devoted to technical considerations, including characteristics of extrinsic fluorescent probes, optical apparatus and data analysis. Finally, recent work using this technique is reviewed.

1.1. Protein rotational motions

Crystallographic studies of proteins trapped in several different configurations analogous to states of the functioning system often lead to hypotheses for the structural basis of the catalytic activity. They also identify likely sites of interaction with upstream and downstream functional partners. Distinct conformations are typically characterized by differences in the relative orientations of nearby compact domains linked by hinges or swivels composed of glycine residues or flexible loops. The implied rotations have direct bearing on the functional output as illustrated here by several examples where large orientation changes have been discovered.

Due to the helical structure of double-stranded DNA (dsDNA), nucleic acid processing enzymes produce a torque when unwinding the helix to gain access to the base sequence. These enzymes catalyze processes such as replication, transcription, repair and recombination.

Topoisomerase II removes accumulated strain in dsDNA and enables separation of chromosomes or untangling of circular DNA after replication. It operates by passing one dsDNA segment through a 'gate' temporarily created in another double-stranded segment. During passage of the intact dsDNA through the gate, quaternary structural changes in the DNA-opening platform take place. Dramatic reorganization of accessory protein domains include rotations of up to 170° (Fig. 1a; Fass et al., 1999).

During protein biosynthesis, two GTP-binding elongation factors (EF-Tu and EF-G) enable the ribosome to accurately translate the genetic code into amino acid sequences and to maintain the codon reading frame. In each elongation cycle of a nascent polypeptide, an aminoacylated transfer RNA (tRNA) binds to the ribosome as a ternary complex with EF-Tu and GTP. If the tRNA's anti-codon matches the codon in the template messenger RNA, EF-Tu hydrolyzes the GTP and dissociates. The amino acid is then transferred from the tRNA to the peptide. Crystal structures show a nearly 90° rotation between domains in EF-Tu accompanying GTP hydrolysis (Fig. 1b; Clark et al., 1999). The tRNA also rotates ~60° relative to the ribosome. EF-G participates in translocation of the mRNA and two tRNA molecules by exactly 3 bases along the synthetic machinery. Whether EF-G functions as a 'motor' and whether it exhibits large domain rotations are controversial (Rodnina et al., 1999).

The F_0–F_1 ATP synthase couples energy from a proton gradient across the inner mitochondrial membrane into phosphorylation of ADP to form ATP. This process is the major end product of mitochondrial respiration. F_0 is thought to be a motor that rotates a protein driveshaft (the γ subunit; Walker, 1994; Boyer, 1997). Torque transmitted to the F_1 domain drives the 'uphill' ATP synthetic reaction. When uncoupled from F_0, F_1 hydrolyzes ATP and rotates the γ subunit in the opposite direction (Yasuda et al., 1998).

Myosin, the motor protein in muscle contraction and in many types of non-muscle motility, translocates along the cytoskeletal filament, actin. In crystal structures of myosin with various nucleotide analogs bound, the angle between the region containing the actin and nucleotide binding sites (the motor domain) and that containing the myosin light chains varies by up to 70° (Fig. 1c; Rayment et al., 1993a,b; Dominguez et al., 1998; Houdusse et al., 1999; Corrie et al., 1999). These studies and other data suggest that during the enzymatic ATPase cycle, a tilting of the light chain domain amplifies sub-nanometer structural changes at the nucleotide binding site into several nanometers of motion (Goldman, 1998).

Voltage-gated ion channels underlie electrical impulses in the surface membranes of excitable cells such as neurons and muscle fibers. Na^+, K^+ and Ca^{2+} channels are all composed of homologous repeated domains that form a membrane-spanning pore. The channels are normally closed when transmembrane voltage is negative inside of the cell relative to the extracellular space, but they open when the potential decreases or reverses. The fourth membrane spanning segment (S4) within each domain contains positively charged residues and is thought to serve as the voltage sensor. X-ray crystallography of ion channels has been more difficult than for soluble proteins, but the structure of a homologous bacterial K^+ channel (without voltage gating) has recently been solved (Doyle et al., 1998). While this structure has resulted in many new insights, the motions associated with gating are still unknown. Experiments using fluorescent probes attached to S4 segments of eukaryotic K^+ channels, however, suggest that the voltage sensor either translates or tilts when the membrane electrical field is altered (Mannuzzu et al., 1996; Yang et al., 1996; Loots and Isacoff, 1998).

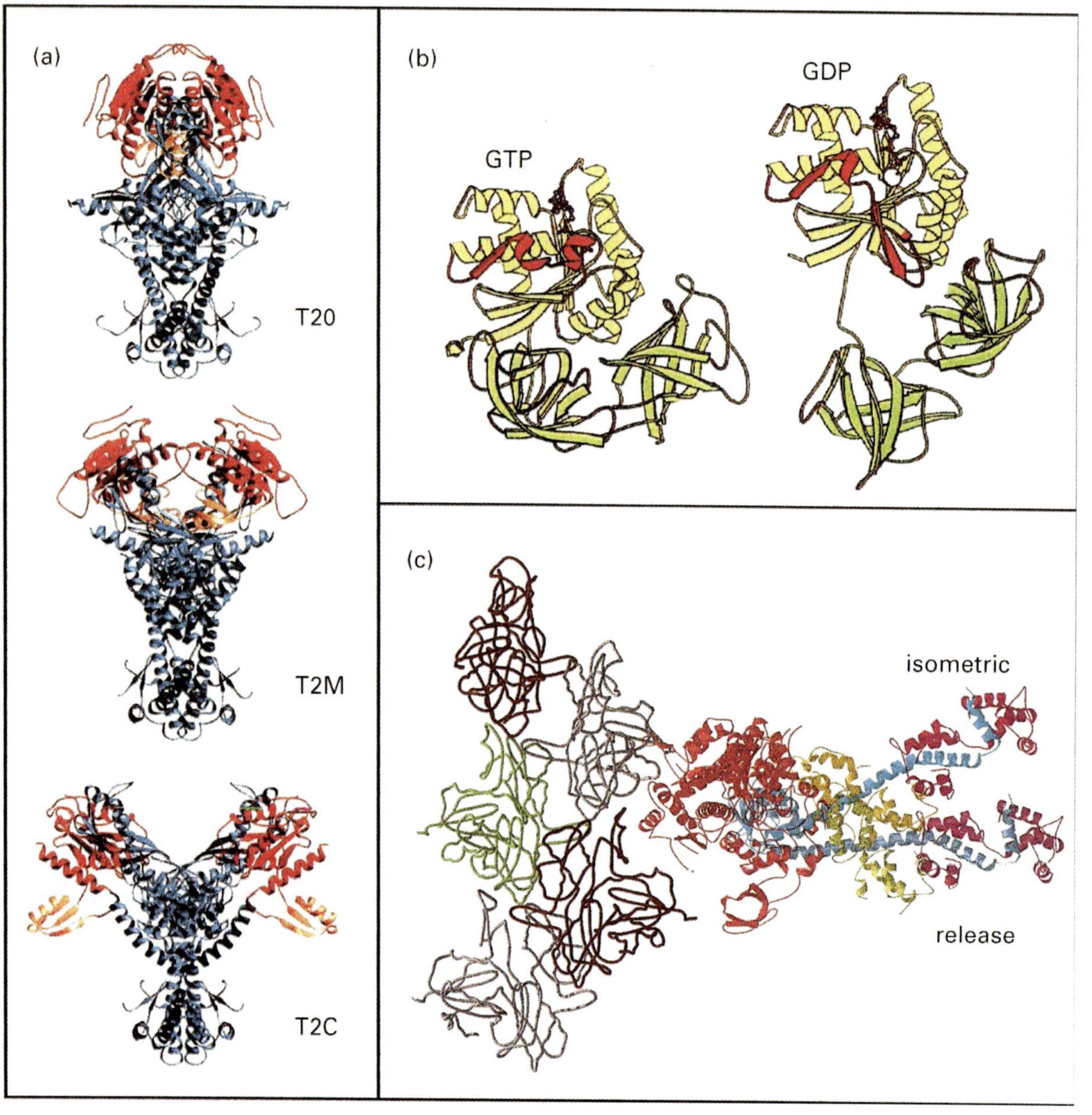

Fig. 1. Examples of crystal structures showing protein domain rotations. (a) Topoisomerase II (T2). Both the Rossman fold (red) and the B′ hook (orange) domains rotate relative to the DNA binding platform (blue) during activity. (Reprinted from with permission from Fass et al., 1999, Structural Biology, Nature 6, 322–326.) (b) Ribosomal elongation factor-Tu undergoes a large nucleotide-hydrolysis-dependent reorientation of domains. Note the relative positions of domain I (yellow) versus domains II and III (green) and the transition from α-helix to β-strand in Switch I (red). (Reprinted from FEBS Letters 452, Clark et al., "Structural information for explaining the molecular mechanism of protein biosynthesis" pp. 41–46 © 1999 with permission from Elsevier Science.) (c) Crystal structure of chicken skeletal muscle myosin subfragment-1 (red, blue, tan, and magenta) docked to actin (gray, brown and green; Rayment et al., 1993a). The angle of the light chain domain has been modified to show the relative orientation change determined from fluorescence polarization experiments with bifunctional rhodamine (see Section 2.1) when a muscle fiber is suddenly allowed to shorten during isometric contraction. (Reprinted from Corrie et al. "Dynamic measurement of myosin light-chain-domain tilt and twist in muscle contraction", Nature 400, 425–430 © 1999 with permission from Nature (http://www.nature.com/).

In all of these systems, plausible relationships have been proposed between the observed structural changes and the functional output. Many aspects, though, are unresolved. These include the kinetics of the events, the relationship to ATP or GTP hydrolysis and how each protein is controlled by the chemical, mechanical and topological states of its functional partners. Thus testing the structurally derived hypotheses requires detection of the timing and extent of the motions during activity.

1.2. Fluorescence polarization

While X-ray crystallography provides remarkable spatial resolution of long-lived, static states, it provides no temporal information on functionally relevant motions except in rare instances (e.g. Schlichting et al., 1990). Conclusions about the mechanism of activity must usually be inferred from the structures of a few trapped states. Furthermore, even these configurations are approximations of the native structures since the environment within a crystal is much different from the cytoplasm or membrane of a cell. In particular, the solvents and intermolecular forces within a crystal are usually unphysiological. Entire macromolecular complexes do not often crystallize, so the structural data are usually limited to subcomplexes, individual subunits or even products of proteolytic cleavage.

Nuclear magnetic resonance (NMR) also yields atomic spatial resolution and may be performed under aqueous conditions, often closer to a native environment. However, this technique is limited to small proteins ($<\sim$30–50 kDa) tumbling freely in solution (Pervushin et al., 1997; Wand et al., 1998). Electron spin resonance (EPR) provides unambiguous angular information and therefore, can also be used to detect domain rotations (Hubbell et al., 1996; Mchaourab et al., 1996). In samples containing multiple states of a reaction pathway, the kinetics of exchange between those states can often be determined by NMR or EPR (Cantor and Schimmel, 1980, Chapter 9). However, direct characterization of consecutive structural states on the time scale of an enzymatic pathway has been achieved with magnetic resonance techniques in only a few cases (e.g. Ostap et al., 1993; Hubbell et al., 1996; Mollaaghababa et al., 2000).

Spectroscopic techniques, including circular dichroism, birefringence, fluorescence resonance energy transfer, fluorescence quenching and Raman scattering, probe structural changes with ample time resolution to monitor functional changes (Cantor and Schimmel, 1980, Chapter 8). These methods often can be used with proteins in native conditions. The structural information provided, however, is usually ambiguous. Structural changes are inferred from a primary signal that is influenced by several parameters including distance between residues, relative orientation and local environmental conditions, such as hydrophobicity or solvent accessibility; sorting out the relative effects of these different factors is often difficult.

Fluorescence polarization provides a technique that bridges the gap between studies with atomic spatial resolution but poor time resolution and those with good time resolution but ambiguous structural information. This technique has been used extensively to detect structural changes in organized systems, such as proteins embedded in cell membranes or muscle fibers, or in isotropic suspensions of soluble proteins or vesicles (Lakowicz, 1999, Chapters 10, 11). In this review we apply the term 'bulk samples' to such experimental systems. Either intrinsic fluorescence is used or the target macromolecule is labeled specifically with an extrinsic fluorescent probe. By exciting the probe with polarized light and by resolving the polarization of the fluorescence emission, the distribution of probe orientations and the time course and extent of probe motions can be determined. The measured polarization ratios are typically sensitive solely to probe orientation. However, a problem in relating this information to the orientation and dynamics of the protein motions in these studies has been that the probe orientations relative to the protein were not known. Newly synthesized probes, discussed in more detail below, which bind to the proteins at predetermined orientations, promise to eliminate this source of uncertainty.

1.3. Single-molecule fluorescence polarization

The single-molecule experiments discussed in this review represent extensions of the fluorescence polarization technique that has been applied to bulk samples. While measurements on macroscopic populations offer the advantage of producing large signals, there are several disadvantages. The average behavior of a population can be measured but some properties of the distribution are lost in the ensemble average. Inhomogenieties are not detected and apparent changes are diminished by asynchrony. By studying single molecules, these problems are avoided.

In bulk measurements, the information accessible about macromolecular orientation is limited by the sample geometry. Intensities of polarized fluorescence, and therefore the derived polarization and anisotropy ratios, depend on the orientational distribution of fluorophores. In suspensions, the rates and extent of probe motions can be interpreted in comparison to expected rates of tumbling, but the isotropic distribution furnishes no average orientation. The fluorescence anisotropy is sensitive to local probe orientation relative to the major hydrodynamic axes of the macromolecule and to internal flexibility of domains and secondary structural elements. In cylindrically symmetrical organized systems such as lipid bilayer membranes (Dale, 1988) and muscle fibers (Mendelson and Wilson, 1987; Hopkins et al., 1998), additional angular information is attainable because the molecules of interest are organized spatially around a symmetry axis. The angular distribution of probes can be measured relative to the axis of symmetry, but azimuthal angles around that axis are unavailable. In contrast, SMFP can completely determine a single fluorophore's mean orientation relative to a laboratory frame of reference which can then be transformed into any convenient coordinate system in the experimental sample.

Asynchrony of molecular transitions also hampers interpretation of bulk measurements. If the sample is undergoing a catalytic cycle, then every state of its reaction pathway will be populated to some extent and a measurement on the stationary system will reflect partitioning of the molecules among these states. In order to observe conformational changes between well-defined states, a perturbation can be imposed on the stationary distribution, such as a temperature or pressure jump, rapid addition of a substrate or photoliberation of a "caged" ligand (Gutfreund, 1995). These perturbations, however, provide a less than complete synchronization of the population, and furthermore, the synchronization degenerates over the same time scale as the processes of interest, due to the stochastic nature of the reactions. In SMFP, no synchronization is necessary. Any conformational state that persists longer than the time required to make a measurement (typically milliseconds) is potentially detectable from changes in the measured intensities. As described in Section 3, motions on time scales both faster and slower than the fluorescence lifetime can also be distinguished with SMFP.

Additional advantages of making fluorescence polarization measurements on individual molecules relate to uncommon or rare events. In the ensemble average from a bulk measurement, signals from rare events are generally overwhelmed by signals from common processes. However, in a single-molecule experiment, rare events are observed directly. Similarly, kinetics in bulk experiments are usually sensitive to only a few 'rate limiting' steps, whereas single-molecule traces potentially reveal the lifetimes of all states persisting longer than the measurement time.

All of these advantages of SMFP follow from detection of the signal from an individual molecule rather than an average from a bulk population. A single molecule is a pure, homogeneous and synchronized sample. Analogous advantages apply to the other single-molecule techniques described in this volume.

1.4. Early single-molecule fluorescence experiments

A brief summary of major developments specifically relevant to SMFP measurements on biological samples is given here. The history of single-molecule detection, in general, has been reviewed extensively (Nie and Zare, 1997; Xie and Trautman, 1998; Moerner and Orrit, 1999; Weiss, 1999).

Some of the first successful efforts to detect individual molecules in solution by fluorescence spectroscopy involved labeling a single macromolecule with many (~100) fluorescent probes (Hirschfeld, 1976; Barak and Webb, 1981; Morikawa and Yanagida, 1981). The multiple dipoles precluded orientation analysis and extensive labeling tends to suppress biological activity. Detection of single fluorophores in aqueous solution was first demonstrated by Shera et al. (1990) using laser-induced fluorescence by illuminating a small volume of solution traveling through a flow cell and using time-gated detection to distinguish fluorescence from Raman scattering. Autocorrelation analysis produced convincing evidence that the signals were derived from individual molecules of rhodamine-6G.

Many early single-molecule experiments were conducted at very low temperatures and in solid substrates where reduction of thermal motion simplified detection and analysis. Under these conditions, single pentacene impurities in *p*-terphenyl crystals were detected using frequency modulated absorption spectroscopy (Moerner and Kador, 1989), and later by fluorescence spectroscopy (Orrit and Bernard, 1990; Ambrose and Moerner, 1991; Orrit et al., 1993). Measurements of the fluorescence intensity of similar crystals, while varying the excitation polarization, yielded discrete orientations of the individual pentacene dipoles projected onto the crystal plane (Güttler et al., 1993, 1996). This work resulted in assignment of the observed spectral properties to crystal lattice states.

Near-field scanning optical microscopy (NSOM) enabled investigation of single molecules at ambient temperatures (Betzig and Trautman, 1992; Betzig and Chichester, 1993; Ambrose et al., 1994; Trautman et al., 1994). In this technique, a sub-wavelength diameter light source is positioned very close to the specimen (Fig. 2a) on a microscope slide. The source is raster-scanned across the slide. By limiting the excitation volume, background fluorescence and scattering are kept much lower than the signal from a single molecule. Betzig and Chichester (1993) measured the orientation of carbocyanine dye molecules ($DiIC_{12}$) on polymethylmethacrylate (PMMA) coated slides. These experiments took advantage of the spatial distribution of optical polarization at the tip of the scanning probe to determine the three-dimensional orientation of single fluorophores, as shown in Fig. 2b. Using identical samples, Trautman and Macklin (1996) and Macklin et al. (1996) compared measurements made with NSOM and spot-confocal microscopy,

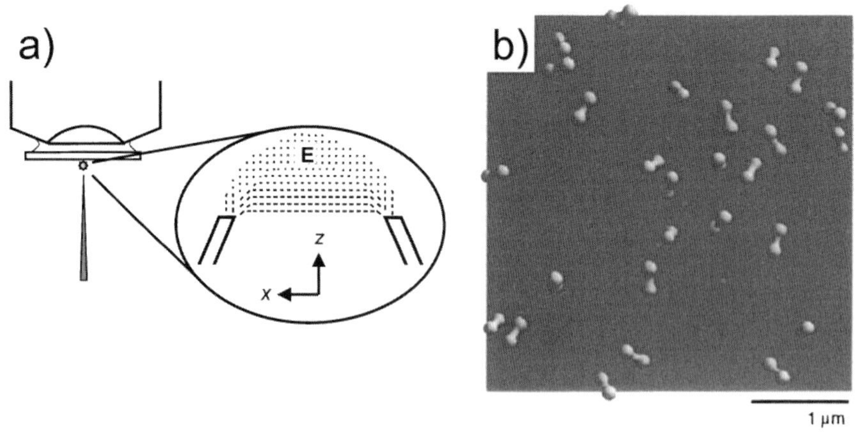

Fig. 2. (a) Near-field scanning optical microscopy utilizes a probe with diameter smaller than the wavelength of light. The probe is raster scanned across the sample. The electric field vector in the evanescent wave at the probe tip is indicated by the bars in the expanded view. (b) Dipole orientations of individual carbocyanine dye molecules determined from Near-field scanning optical fluorescence microscopy. (Reprinted with permission from Betzig and Chichester, "Single molecules observed by near-field scanning optical microscopy", Science 262, 1423–1424. Copyright (1993) The American Association for the Advancement of Science).

a far-field technique. These studies showed that NSOM provided better spatial resolution, as expected, as well as more complete orientation information. However, the much simpler far-field spot-confocal technique furnished more accurate measurements of excited state lifetime and higher temporal resolution. These experiments also demonstrated the orientation sensitivity of the far-field technique by exciting with two orthogonal polarizations. Using analysis procedures described below, the two resulting fluorescence intensities were combined to determine the orientation of the absorption dipole projected onto the plane of the microscope stage.

Refinements to epifluorescence microscopes using laser excitation sources enabled two groups to detect single labeled proteins under aqueous conditions appropriate for functional activity (Funatsu et al., 1995; Sase et al., 1995). Funatsu et al. (1995) also used the evanescent wave generated by total internal reflection as an excitation source and obtained an additional reduction of background compared to laser epifluorescence. They showed that kinetics of single enzymes can be studied by measuring interaction of a fluorescent ATP analog with individual fluorescently labeled myosin molecules. Sase et al. (1995) recorded continuous images of actin filaments, sparsely labeled with tetramethylrhodamine, as they were propelled by myosin in an in vitro motility assay. These pioneering studies inspired many of the SMFP experiments summarized later in this review by establishing that single functioning protein molecules could be monitored in real time.

2. Experimental methods

This section presents an overview of methods used for SMFP experiments. Many suitable fluorescent probes are available for labeling the protein (or macromolecular system) under investigation, the choice being dictated by optical and biological considerations. A variety of optical arrangements with low background and high sensitivity have been developed to detect individual probes. The excitation methods generally minimize background signal levels by illuminating a very small volume. The detectors possess high sensitivity and low intrinsic noise. The discussion here is limited to details of particular relevance to SMFP. More general descriptions of methods for single-molecule microscopy may be found in several recent reviews (Basché et al., 1997; Nie and Zare, 1997; Xie and Trautman, 1998) and other chapters in this volume.

2.1. Fluorescent probes

The use of intrinsic fluorescence and the available extrinsic fluorescent probes have been reviewed (Haugland, 1996, Chapter 1; Johnson, 1998; Lakowicz, 1999, Chapter 3). Rhodamines, Cy3, Cy5, Green Fluorescent Protein (GFP), Texas Red, pentacene and $DiIC_{12}$ have all been used in single-molecule investigations and others may be applicable. High quantum yield for fluorescence emission (Q_f) and photostability are especially desirable characteristics for SMFP due to the low signal levels. The total number of photons emitted by an individual molecule before it is irreversibly photobleached must be high enough to obtain sufficient signal levels over a useful time period (typically 0.1–10 s for protein studies). Total photon emission is approximately equal to the ratio of Q_f to the quantum yield for photobleaching (Q_b). Rhodamine and Cy3 are examples of probes with high Q_f/Q_b ratios. When photobleaching is minimized by eliminating reactive oxygen in the medium, $\sim 10^6$ fluorescent photons are emitted by these probes before irreversible photodestruction (Soper et al., 1991; Nie and Zare, 1997).

In fluorescence polarization studies, close alignment of the probe's absorption and emission transition dipole moments is advantageous since it simplifies data analysis. This characteristic is quantified by r_o, the fluorescence anisotropy of a sample containing randomly oriented immobilized probes or the zero-time anisotropy in nanosecond time-resolved measurements of freely rotating probes. r_o equals 0.4 for probes with perfectly aligned absorption and emission dipoles. For instance, for the major visible absorption transition in rhodamine and other xanthene derivative fluorophores (e.g. fluorescein, eosin, etc.), $r_o \cong 0.38$ (Chen and Bowman, 1965; VanderMeulen et al., 1990; van der Heide et al., 1992) indicating that the two dipoles are aligned to within $\sim 10°$. In this case, the reasonable assumption is commonly made that the absorption and emission dipoles are colinear.

In experiments using extrinsic probes, the extent and specificity of labeling should be considered and functional characteristics of the labeled protein should be compared to those in the native state. Probes are available with reactive linkers that target specific residues which can be engineered into the polypeptide sequence to place probes at particular locations. Many examples of site-specific labeling of surface residues at high stoichiometry and with relatively little perturbation of activity have been reported (Maxfield, 1989; Wang, 1989; Waggoner, 1995; Mchaourab et al., 1996; Sabido-David et al., 1998).

Green fluorescent protein (GFP) and its derivatives (Cubitt et al., 1995; Heim and Tsien, 1996; Tsien, 1998; Yang et al., 1998) have been used widely in cell biological studies because they can be easily linked to many proteins and introduced into the native environment. GFP is well polarized ($r_o = 0.32$; Swaminathan et al., 1997). However, GFP has not been effectively utilized in SMFP studies because its emission is not stable. It exhibits transient periods of zero fluorescence — a phenomenon that has been termed 'blinking' (Dickson et al., 1997; Pierce et al., 1997). Instability of emission causes intensity changes unrelated to probe orientation and, therefore, fluorescence polarization data from GFP are difficult to analyze. Blinking has also been observed with other fluorophores (e.g. Ha et al., 1996).

The fluorophore orientation that gives rise to experimental fluorescence polarization signals reflects both the protein orientation and the angular position of the probe relative to the protein. Typically, the latter angle is not known and may even be inhomogeneous. Given these uncertainties, it is generally not possible to determine the orientation of a protein domain from the measured orientation of the probe dipoles, a major limitation of fluorescence polarization studies. Recent developments, however, circumvent this difficulty by fixing the probe to two or more amino acids, thereby defining the local orientation.

A bifunctional rhodamine (BR) (Fig. 3a and b) was synthesized by Corrie et al. (1998) to bind to two appropriately spaced cysteine residues engineered into a protein. The absorption and

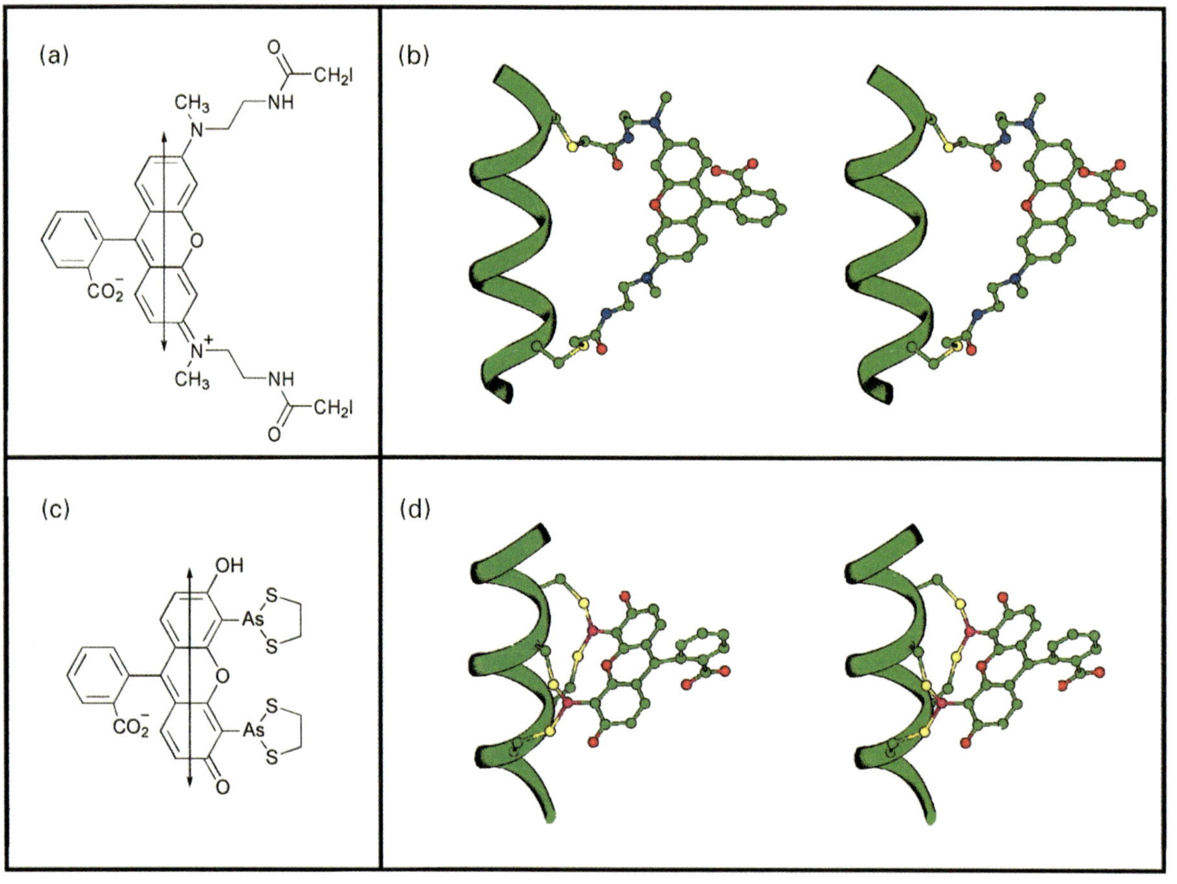

Fig. 3. New probes which can be bound with known location and orientation to an engineered protein. (a) Bisiodoacetamidorhodamine (BR-I_2), (b) stereo view of bifunctional rhodamine (BR) crosslinked to two cysteines in an α-helix, (c) 4′,5′-bis(1,3,2-dithioarsolan-2-yl)fluorescein (FlAsH-EDT_2), (d) stereo view of FlAsH bound to four cysteines in an α-helix. The FlAsH chromophore is depicted in one of two possible orientations (see text). The absorption and emission transition dipole moments are shown as double-headed arrows in (a) and (c).

emission dipoles in rhodamine are aligned with the long axis of the aromatic ring system that forms the chromophore (double-headed arrow in Fig. 3a). The two iodoacetamide linkers on the ends of the chromophore can bind covalently to accessible cysteine residues suitably spaced in the protein. The orientation of the dipoles relative to the protein is then expected to be along a line between the two cysteine residues which is known from the atomic structure. In order to detect rotations about different protein axes, various relative orientations can be obtained by replacing different pairs of native amino acids with cysteines.

Bifunctional rhodamine has been applied to the study of the myosin regulatory light chain (RLC) orientation in muscle fibers (Corrie et al., 1999). Four mutants were constructed, each containing a pair of cysteine residues 1.0–1.6 nm apart. These mutant proteins were reacted with the bifunctional rhodamine and the 1 : 1, two-site labeled protein was isolated from the reaction mixture by FPLC. The probe was shown to bind specifically and cross-link the pair of engineered cysteine residues in each mutant. In combination, the data from the four mutants enabled determination of the motions of the RLC relative to the muscle fiber axis in steady-state experiments and in dynamic measurements with 50 μs time resolution. BR may find many applications in single-molecule studies of myosin and other systems.

Another probe designed to provide a stable, uniform orientation relative to the protein is 4′, 5′-bis(1,3,2-dithioarsolan-2-yl)fluorescein (FlAsH-EDT_2; Fig. 3c; Griffin et al., 1998). FlAsH is an organoarsenical compound that binds tightly to a protein α-helix containing cysteines at positions i, $i+1$, $i+4$ and $i+5$ (Fig. 3d). Native residues are replaced by cysteines to position this amino acid motif at a known orientation with respect to the crystal structure. Whether the chromophore is oriented along the $i \leftrightarrow i+4$ direction or $i \leftrightarrow i+1$ (or both) is not currently known, but its rotational mobility is restricted when it interacts with calmodulin containing the binding motif (S.R. Adams, pers. comm.). The FlAsH reagent itself is virtually non-fluorescent but becomes fluorescent upon binding to the four cysteine residues of the target protein. Thus, appearance of fluorescence is an indicator of specific labeling. Excess ethanedithiol (EDT) can reverse this reaction, and dissociate FlAsH from the protein. These features enable FlAsH to be used both in vitro and inside live cells.

The two fluorescent probes with multiple linkers mentioned here have been synthesized in academic laboratories. Other bifunctional fluorescent compounds have recently become available commercially (Ehrhardt et al., 1998). As the value of this new type of probe becomes more evident, other bifunctional or motif-specific labels should become available.

2.2. *Instrumentation*

In most SMFP experiments the sample under study is immobilized on a surface or in a polymer matrix (Dickson et al., 1996; Lu et al., 1998) to constrain individual molecules within the field of view long enough for their reaction processes to be observed. Either the sample sparsely occupies the detected volume or it is sparsely labeled with probes so that individual fluorophores are spatially separated and distinguishable. The surface interaction may also provide a fixed orientation of the attached domain. Rotational motions are then referred to that stable platform. Samples can be attached to glass (Howard et al., 1993, Unger et al., 1999), fused silica (Iler, 1979, Chapters 6, 7; Funatsu et al., 1995; Sase et al., 1995) or freshly cleaved mica surfaces (Bopp et al., 1997; Sytnik et al., 1999). Methods for attaching motor and cytoskeletal proteins to surfaces have been discussed by Scholey (1993). Nitrocellulose, a surface coating commonly used to attach myosin to glass coverslips, exhibits fluorescence, presumably due to contaminants (Conibear et al., 1998), that limits its usefulness in SMFP. Other methods for sample attachment to surfaces are described in the studies cited below and in other chapters in this volume.

The sample slide is placed on the stage of an upright or inverted microscope fitted with polarizing optics and a sensitive detector. The input optics are arranged to limit the excitation electric field to a very small volume (Nie and Zare, 1997). This minimizes background fluorescence from contaminants and from labeled proteins elsewhere in the sample chamber. It also reduces the amount of Rayleigh and Raman scattering from the buffer itself. A high numerical aperture (NA) microscope objective is used to collect as large a fraction of the emitted fluorescence as possible and to project the emission through bandpass or long-pass spectral filters and polarizing optics to the detector.

Lasers are ideal excitation sources for SMFP measurements because of their spectral purity, coherence, and high degree of polarization. Emission from a standard polarized laboratory laser generally exhibits a high extinction ratio (typically on the order of 100 : 1). A 'clean-up' polarizer can improve the linear polarization to achieve an extinction ratio of $10^4:1$ or better (for a technical discussion, see e.g. catalogs of Melles Griot Corp., 1999, Chapter 1, or Newport Corp., 1999, Chapters 8, 11). Comparing emission intensities for different excitation polarizations enables determination of the orientation of the probe absorption dipole as explained below in Section 3. The polarization angle can be altered by mechanically switching linear polarizers or by rotating a retardation plate. Faster modulation of the polarization (in the μs range) is achieved using electronic variable retarders (Jenkins and White, 1976, Chapter 32).

Power of the excitation light influences the intensity of fluorescence, as well as the photobleaching rate. An increase in excitation power will increase the emitted photon flux proportionally, and therefore, increase the signal-to-noise ratio (Inoué and Spring, 1997, Chapter 6). However, because the number of photons emitted by a fluorophore before photobleaching is limited, such an increase in excitation power will also accelerate photobleaching. Therefore, a tradeoff exists between signal-to-noise ratio for individual measurements and total recording time before photobleaching. Also, the excitation power should be kept low enough so that the absorption probability is linear in intensity (i.e. the fluorophore should not be saturated; Yariv, 1975, Chapter 8). The absolute intensity is difficult to quantify, so orientation measurements are usually made by comparing two or more emission intensities with different excitation and/or detection polarizations (Lakowicz, 1999, Chapter 10). Reliable determination of calibration factors that quantify the relative excitation power and detection sensitivity at different polarizations is crucial for this type of experiment. Measurements on a sample with known polarization properties are used to determine these factors. This is similar to the determination of the instrument factors in a conventional spectrofluorimeter.

2.2.1. Far-field excitation

With relatively minor modifications, a standard research epifluorescence microscope may be used to perform SMFP experiments (Funatsu et al., 1995; Sase et al., 1995; Unger et al., 1999). Usually, a laser excitation source replaces the arc lamp illuminator. The objective and immersion fluid are selected for low autofluorescence. The laser is either focussed or collimated at the back focal plane of the objective, leading to wide-field epiillumination or spot-confocal excitation

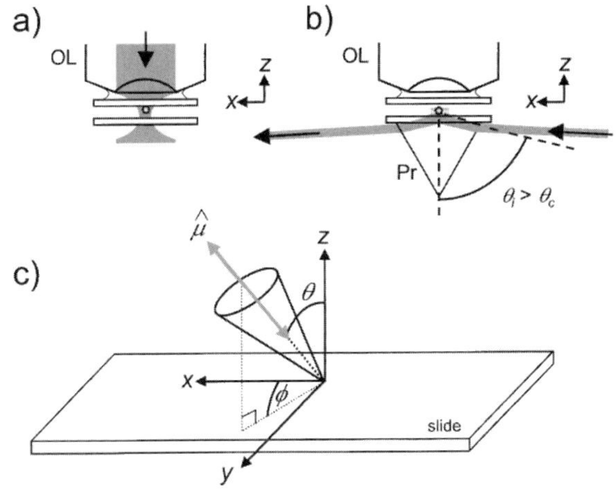

Fig. 4. Fluorescence excitation schemes. (a) With laser spot-confocal excitation, a laser beam is focused through the objective lens (OL) to a diffraction limited spot. (b) Prism (Pr)-type total internal reflection produces an evanescent wave near the microscope slide/buffer interface. (c) Cartesian coordinate system used in Section 3 and Appendix A. The origin is on the microscope slide. The x- and y-axis are in the plane of the slide. The z-axis coincides with the optical axis of the microscope and points towards the objective. The orientation of a transition dipole moment, $\hat{\mu}$, is described by its polar angles θ (axial) and ϕ (azimuthal).

(Fig. 4a) at the slide surface, respectively. Diffraction limited illumination and detection of emission from the same small volume in laser spot-confocal microscopy provides high spatial resolution and minimizes fluorescence from other regions of the sample (Pawley, 1995). The sample is scanned in order to obtain a two-dimensional image of the slide surface. Wide-field epiillumination avoids the necessity for lateral scanning of the excitation at the expense of higher background.

For either of these far-field excitation techniques, using a Gaussian mode laser, the excitation beam at the slide surface propagates predominantly along the optical axis (z-axis) of the microscope. The optical polarization (direction of the oscillating electric field) is thus predominately in the plane of the slide (x- and y-axis; Fig. 4c; see Xie and Trautman, 1998, for an exception). As will be discussed in more detail in Section 3, this characteristic of far-field propagating excitation makes it difficult to use for measuring the axial polar angle, θ, of the molecular orientation with respect to the z-axis.

2.2.2. *Total internal reflection excitation*

Total internal reflection (TIR) microscopy (Fig. 4b) makes use of the evanescent wave formed beyond a reflecting surface under glancing illumination. The input laser beam is directed through a glass or fused silica microscope slide toward the interface with the aqueous medium containing the experimental sample (see Axelrod (1989a) and Ambrose et al. (1999) for other possible geometries). The index of refraction (n_1) of the slide is greater than that of the medium (n_2). If the incident angle (θ_i with respect to the slide normal) is greater than the critical angle defined by $\theta_c = \sin^{-1}(n_2/n_1)$, then all of the energy is reflected back into the slide. The electric field, however, does not vanish abruptly at the surface. An oscillating electromagnetic field, termed an evanescent wave, is present on the aqueous side of the interface (Axelrod, 1989a). The evanescent wave is capable of exciting fluorophores (Carniglia et al., 1972). Its amplitude decays exponentially away from the interface and, under conditions applicable to TIR microscopy, the $1/e$ decay constant is only a few hundred nm ensuring that only fluorophores at or near the surface are excited. Thus, contaminants and sample molecules in the bulk medium are not excited, reducing background several orders of magnitude as compared to conventional epifluorescence microscopy (Funatsu et al., 1995).

The evanescent wave has unusual but predictable polarization properties. Consider the geometry in which the scattering plane, which is defined by the incident and reflected beams, is the x–z plane (see Fig. 4b). If the incident beam is polarized perpendicular to the scattering plane (s-polarization), then the evanescent wave is polarized in the y direction (see Fig. 4c). If the incident beam is polarized parallel to the scattering plane (p-polarization) then the evanescent wave is polarized predominantly along the z direction, with a small component along the x direction (Axelrod et al., 1984). The relative magnitudes of the z- and x-component of polarization depend on n_1/n_2 and θ_i. For a typical setup with p-polarized excitation, the ratio of intensities (x-component/z-component) is approximately 0.05. As discussed further in Section 3.3, the robust z-component of polarization in the evanescent wave allows sensitive detection of the fluorophore orientation (θ) relative to the optical axis.

2.2.3. Near-field optical probe excitation

Near-field scanning optical microscopy is the optical equivalent of atomic force microscopy. A tapered optical fiber, coated with metal, serves as a near-field optical probe. Even though the probe aperture is smaller than the wavelength of the excitation light, a non-propagating oscillating electric field extends several hundred nanometers beyond the tip (Fig. 2a). As in TIR microscopy, this evanescent wave has predictable spatial extent and polarization. The shape of the electric field distribution provides various polarizations, including z-axis components, as the probe is scanned near an individual fluorophore (Fig. 2a; Dürig et al., 1986; Betzig and Chichester, 1993). The orientation of the electric field can also be altered by changing the polarization of the input light (Xie and Dunn, 1994; Ruiter et al., 1997).

A unique advantage of NSOM is that spectroscopic and topographic information can be measured simultaneously. Disadvantages are low-power throughput, poor reproducibility of the field distribution at the tip and the possibility of mechanical or electromagnetic perturbation of the sample by the probe (Macklin et al., 1996; Trautman and Macklin, 1996). Time resolution depends on the scanning area and signal-to-noise ratio, but is typically lower than with the spot-confocal, wide-field epiillumination or TIR methods.

2.2.4. Microscope objectives and emission optics

Because the signal-to-noise ratio in SMFP is limited by the number of detected photons, collection efficiency of the microscope objective is often more important than spatial resolution, flatness of field or correction of optical aberrations. Thus the numerical aperture and transmittance of the objective are the key specifications. When the excitation laser passes through the objective, a further requirement is very low autofluorescence.

The large solid angle subtended by high NA objectives complicates the data analysis somewhat due to the following effect. An individual fluorophore emits over a wide angle. When this light is collected and collimated by the objective, it is redirected to propagate along the optical axis (i.e.

the z-axis). Because the electric field remains normal to the direction of propagation for each ray, the polarization is also redirected by the objective. The effect of this refraction in regards to SMFP is that polarization originally along one axis of the microscope coordinate system is transferred partly onto the other two axes (Axelrod, 1989b). This depolarization must be taken into account as explained in Section 3.

The light collected by the objective lens is filtered in order to reject wavelengths outside of the fluorescence emission band. Stray light from the excitation beam as well as Raman and Rayleigh scattering are the greatest sources of background contamination (Nie and Zare, 1997). High-quality optical long-pass and band-pass filters are commercially available. The excitation wavelength may be further suppressed with sharp holographic notch filters which have recently become available.

Dichroic beamsplitters, filters, or mirrors are often used to redirect the emitted light. Care must be taken to account for polarization-dependent transmissions, reflections, and phase shifts (Born and Wolf, 1964, Section 1.5; see also Newport Corp. catalog, 1999, p. 11.14). This is particularly important if the light is not collimated, or when the detected polarization is not parallel or perpendicular to the reflection plane defined by the propagation direction of the incident light and the normal to the mirror. In these cases mixing between the detected polarization signals is possible. Similar precautions apply when reflecting surfaces are used in the excitation pathway.

Either total fluorescence is detected or orthogonal components of polarization are selected by a polarizer (analyzer). Hereafter, the term detector polarization will be used to denote the polarization transmitted by the analyzer. Simultaneous detection of both x- and y-polarization by using a polarizing beam splitting prism (e.g. a Wollaston or Thompson prism) and directing the polarized components onto two separate detectors is a common arrangement (Fig. 5a) because this scheme utilizes all of the collected light. A variant of this arrangement is the 'dual-view microscope' (Kinosita et al., 1991; Sase et al., 1997). The two polarization components are projected onto two-halves of the detector surface in a single video camera to simultaneously obtain spatial resolution and polarized fluorescence (Fig. 5b).

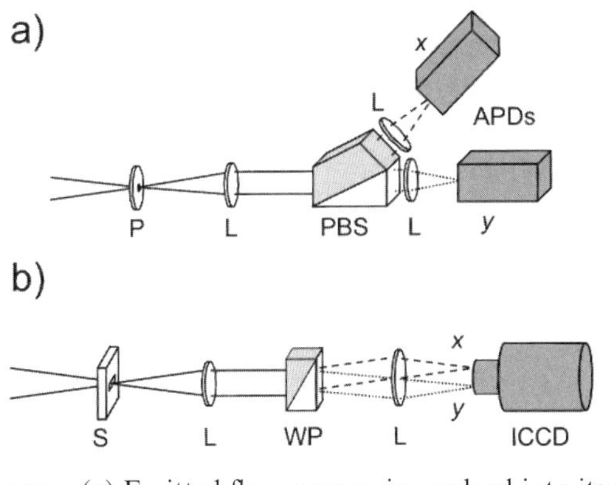

Fig. 5. Fluorescence detection schemes. (a) Emitted fluorescence is resolved into its two orthogonal components (here, x and y) with a polarizing beam splitter (PBS) and focused onto two separate avalanche photodiodes (APDs). (b) Emitted fluorescence is split by a Wollaston prism (WP), and imaged onto two halves of the detector area of an ICCD camera (Kinosita et al., 1991). L, lens; P, pinhole; S, slit.

2.2.5. Detectors

The light collected from individual fluorophores in SMFP experiments is ~10,000 photons per second or less, thus necessitating highly sensitive detectors. For single-molecule experiments in which localization of the molecule is the primary goal, a signal-to-noise ratio of 2 may be sufficient. However, for SMFP experiments, the accuracy of the deduced orientations is directly related to the measurement precision of the various polarized intensities and thus to the signal-to-noise ratio. The specific characteristics of a detector that govern the signal-to-noise ratio include the quantum efficiency, intrinsic background noise, and the fidelity of the amplification and readout processes. The time resolution of the detector is also important in dynamic experiments. Either solid-state (CCD) cameras or avalanche photodiodes have been used in most SMFP experiments.

Cameras offer the obvious advantage of simultaneously collecting data from a two-dimensional field on the microscope slide (Inoué and Spring, 1997, Chapters 7, 8). Slow scan, digital readout, back illuminated CCD cameras are available with very high quantum efficiencies (as high as 90% at 600 nm) and low combined readout and background noise (e.g. less than 10 counts per pixel per image). However, these cameras usually require several seconds to transfer data from an entire image. While data acquisition rates can be increased by binning pixels, or using on-chip frame transfer technology, slow-scan cameras are generally not fast enough to observe biologically relevant events in real time. CCD cameras coupled to image intensifiers offer considerably higher readout rates with transfer times for an entire frame as short as 30 ms and even faster subframe transfer times. The tradeoff for improved time resolution is a lower quantum efficiency (typically peaking at 40% near 550 nm) and higher measurement uncertainty caused by the statistical nature of the intensifier gain.

In SMFP, measurements of polarized fluorescence from individual stationary molecules have been made almost exclusively with avalanche photodiode (APD) detectors. These photon counting devices have an active area a few hundred μm in diameter, peak photon detection efficiency (including quantum efficiency) of ~70% at 700 nm, dark noise as low as 25 counts per second and maximum counting rates up to 10 MHz (Dautet et al., 1993) and manufacturer data sheets cited there. Thus, they impose no restriction on the measurement time resolution, which is then limited only by photon statistics.

3. Analysis of single-molecule fluorescence polarization data

Single-molecule fluorescence polarization data consist of a number of intensity measurements made with different excitation and detector polarizations. To analyze these data, the dependence of the intensities on molecular orientation and motions must be considered. Relationships of polarized fluorescence intensities to fluorophore orientation and mobility appropriate for SMFP data are presented here, in the appendix, and the references therein. The equations assume that a fluorescent molecule with non-degenerate, linear, absorption and emission transition dipole moments is excited by an electromagnetic field with known polarization, and that fluorescence is collected by a high NA objective lens. The excitation is assumed to be non-saturating so that the signal is linearly proportional to excitation intensity. The equations assume that the data have been corrected for background light and for polarization-dependent variations in excitation intensity and detector sensitivity (instrumental artifacts).

The analysis given here does not account for Fresnel reflections at optical interfaces or for non-propagating components of dipole emission. These effects introduce only small perturbations for molecules more than a few wavelengths from the cover slip on the objective side of the sample chamber. However, the corrections may be larger for molecules on the surface of the cover slip if an oil immersion objective is used. In this case, non-propagating emission toward the objective may be converted into propagating radiation in the cover slip and contribute to the measured intensity. These effects, which are discussed by Hellen and Axelrod (1987), may change the quantitative analysis, but do not alter the types of information that can be obtained.

For the purposes of this discussion, it will be useful to define a number of distinct time scales of possible molecular motion. These are defined relative to the fluorophore's excited state lifetime, typically 1–10 ns, and relative to the time required for a single intensity measurement, typically 1–10 ms for photon counting experiments, or 33 ms for video imaging. Changes in molecular orientation that occur on a time scale much longer than the measurement time will lead to changes in measured intensity between successive data samples. Functionally related conformational changes may cause this behavior. A molecule undergoing such motion may be considered static for analyzing each individual measurement. The analysis for this case is the most straightforward, and will be considered first.

Following that, molecular motion with a time scale shorter than the measurement time will be considered. While a single equation may, in principle, account for any motion of this type, two limiting cases are of particular interest. "Fast wobble" will refer to the case of motion with a time scale much shorter than the fluorescence lifetime. Independent thermal vibrations of the probe may occur in this time regime. "Slow wobble" will refer to motions with a time scale much longer than the fluorescence lifetime, but still much shorter than the measurement time. Such rotations are likely to result from thermal motions of protein domains or from steps in the enzymatic reaction pathway. Motions in these characteristic time regimes confer distinct characteristics on the measured intensities. Furthermore, these cases are useful because often a small fluorescent probe will undergo fast wobble relative to its attachment point on a protein, while the protein (and attached probe) may exhibit slow wobble relative to the laboratory frame. While not treated explicitly here, the equations for slow wobble and for fast wobble may be combined to take into account motions on both time scales.

3.1. Stationary fluorophores

If the fluorophore is effectively stationary during the measurement time, the analysis provides the polar angles, θ and ϕ, of the absorption and/or emission dipole, as defined by the coordinate system shown in Fig. 4c.

3.1.1. Absorption polarization ratios

Switching between two excitation polarizations parallel to the x- and y-axis, and collecting the emitted fluorescence with a high NA objective lens, results in two measured intensities. As shown in the appendix, these measured intensities are proportional to the projections of the absorption transition dipole moment onto the directions of polarization (i.e. along the x- and y-axis):

$$\begin{aligned} {}_xI &\propto \sin^2\theta_a \cos^2\phi_a \\ {}_yI &\propto \sin^2\theta_a \sin^2\phi_a \end{aligned} \tag{1}$$

where θ_a and ϕ_a are the polar angles of the absorption dipole, and the subscripts preceding I indicate the polarizations of the excitation light. In order to eliminate unknown multiplicative factors which are common to both measured intensities, including laser power, absorption cross section, objective collection efficiency, etc., an absorption polarization ratio, Q, may be calculated:

$$Q \equiv \frac{{}_yI - {}_xI}{{}_yI + {}_xI} = \frac{\sin^2\theta_a(\sin^2\phi_a - \cos^2\phi_a)}{\sin^2\theta_a(\sin^2\phi_a + \cos^2\phi_a)} = 1 - 2\cos^2\phi_a. \tag{2}$$

ϕ_a may then be determined from Q, which ranges from -1 to $+1$ as ϕ_a ranges from 0 to $\pi/2$. Note that due to the mirror symmetry of the transition dipole, optical techniques cannot differentiate between probe orientations (θ_a, ϕ_a) and $(\pi - \theta_a, \pi + \phi_a)$. For measurements using only linear polarizations along the x-, y- and z-axis, a further four-fold ambiguity arises in the determination of orientation due to reflection symmetries.

In order to increase accuracy, and to partially resolve this symmetry problem, measurements may be made at additional excitation polarizations. For an excitation laser with polarization in the plane of the stage and at an angle ζ with respect to the x-axis, the fluorescence intensity is given by (see Appendix A)

$${}_\zeta I \propto \sin^2\theta_a(\cos\phi_a \cos\zeta + \sin\phi_a \sin\zeta)^2 = \sin^2\theta_a \cos^2(\zeta - \phi_a). \tag{3}$$

As ζ is varied continuously, the phase of the intensity change gives ϕ_a.

3.1.2. Emission polarization ratios

The orientation of the emission dipole of a static molecule may be interrogated by resolving the polarization of the fluorescence. In analogy to the excitation case, the intensity of the emitted light detected through the analyzer is dependent mainly on the projection of the emission dipole onto the detector polarization axis. However, due to collimation of the light collected over the large solid angle subtended by a high NA objective, the detected intensity will also have some smaller contributions from components of the emission dipole perpendicular to the detector polarization. The extent of mixing of contributions from orthogonal components of the emission dipole depends on the numerical aperture of the objective as described in Appendix A. The intensities of detected fluorescence with polarizations along the x- and y-axis are given by

$$\begin{aligned} I_x &\propto C_1\sin^2\theta_e\cos^2\phi_e + C_2\sin^2\theta_e\sin^2\phi_e + C_3\cos^2\theta_e, \\ I_y &\propto C_2\sin^2\theta_e\cos^2\phi_e + C_1\sin^2\theta_e\sin^2\phi_e + C_3\cos^2\theta_e, \end{aligned} \tag{4}$$

where the subscript after each I indicates the polarization component detected, and θ_e and ϕ_e are the polar angles of the emission dipole. The coefficients C_1, C_2 and C_3 take into account the effects of the numerical aperture (see Appendix A and Axelrod, 1989b). For a water immersion objective with a typical numerical aperture of 1.2, $C_1 = 0.76$, $C_2 = 0.01$, and $C_3 = 0.23$. Calculating the emission polarization ratio, P, from the intensities in Eq. (4) yields

$$P \equiv \frac{I_y - I_x}{I_y + I_x} = \frac{(C_1 - C_2)(\sin^2\phi_e - \cos^2\phi_e)\sin^2\theta_e}{(C_1 + C_2)\sin^2\theta_e + 2C_3\cos^2\theta_e}.$$

$$= \frac{(C_1 - C_2)(\sin^2\phi_e - \cos^2\phi_e)}{(C_1 + C_2) + 2C_3\cot^2\theta_e}. \qquad (5)$$

P is sensitive not only to ϕ_e, but also to θ_e due to the high numerical aperture of the objective. The extent of the ratio's dependence on θ_e may be considered by comparing the relative magnitudes of $2C_3$ and $(C_1 + C_2)$. For the objective mentioned above, $2C_3 = 0.46$, while $(C_1 + C_2) = 0.77$. In principle, this polarization ratio may be combined with another measurement to determine both polar angles. For example, if the absorption and emission dipoles are colinear with polar angles $(\theta_a, \phi_a) = (\theta_e, \phi_e) \equiv (\theta, \phi)$, then the absorption polarization ratio in Eq. (2) will give the azimuthal angle, ϕ, and the emission polarization ratio can then be solved for the axial angle, θ.

Note that for fluorescence from a single stationary molecule, absorption polarization ratios are independent of the detector polarization, as indicated by Eq. (2). Emission polarization ratios (e.g. Eq. (5)) are independent of the excitation polarization. This is in striking contrast to the analogous measurements on bulk samples where, even if the molecules are stationary, the absorption polarization ratio depends on the detector polarization and the emission polarization ratio depends on the excitation polarization. For example, the excitation source preferentially excites a subpopulation of molecules aligned along its polarization. This photoselection biases the polarization of the emitted light.

For a stationary single molecule the measured fluorescence *intensities* (corrected for instrumental artifacts) depend on both excitation and detection polarization; however, the *ratios* do not. Regardless of how the molecule is excited, polarization of the emitted light is determined solely by the orientation of the emission dipole. Analogously, the probability of excitation depends only on the angle between the excitation polarization and the molecular absorption dipole. Thus, absorption polarization ratios do not depend on the detector polarization.

3.2. Non-stationary molecules

The general form of the equation describing polarized fluorescence intensity from a molecule undergoing rotation during an individual measurement is

$${}_{\varepsilon}I_{\alpha} = K\int_0^{\infty}\iint \rho(\theta_a, \phi_a, \theta_e, \phi_e, t)(1/\tau)e^{-(t/\tau)}P_a(\theta_a, \phi_a, \hat{\varepsilon})P_e(\theta_e, \phi_e, \hat{\alpha})\, d\Omega_a\, d\Omega_e\, dt \qquad (6)$$

(see also Cantor and Schimmel, 1980, Section 8-2). Here, $\rho(\theta_a, \phi_a, \theta_e, \phi_e, t)$ is a correlation function which describes the probability density of finding the absorption dipole at orientation (θ_a, ϕ_a) and the emission dipole at orientation (θ_e, ϕ_e) some time t later. Integrations are taken over all possible orientations of absorption and emission dipole, and over time. $d\Omega_a = \sin\theta_a\, d\theta_a\, d\phi_a$; $d\Omega_e = \sin\theta_e\, d\theta_e\, d\phi_e$. $P_a(\theta_a, \phi_a, \hat{\varepsilon})$ is the relative probability of absorbing a photon, given absorption dipole orientation (θ_a, ϕ_a) and electric field polarization $\hat{\varepsilon}$. $P_e(\theta_e, \phi_e, \hat{\alpha})$ is the relative probability of detecting an emitted photon, given emission dipole orientation (θ_e, ϕ_e) and analyzer aligned along $\hat{\alpha}$. Expressions for $P_a(\theta_a, \phi_a, \hat{\varepsilon})$ and $P_e(\theta_e, \phi_e, \hat{\alpha})$ are given in Appendix A. The factor $(1/\tau)e^{-t/\tau}$ is a weighting function that expresses the normalized probability of emitting a photon at time t, after excitation, assuming a single fluorescence lifetime, τ. The infinite time integral is closely approximated when the measurement time is much greater than τ. K is a constant of proportionality that is independent of excitation/detector polarization and molecular orientation.

As described in the Applications Section 4 below, Ha et al. (1998, 1999a) have used a form of Eq. (6) to analyze SMFP data and obtained average azimuthal angles, extent of motion and rotational diffusion coefficients of their probes. It is useful to simplify Eq. (6) for the limiting cases of fast and slow wobble.

3.2.1. Fast wobble

In the case of fast wobble, the fluorophore fully samples a distribution of angles, within a region restricted by its local environment, over a time period much faster than the fluorescence lifetime. Therefore, all "memory" of the probe orientation (within its restricted range) at the instant of excitation is lost before the instant of emission. The orientation of the emission dipole, within its restricted range, at emission is independent of the orientation of the absorption dipole at excitation. This property implies that, $\rho(\theta_a, \phi_a, \theta_e, \phi_e, t)$ can be written as the product of two time-independent distribution functions, $\rho_{fa}(\theta_a, \phi_a)$ and $\rho_{fe}(\theta_e, \phi_e)$ which describe the independent probabilities of finding the absorption and emission dipoles in particular orientations. Eq. (6) then simplifies to the product of two terms that describe the absorption and emission separately

$$
{}_{\varepsilon}I_{\alpha} = K\left[\int \rho_{fa}(\theta_a, \phi_a)P_a(\theta_a, \phi_a, \hat{\varepsilon})d\Omega_a\right]\left[\int \rho_{fe}(\theta_e, \phi_e)P_e(\theta_e, \phi_e, \hat{\alpha})d\Omega_e\right]. \quad (7)
$$

Polarization ratios calculated from measurements taken with different values of $\hat{\varepsilon}$ and/or $\hat{\alpha}$ may be considered as in the stationary molecule case. For a set of intensity measurements that vary only in excitation polarization, $\hat{\varepsilon}$, the second integral term in Eq. (7) will be constant. Therefore, this term will cancel when an absorption polarization ratio is calculated from these intensities. Analogously, an emission polarization ratio will be independent of the first integral term in Eq. (7). Therefore, as in the case of a static fluorophore, absorption polarization ratios are independent of the detection polarization, and emission polarization ratios are independent of the excitation polarization. For example, if a molecule is alternately excited with x and y polarized light, the resulting absorption polarization ratio, using expressions for $P_a(\theta_a, \phi_a)$ and $P_e(\theta_e, \phi_e)$ given in the Appendix, is

$$
Q = \frac{{}_{y}I_{\alpha} - {}_{x}I_{\alpha}}{{}_{y}I_{\alpha} + {}_{x}I_{\alpha}} = \frac{\int \rho_{fa}(\theta_a, \phi_a)\sin^2\theta_a(\sin^2\phi_a - \cos^2\phi_a)d\Omega_a}{\int \rho_{fa}(\theta_a, \phi_a)\sin^2\theta_a d\Omega_a}. \quad (8)
$$

Note that this equation is similar to that for the static case (Eq. (2)) except that the intensities are now averaged over the distribution of dipole orientations produced by the motion. Adding fast wobble reduces the range of possible polarization ratios, since the fluorophore is not aligned along any precise direction for the entire duration of a measurement. In effect, fast wobble, while still preserving the independence of the absorption and emission polarizations from each other, causes a reduction in the magnitude of polarization ratios.

3.2.2. *Slow wobble*

If the characteristic time scale of motion is much longer than the fluorescence lifetime, but still shorter than the measurement time, then the molecule is effectively static between corresponding absorption and emission events. Therefore, during the time between absorption and emission (i.e. over the time integral of Eq. (6)) $\rho(\theta_a, \phi_a, \theta_e, \phi_e, t)$ is constant and may be approximated by the time-independent probability, $\rho_s(\theta_a, \phi_a, \theta_e, \phi_e)$, of finding the absorption and emission dipoles oriented in the directions (θ_a, ϕ_a) and (θ_e, ϕ_e), respectively. Eq. (6) then simplifies to

$$_{\varepsilon}I_{\alpha} = K \iint \rho_s(\theta_a, \phi_a, \theta_e, \phi_e) P_a(\theta_a, \phi_a, \hat{\varepsilon}) P_e(\theta_e, \phi_e, \hat{\alpha}) \mathrm{d}\Omega_a \mathrm{d}\Omega_e. \tag{9}$$

In this case the integrations over $\mathrm{d}\Omega_a$ and $\mathrm{d}\Omega_e$ are not separable. Therefore, when absorption polarization ratios are calculated, terms with $P_e(\theta_e, \phi_e, \hat{\alpha})$ do not cancel as in the static and fast wobble cases. Analogously, when emission polarization ratios are calculated, terms with $P_a(\theta_a, \phi_a, \hat{\varepsilon})$ do not cancel. In contrast to the static fluorophore and fast wobble cases (Eqs. (2) and (8)), the absorption polarization ratios for slow wobble depend on the detector polarization, and the emission polarization ratios depend on the excitation polarization. This property, and Eq. (9), also describe fluorescence polarization data for a static distribution of many fluorophores. Slow wobble thus makes a single molecule appear to have polarization properties similar to those of a bulk sample.

Again, taking x and y polarized excitations as an example, the absorption polarization ratio, analogous to that in Eqs. (2) and (8), is given by

$$\begin{aligned} Q_{\alpha} &= \frac{{}_{y}I_{\alpha} - {}_{x}I_{\alpha}}{{}_{y}I_{\alpha} + {}_{x}I_{\alpha}} \\ &= \frac{\iint \rho_s(\theta_a, \phi_a, \theta_e, \phi_e)(\sin^2\theta_a)(\sin^2\phi_a - \cos^2\phi_a) P_e(\theta_e, \phi_e, \hat{\alpha}) \mathrm{d}\Omega_a \, \mathrm{d}\Omega_e}{\iint \rho_s(\theta_a, \phi_a, \theta_e, \phi_e)(\sin^2\theta_a) P_e(\theta_e, \phi_e, \hat{\alpha}) \, \mathrm{d}\Omega_a \, \mathrm{d}\Omega_e}. \end{aligned} \tag{10}$$

Note that here, the subscript, α, has been added to the polarization ratio, Q. This is necessary since, as explained above, the value of this *absorption* polarization ratio, is dependent on the orientation, $\hat{\alpha}$, of the *detector* polarizer, which determines $P_e(\theta_e, \phi_e, \hat{\alpha})$. Q_x measured from x-polarized detection does not equal Q_y measured from y-polarized detection. A simple check, then, to evaluate the extent of slow motion in a single molecule system, which follows from this analysis, is to measure a pair of absorption or emission polarization ratios (e.g. $Q_x$and Q_y or ${}_xP$ and ${}_yP$). The pair will differ only if slow wobble is present.

3.3. *Measurement of the axial angle, θ*

Wide-field epiillumination and laser spot-confocal excitation cannot readily be used to determine the axial angle of the absorption transition dipole moment, θ_a. The difficulty is that for an excitation laser beam propagating along the optical axis of the microscope, the polarization at the center of the beam is in the plane of the microscope stage (Axelrod, 1989b). The intensities in Eqs. (1) and (3) are sensitive to $\sin^2\theta_a$ but they are also proportional to a number of other physical parameters, such as intensity of the laser, absorption cross section, etc. Because these factors are difficult to quantify, θ_a cannot be inferred directly. The excitation polarization ratio (Eq. (2)) is completely independent of θ_a. The use of a high numerical aperture objective results in some dependence of emission polarization ratios on θ_e, as shown in Eq. (5). Introduction of motion may also introduce some θ_a or θ_e dependence of absorption and emission polarization ratios. However with axially propagating excitation, the sensitivities of polarized fluorescence ratios to θ_a and θ_e are low.

To get around this problem, some studies have made use of specialized fluorophores with multiple degenerate transition dipole moments distributed isotropically in a plane perpendicular to a symmetry axis (Bopp et al., 1997, 1999; Empedocles et al., 1999). With these unusual fluorophores, both axial and azimuthal angles of the symmetry axis can be determined with wide-field epiillumination or spot-confocal excitation.

A more direct solution, that applies to conventional fluorophores with linear, non-degenerate, transition dipole moments, is to utilize excitation with a large component of electric field along the optical axis of the microscope (i.e. along the z-axis; Betzig and Chichester, 1993; Dickson et al., 1998; Forkey et al., 1999). The evanescent waves generated by total internal reflection and near-field scanning optical probes have polarization with large z-components. For a static molecule excited with a z-polarized electric field, the intensity expression, analogous to Eq. (1) is

$$ {}_zI \propto \cos^2\theta_a \tag{11} $$

and an absorption polarization ratio, analogous to that in Eq. (2) can be calculated. For example

$$ Q' \equiv \frac{{}_zI - {}_xI}{{}_zI + {}_xI} = \frac{\cos^2\theta_a - \sin^2\theta_a\cos^2\phi_a}{\cos^2\theta_a + \sin^2\theta_a\cos^2\phi_a} = \frac{1 - \tan^2\theta_a\cos^2\phi_a}{1 + \tan^2\theta_a\cos^2\phi_a}. \tag{12} $$

This polarization ratio is dependent on the axial angle, θ_a, and, therefore, may be used in conjunction with another measurement, such as Q in Eq. (2) to determine both polar angles of the absorption dipole.

The evanescent wave generated by total internal reflection is an excellent source for this type of measurement. Expressions for the evanescent wave electric fields generated by total internal reflection of p- and s-polarized laser beams (see Section 2.2.2) at the interface between a microscope slide and an aqueous sample chamber are (Axelrod et al., 1984; Thompson et al., 1984):

$$ \mathbf{E}_p = \begin{bmatrix} \hat{x}2A_p\cos\theta_i\sin\delta_p\exp(-i\frac{\pi}{2})\exp(-i\delta_p) \\ +\hat{z}2A_p(1/\xi^2)\sin\theta_i\cos\delta_p\exp(-i\delta_p) \end{bmatrix} \exp(-z/d)\exp(-i(\omega t - n_1k_0x\sin\theta_i)) $$

$$ \mathbf{E}_s = [\hat{y}2A_s\cos\delta_s\exp(-i\delta_s)]\exp(-z/d)\exp(-i(\omega t - n_1k_0x\sin\theta_i)) $$

$$ \xi = \frac{n_2}{n_1}, \quad d = \left(n_1k_0\sqrt{\sin^2\theta_i - \xi^2}\right)^{-1}, \quad \theta_i > \theta_c \equiv \sin^{-1}(\xi), \tag{13} $$

$$ \delta_p = \tan^{-1}\left[\frac{\sqrt{\sin^2\theta_i - \xi^2}}{\xi^2\cos\theta_i}\right], \quad \delta_s = \tan^{-1}\left[\frac{\sqrt{\sin^2\theta_i - \xi^2}}{\cos\theta_i}\right], $$

where the actual electric fields are given by the real parts of the expressions for $\mathbf{E}_p$ and $\mathbf{E}_s$. These equations assume that the scattering plane is the x–z plane (see Fig. 4b). A_p and A_s are the electric field amplitudes of the p- and s-wave laser beams, respectively. θ_i is the angle of the incident laser beam relative to the direction normal to the interface and θ_c is the critical angle, beyond which total internal reflection occurs. n_1 is the index of refraction on the side of the interface with the propagating laser beam, and n_2 is the refractive index on the side of the interface with the evanescent wave. d represents the $1/e$ decay distance of the amplitude of the evanescent wave. ω is the radial frequency, and k_o the vacuum wave number of the excitation laser beam.

For a quartz slide ($n_1 = 1.46$)/water ($n_2 = 1.33$) interface, total internal reflection of an argon ion laser beam with wavelength 514 nm occurs for incident angles greater than the critical angle of $\sim 66°$. For an incident angle, θ_i, of 70° an evanescent wave is generated with amplitude decay distance, d, of 240 nm (intensity decay distance, $d/2 = 120$ nm). The evanescent wave generated by an s-polarized incident beam is linearly polarized along the y-direction and has intensity $|\mathbf{E}_s \cdot \hat{y}|^2 = 2.7A_s^2$ at the slide surface; the evanescent wave generated by a p-polarized incident beam is elliptically polarized in the x–z plane with $|\mathbf{E}_p \cdot \hat{z}|^2 = 3.1A_p^2$ and $|\mathbf{E}_p \cdot \hat{x}|^2 = 0.2A_p^2$ at the slide surface. For the p-generated wave, the component of the electric field along the x-direction can easily be taken into account in expressions for the polarization ratios. However, the x-component of the electric field is so much smaller than the z-component, that the excitation probability is dominated by the z-component of the absorption dipole. This makes total internal reflection a particularly useful excitation source for determining the full three-dimensional orientation of single fluorescent molecules.

4. Applications

The studies summarized in this section make use of the experimental and analytical methods described above. In contrast to many of the early experiments, those described here apply to functioning biophysical systems in aqueous media.

Using a laser spot-confocal apparatus, Ha et al. (1996) determined the orientation of single DNA-bond fluorophore dipoles projected onto the x–y plane. With data averaged over ~ 4 s they achieved an impressive angular resolution of 0.2°. The data enabled unambiguous differentiation between molecular rotations, and spectral jumps or transitions to non-fluorescent states. While these initial measurements were made at a slide/air interface, subsequent studies incorporated an aqueous medium. Ha et al. (1998) resolved the orientation of the transition dipole moment of single Texas Red probes covalently linked to short single-stranded DNA (ssDNA) bound to silanized glass. The DNA tethered the fluorophore in the aqueous solution, close to the slide while allowing rotation. The linear polarization angle of the fluorescence excitation was swept through a 120° range in the plane of the microscope stage during a period of ~ 50 ms, and the fluorescence emission was resolved into orthogonal components by a polarizing beam splitter. Averages over several angle scans (Fig. 6) were analyzed using a limited diffusion model (similar to Eq. (6) above) and assuming that the absorption and emission dipoles were colinear. Analysis of the experiments yielded the average orientation and extent of motion projected onto the x–y plane, as well as the rotational diffusion rate. Individual fluorophores were observed to "jump" repeatedly from a rotating state to a fixed orientation and vice versa, suggesting that individual probes repeatedly adsorbed to the glass substrate and were released from it. Some molecules which exhibited multiple jumps returned to fixed orientations, indicating preferential binding sites on the substrate. Further experiments with DNA may reveal new aspects

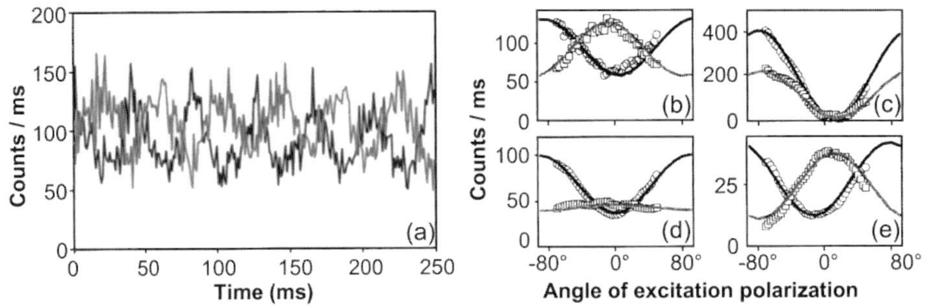

Fig. 6. SMFP experiments with high angular resolution. Fluorescence from single Texas Red molecules was excited by polarized spot-confocal excitation. The linear polarization of the excitation beam was repeatedly rotated through 120° in the *x–y* plane. Fluorescence was split into its two orthogonal polarization components by a polarizing beam splitter. (a) Raw data shows the two intensities (indicated by black and gray traces) changing with excitation polarization. The orthogonally polarized signals are out of phase as expected for a freely rotating molecule. (b–e) Time-averaged data (symbols) and fitted curves based on a limited rotational diffusion model. Amplitudes and relative phases indicate mean probe orientation and degree of rotational freedom. (Modified with permission from Ha et al., 1998.).

of its physical properties and its interactions with other macromolecules. Using a similar technique, structural changes of staphylococcal nuclease upon binding of an inhibitor, have also been detected (Ha et al., 1999b).

Schmidt et al. (1995) interrogated motions of individual tetramethylrhodamine (TMR)-labeled 1-2-dihexadecanoyl-*sn*-glycero-3-phosphoethanolamine (DHPE) molecules in bilayer membranes. They demonstrated tracking, with 10 ms time resolution, of single DHPE molecules diffusing in 1-palmitoyl-2-oleoyl-*sn*-glycero-3-phosphocholine (POPC) membranes. In further experiments, SMFP was used to detect rotational mobility (Schütz et al., 1997). Polarization of the laser excitation beam was switched at 30 Hz, by an electro-optic modulator, between two discrete orientations in the sample plane and absorption polarization ratios (Q given in Eqs. (2), (8) and (10)) were calculated. For TMR-labeled DHPE in POPC membranes, the average absolute value of Q ($|\bar{Q}|$) was 0.14. This value indicates virtually free rotational mobility on the submillisecond time scale. In contrast, DHPE embedded in a solid membrane composed of 1,2-dipalmitoyl-*sn*-glycero-3-phosphocholine (DPPC) gave $|\bar{Q}| = 0.66$, consistent with immobile fluorophores randomly oriented in the plane of the bilayer. Similar experiments with TMR- and Cy7-labeled lipid molecules in DPPC membranes allowed sequential measurements of orientation to determine the trajectory of slow motion (Harms et al., 1999). The extent of rotational mobility of the probe molecules agrees with earlier studies on bulk samples, thus indicating the potential for further investigations of individual macromolecules embedded in membranes.

The study of motor proteins is a natural application of SMFP because rotational motions are prevalent and in vitro assays of their function have already been well characterized (Scholey, 1993; Vallee, 1998). Sase et al. (1997) monitored the emission polarization ratio (P, Eq. (5)) from single TMR molecules conjugated to actin filaments. A 30 μm diameter region of the microscope slide was illuminated by circularly polarized wide-field epiillumination. As the filament was actively propelled by myosin, P oscillated between ca. –0.5 and +0.5 on a $\frac{1}{2}$ s timescale (Fig. 7). This experiment elegantly demonstrated rotation of the actin filament about its axis during translation. Because the observed pitch of rotation was more than 10-fold greater than that of the actin helix,

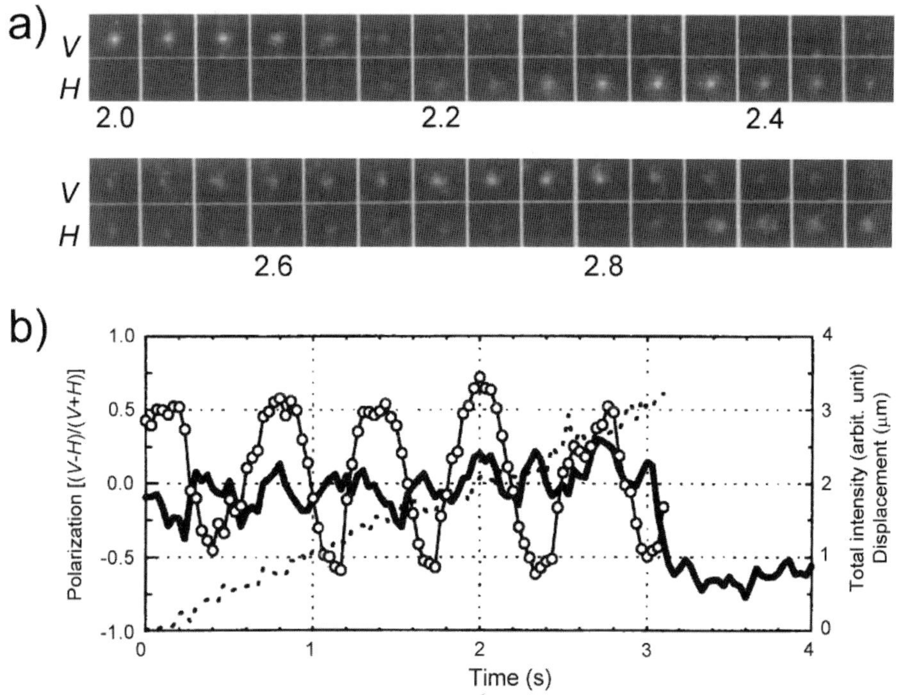

Fig. 7. Twisting of actin during translation in an in vitro actomyosin motility assay. (a) Polarized fluorescence images of an individual fluorophore attached to actin. The experiments used wide-field laser epiillumination with circularly polarized excitation, and the imaging system shown in Fig. 5b. V and H indicate vertically and horizontally polarized emission. Images were collected at 33-ms intervals. Alternating intensity levels in the two sequences of polarized images demonstrate that the actin filament rotated about its axis as it translated. (b) Total intensity of a spot during translocation ($I = V + H$, solid line); emission polarization ratio ($P = (V - H)/(V + H)$, open circles); displacement of the spot from its position at $t = 0$ (broken line). (Reproduced with permission from Sase et al., "Axial rotation of sliding actin filaments revealed by single-fluorophore imaging", Proceedings of the National Academy of Sciences, USA 94, 5646–5650. Copyright (1994) National Academy of Sciences, USA).

these results indicate, in agreement with other studies, that the myosin is not rigidly constrained along the actin helix.

Warshaw et al. (1998) detected motions of TMR bound to myosin RLC (see Fig. 1c) in an in vitro motility assay. They excited the sample with circularly polarized light in a laser spot-confocal microscope, resolved emission from the probes into orthogonal components with a polarizing beam splitter and collected the fluorescence with two APDs at 1–10 ms time resolution. The intensity signals exhibited discrete, reversible transitions in the presence of ATP and actin. The duration of these events versus ATP concentration indicated that they were due to actomyosin association and dissociation during the force-generating enzymatic cycle. An emission polarization ratio similar to the P ratio (Eq. (5)) indicated extensive mobility of the RLC when the myosin was dissociated from actin and discrete orientations during association. The polarization ratio was constant during an event. This result indicates that attached 'pre-force' states are either disordered like detached ones or are occupied for a period shorter than the time resolution.

In a similar experiment, Quinlan et al. (1999) used myosin RLC labeled with bifunctional rhodamine (see Section 3.1.2 on Fluorescent Probes above and Fig. 2a). Evanescent waves with polarizations along the y-axis and predominantly along the z-axis were generated by total internal reflection. The polarization of the input laser beam was switched by an electro-optic modulator at 50 Hz. The fluorescent emission was resolved into its components polarized along the x- and y-axis by a beam splitting prism and detected by two APDs (Forkey et al., 1999). As discussed above, this excitation scheme enables resolution of probe orientation in the x–y plane (ϕ) and along the z-axis (θ, Fig. 4c). Discrete changes of intensities were observed, often in opposite directions, indicating sudden changes of orientation and mobility of the RLC during actomyosin motility (Fig. 8). Coumarin probes attached to the actin filaments were also excited by an ultraviolet evanescent wave to select myosins colocalized with actin and to provide an angular reference axis. The bifunctional probe potentially enables protein orientation to be quantified

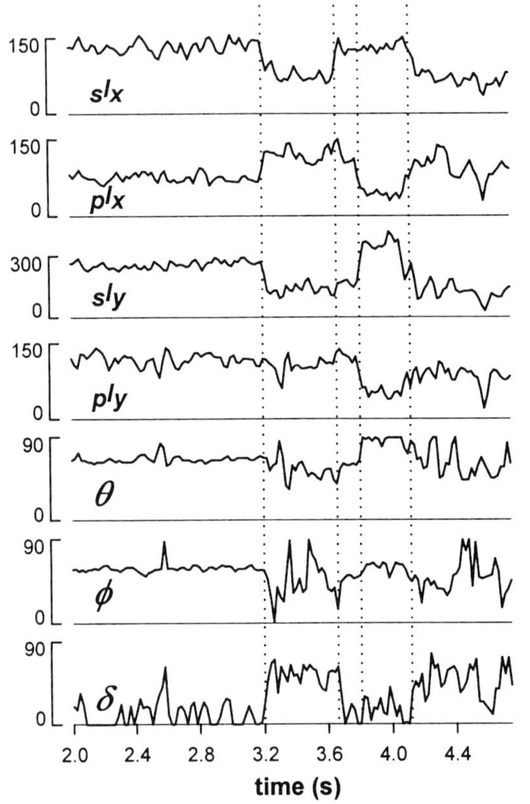

Fig. 8. Polarized fluorescence intensities and fitted orientation parameters for a single bifunctional rhodamine probe. The fluorophore was bound to the regulatory light chain of rabbit skeletal muscle myosin in an in vitro motility assay. The excitation source was an evanescent wave generated by total internal reflection (Fig. 4b). Time multiplexed s- and p-polarized excitation (see text), and two detector polarizations (Fig. 5a) yielded four independent intensities: ${}_{s}I_{x}$, ${}_{p}I_{x}$, ${}_{s}I_{y}$, and ${}_{p}I_{y}$. Intensities are plotted as number of photons detected during two 10 ms periods with constant excitation polarization. Axial angle (θ), azimuthal angle (ϕ) and a mobility parameter (δ) are given in degrees. θ is the half-angle of a cone describing the extent of mobility on the 'slow wobble' time scale (see Section 3). Time is indicated in seconds relative to the start of excitation. Values of the measured parameters between 0 and 2 s were approximately constant. The abrupt changes in intensities, highlighted by the dotted lines, indicate changes in orientation and extent of mobility of the myosin light chain domain.

from the probe orientation in relation to this reference axis as was reported in bulk experiments with the actomyosin system (Corrie et al., 1999).

A third research group applying SMFP to myosin-based motility used evanescent wave excitation through a high numerical aperture microscope objective (Saito et al., 1997). In this case, the propagation direction of a totally reflected, s-polarized incident beam was switched between x–z and y–z scattering planes by alternating (at 10 Hz) its position at the objective back focal plane. These two excitation modes generated evanescent wave polarizations along the y- and x-axis, respectively. Total fluorescence intensity was monitored from single probe molecules bound to the motor domain of myosin (see Fig. 1c) or to the RLC. Absorption polarization ratios indicated that fluorophores bound to the motor domain were less mobile than those bound to the RLC.

While the studies described thus far all made use of extrinsic fluorophores attached to the biological molecule of interest, intrinsic fluorophores may also be used. This was demonstrated by Bopp et al. (1997) who measured the molecular orientation of the light harvesting complex (LH-2) from the purple bacterium, *Rhodopseudomonas acidophila*, using a spot-confocal microscope. Instead of a discrete absorption dipole as in the previously discussed systems, LH-2 contains a set of fluorophores arranged around a molecular symmetry axis. Measurements by Bopp et al. using two orthogonal excitation polarizations in the plane of the microscope slide, were consistent with a homogeneous surface distribution of cyllindrically symmetric LH-2 complexes with their axes tilted 20–30° relative to the substrate normal. More recently, the excitation polarization was swept through a range of orientations on a time scale faster than 100 ms (Bopp et al., 1999). These measurements indicated that the molecules are not circularly symmetric, but can be described as elliptic absorbers. Discrete jumps in ellipticity and orientation of the major axes found in the 10 s time domain indicated structural deformations.

5. Conclusions

The technology, components and analytical tools are now readily available to perform biologically relevant SMFP experiments. Nevertheless, the method has not yet been widely applied partly because it is new, but also because significant hurdles limit the preparation and characterization of the experimental materials. Functioning samples that can be integrated into SMFP instruments have been designed for several macromolecular systems. Most of these involve attachment of the enzyme complex or lipid bilayer to the surface of the microscope slide necessitating careful evaluation of functional integrity under these conditions. Measurements of fluorescence polarization from bound probes reveal the mobility of the labeled component in various occupied states and kinetics of transitions between them. Quantifying domain orientation from the probe angles, however, requires knowledge of the local probe orientation within the macromolecule. Determination of these angles has been accomplished in only a few cases. Availability of new probes capable of labeling proteins with predetermined orientation, improved methods for attachment without compromising activity and innovations in the assay methods should promote application of SMFP to studies of many more biophysical systems.

Further development of SMFP will increase the accessible information. For example, the use of multiple, spectrally distinct fluorophores would allow simultaneous orientation measurements on several domains. Also, continued advancement of video cameras may enhance polarization measurements on spatially separated molecules. Integration of SMFP with other single molecule techniques will enable direct correlation of structural changes with other processes. Combination with optical traps would enable correlation between structural changes and mechanical events such as force production or movement. Combination of SMFP with detection of ligand binding and dissociation would provide the timing of conformational changes relative to these chemical reactions. With continued development and implementation, SMFP is expected to play a major role in elucidating the dynamic relationships between structural changes in macromolecular machines and their functional output.

Acknowledgements

This work was supported by the NIH, MDA, AHA and HHMI. We thank Drs. John E. T. Corrie, Enrique M. De La Cruz, Carol J. Deutsch, Dennis E. Discher, Uulke van der Heide, Michael Ostap, Brian M. Salzberg and Mr. Francesco Vanzi for helpful discussions and comments on the manuscript.

Appendix A. Dependence of polarized fluorescence intensity on molecular orientation

In this appendix, equations are presented that show the dependence on molecular orientation of single-molecule polarized fluorescence intensities. The molecule is presumed to have well defined, linear, absorption and emission transition dipole moments. The fluorescence from a single molecule on a microscope slide is collected and collimated by a microscope objective having a high numerical aperture. The right handed Cartesian coordinate system, described by the unit vectors $\hat{x}, \hat{y}$ and $\hat{z}$ (Fig. 4c), has its origin on the slide surface. The positive z-axis coincides with the optical axis of the microscope, and points toward the objective.

A.1. Excitation of a single fluorophore

The oscillating electric field $\mathbf{E}$ at a given point for any of the excitation schemes used in single-molecule experiments may be written in the form

$$\mathbf{E} = E\hat{\varepsilon}\mathrm{e}^{-\mathrm{i}\omega t},$$

$$\hat{\varepsilon} = \left[\hat{x}\,\varepsilon_x \mathrm{e}^{-\mathrm{i}\delta_x} + \hat{y}\varepsilon_y \mathrm{e}^{-\mathrm{i}\delta_y} + \hat{z}\varepsilon_z \mathrm{e}^{-\mathrm{i}\delta_z}\right], \tag{A.1}$$

where the actual electric field is given by the real part of this complex expression. Here, E is the magnitude of the electric field, ω is the radial frequency of the field, and t is time. $\hat{\varepsilon}$ describes the polarization. ε_j and δ_j represent the relative magnitude and phase, respectively, of the jth component of the electric field and the ε_j values have been normalized so that $|\hat{\varepsilon}| = \sqrt{\varepsilon_x^2 + \varepsilon_y^2 + \varepsilon_z^2} = 1$.

For an individual molecule, the orientation of its absorption transition dipole moment may be described either by axial and azimuthal polar angles, θ_a and ϕ_a, respectively, or equivalently by the unit vector $\hat{\mu}_a = \hat{x}\mu_{ax} + \hat{y}\mu_{ay} + \hat{z}\mu_{az}$, where $\mu_{ax} = \sin\theta_a \cos\phi_a$, $\mu_{ay} = \sin\theta_a \sin\phi_a$, and $\mu_{az} = \cos\theta_a$. The probability per unit time of exciting this molecule is proportional to the square of the magnitude of the projection of the electric field along the direction of the absorption dipole. In particular, the dependence of the excitation probability on the electric field polarization, $\hat{\varepsilon}$, and molecular orientation, $\hat{\mu}_a$, is given by $P_a(\hat{\mu}_a, \hat{\varepsilon})$ which is defined as

$$
\begin{aligned}
P_a(\hat{\mu}_a, \hat{\varepsilon}) &\equiv |\hat{\mu}_a \cdot \hat{\varepsilon}|^2 \\
&= \left(\varepsilon_x \mu_{ax} \cos\delta_x + \varepsilon_y \mu_{ay} \cos\delta_y + \varepsilon_z\, \mu_{az} \cos\delta_z\right)^2 \\
&\quad + \left(\varepsilon_x \mu_{ax} \sin\delta_x + \varepsilon_y \mu_{ay} \sin\delta_y + \varepsilon_z\, \mu_{az} \sin\delta_z\right)^2,
\end{aligned}
\tag{A.2}
$$

where the dependence of μ_{ax}, μ_{ay} and μ_{az} on (θ_a, ϕ_a) has been given above.

Eq. (A.2) may often be reduced to a relatively simple expression depending on the particular mode of excitation. A number of examples are:

Epifluorescence (wide-field or spot-confocal) with electric field linearly polarized along the x-axis:

$$\varepsilon_x = 1,\ \varepsilon_y = \varepsilon_z = 0,$$

$$P_a(\hat{\mu}_a, \hat{\varepsilon}) = \mu_{ax}^2 = \sin^2\theta_a \cos^2\phi_a;$$

Epifluorescence with electric field linearly polarized along the y-axis:

$$\varepsilon_y = 1, \quad \varepsilon_x = \varepsilon_z = 0,$$

$$P_a(\hat{\mu}_a, \hat{\varepsilon}) = \mu_{ay}^2 = \sin^2\theta_a \sin^2\phi_a;$$

Epifluorescence with electric field linearly polarized in the x–y plane and making an angle ζ relative to the x-axis:

$$\varepsilon_x = \cos\zeta, \quad \varepsilon_y = \sin\zeta, \quad \varepsilon_z = 0, \quad \delta_x = \delta_y,$$

$$
\begin{aligned}
P_a(\hat{\mu}_a, \hat{\varepsilon}) &= \left(\mu_{ax}\cos\zeta + \mu_{ay}\sin\zeta\right)^2 \\
&= (\sin\theta_a \cos\phi_a \cos\zeta + \sin\theta_a \sin\phi_a \sin\zeta)^2 = \sin^2\theta_a \cos^2(\zeta - \phi_a);
\end{aligned}
$$

Epifluorescence with a circularly polarized electric field in the x–y plane:

$$\varepsilon_x = \varepsilon_y = 1/\sqrt{2}, \quad \varepsilon_z = 0, \quad \delta_y = \delta_x \pm \pi/2$$

$$P_a(\hat{\mu}_a, \hat{\varepsilon}) = \tfrac{1}{2}\left(\mu_{ax}^2 + \mu_{ay}^2\right) = \tfrac{1}{2}\sin^2\theta_a;$$

Evanescent wave generated by total internal reflection of an s-polarized excitation laser beam propagating in the x–z plane (see Section 3.3):

$$\varepsilon_x = \varepsilon_z = 0, \quad \varepsilon_y = 1,$$

$$P_a(\hat{\mu}_a, \hat{\varepsilon}) = \mu_{ay}^2 = \sin^2\theta_a \sin^2\phi_a;$$

Evanescent wave generated by total internal reflection of a p-polarized excitation laser beam propagating in the x–z plane (see Section 3.3):

$$\varepsilon_x = \sqrt{\kappa},\ \varepsilon_y = 0,\ \varepsilon_z = \sqrt{1-\kappa},\ \delta_x = \delta_z + \pi/2\ \ (0<\kappa \ll 1).$$

$$P_{\mathrm{a}}(\hat{\mu}_{\mathrm{a}}, \hat{\varepsilon}) = \kappa\mu_{\mathrm{ax}}^2 + (1-\kappa)\mu_{\mathrm{az}}^2 = \kappa \sin^2\theta_{\mathrm{a}} \cos^2\phi_{\mathrm{a}} + (1-\kappa)\cos^2\theta_{\mathrm{a}}.$$

A.2. Fluorescence from a single fluorophore

When a molecule fluoresces, the probability of any emitted photon propagating in some direction, $\hat{k}$, is proportional to the square of the magnitude of the induced electric field propagating in that direction. At locations much farther from the molecule than the wavelength (i.e. far field), this electric field is given by

$$\mathbf{E}_{\mathrm{dipole}} = K'\left[\hat{\mu}_{\mathrm{e}} - \left(\hat{\mu}_{\mathrm{e}} \cdot \hat{k}\right)\hat{k}\right]\mathrm{e}^{-\mathrm{i}\omega t}, \tag{A.3}$$

where $\hat{\mu}_{\mathrm{e}} = \hat{x}\mu_{\mathrm{ex}} + \hat{y}\mu_{\mathrm{ey}} + \hat{z}\mu_{\mathrm{ez}}$ is a unit vector in the direction of the molecule's emission transition dipole moment (Born and Wolf, 1964, Section 2.2). K' is a scale factor that is independent of molecular orientation but depends on other physical parameters such as excitation rate and fluorescence quantum efficiency. The magnitude squared of this field is proportional to $1-(\hat{\mu}_{\mathrm{e}} \cdot \hat{k})^2$ which equals $\sin^2\chi$, where χ is the angle between $\hat{\mu}_{\mathrm{e}}$, and the propagation direction, $\hat{k}$.

Because the fluorescence in this far-field region is a transverse propagating wave, it is polarized in the plane perpendicular to $\hat{k}$. The intensity of the fluorescence with polarization along some specific direction, $\hat{A}$, perpendicular to $\hat{k}$ (i.e. $\hat{A} \cdot \hat{k} = 0$), is proportional to the square of the projection of $\mathbf{E}_{\mathrm{dipole}}$ onto $\hat{A}$, as given by

$$P'_{\mathrm{e}}(\hat{\mu}_{\mathrm{e}}, \hat{A}) \equiv \left|\left[\hat{\mu}_{\mathrm{e}} - \left(\hat{\mu}_{\mathrm{e}} \cdot \hat{k}\right)\hat{k}\right] \cdot \hat{A}\right|^2 = \left|\hat{\mu}_{\mathrm{e}} \cdot \hat{A}\right|^2. \tag{A.4}$$

For a system that uses a high numerical aperture objective lens, the polarization analyzer is placed after the objective, where the collected light is approximately collimated along the z-axis. Integrating Eq. (A.4) over all $\hat{k}$ directions subtended by the objective, and taking into account the collimation of the fluorescence, yields the following probability function for detecting a photon after the objective, with the analyzer along $\hat{\alpha}$, in the x–y plane:

$$P_{\mathrm{e}}(\hat{\mu}_{\mathrm{e}}, \hat{\alpha}) = C_1\left(\mu_{\mathrm{ex}}\cos\eta + \mu_{\mathrm{ey}}\sin\eta\right)^2 + C_2\left(\mu_{\mathrm{ex}}\sin\eta - \mu_{\mathrm{ey}}\cos\eta\right)^2 + C_3\mu_{\mathrm{ez}}^2, \tag{A.5}$$

where η is the angle between $\hat{\alpha}$ and the x-axis, defined to be positive for rotation from the positive x-axis towards the positive y-axis. The formulas for C_1, C_2 and C_3 that express mixing of polarizations between the orthogonal directions due to the high numerical aperture have been derived elsewhere for use in the analysis of polarization microscopy measurements on bulk samples (Axelrod, 1989b):

$$C_1 = (1/8)\left[5 - 3\cos(\delta_{\mathrm{c}}) - \cos^2(\delta_{\mathrm{c}}) - \cos^3(\delta_{\mathrm{c}})\right]\left[1-\cos(\delta_{\mathrm{c}})\right]^{-1},$$

$$C_2 = (1/24)\left[1 - 3\cos(\delta_{\mathrm{c}}) + 3\cos^2(\delta_{\mathrm{c}}) - \cos^3(\delta_{\mathrm{c}})\right]\left[1-\cos(\delta_{\mathrm{c}})\right]^{-1},$$

$$C_3 = (1/6)\left[2 - 3\cos(\delta_{\mathrm{c}}) + \cos^3(\delta_{\mathrm{c}})\right]\left[1-\cos(\delta_{\mathrm{c}})\right]^{-1}. \tag{A.6}$$

Evanescent wave generated by total internal reflection of a p-polarized excitation laser beam propagating in the x–z plane (see Section 3.3):

$$\varepsilon_x = \sqrt{\kappa},\ \varepsilon_y = 0,\ \varepsilon_z = \sqrt{1-\kappa},\ \delta_x = \delta_z + \pi/2\ \ (0 < \kappa \ll 1).$$

$$P_{\mathrm{a}}(\hat{\mu}_{\mathrm{a}}, \hat{\varepsilon}) = \kappa\mu_{\mathrm{ax}}^2 + (1-\kappa)\mu_{\mathrm{az}}^2 = \kappa \sin^2\theta_{\mathrm{a}} \cos^2\phi_{\mathrm{a}} + (1-\kappa)\cos^2\theta_{\mathrm{a}}.$$

A.2. Fluorescence from a single fluorophore

When a molecule fluoresces, the probability of any emitted photon propagating in some direction, $\hat{k}$, is proportional to the square of the magnitude of the induced electric field propagating in that direction. At locations much farther from the molecule than the wavelength (i.e. far field), this electric field is given by

$$\mathbf{E}_{\mathrm{dipole}} = K'\left[\hat{\mu}_{\mathrm{e}} - \left(\hat{\mu}_{\mathrm{e}}\cdot\hat{k}\right)\hat{k}\right]\mathrm{e}^{-\mathrm{i}\omega t}, \tag{A.3}$$

where $\hat{\mu}_{\mathrm{e}} = \hat{x}\mu_{\mathrm{ex}} + \hat{y}\mu_{\mathrm{ey}} + \hat{z}\mu_{\mathrm{ez}}$ is a unit vector in the direction of the molecule's emission transition dipole moment (Born and Wolf, 1964, Section 2.2). K' is a scale factor that is independent of molecular orientation but depends on other physical parameters such as excitation rate and fluorescence quantum efficiency. The magnitude squared of this field is proportional to $1-(\hat{\mu}_{\mathrm{e}}\cdot\hat{k})^2$ which equals $\sin^2\chi$, where χ is the angle between $\hat{\mu}_{\mathrm{e}}$, and the propagation direction, $\hat{k}$.

Because the fluorescence in this far-field region is a transverse propagating wave, it is polarized in the plane perpendicular to $\hat{k}$. The intensity of the fluorescence with polarization along some specific direction, $\hat{A}$, perpendicular to $\hat{k}$ (i.e. $\hat{A}\cdot\hat{k} = 0$), is proportional to the square of the projection of $\mathbf{E}_{\mathrm{dipole}}$ onto $\hat{A}$, as given by

$$P'_{\mathrm{e}}\left(\hat{\mu}_{\mathrm{e}}, \hat{A}\right) \equiv \left|\left[\hat{\mu}_{\mathrm{e}} - \left(\hat{\mu}_{\mathrm{e}}\cdot\hat{k}\right)\hat{k}\right]\cdot\hat{A}\right|^2 = \left|\hat{\mu}_{\mathrm{e}}\cdot\hat{A}\right|^2. \tag{A.4}$$

For a system that uses a high numerical aperture objective lens, the polarization analyzer is placed after the objective, where the collected light is approximately collimated along the z-axis. Integrating Eq. (A.4) over all $\hat{k}$ directions subtended by the objective, and taking into account the collimation of the fluorescence, yields the following probability function for detecting a photon after the objective, with the analyzer along $\hat{\alpha}$, in the x–y plane:

$$P_{\mathrm{e}}(\hat{\mu}_{\mathrm{e}}, \hat{\alpha}) = C_1\left(\mu_{\mathrm{ex}}\cos\eta + \mu_{\mathrm{ey}}\sin\eta\right)^2 + C_2\left(\mu_{\mathrm{ex}}\sin\eta - \mu_{\mathrm{ey}}\cos\eta\right)^2 + C_3\mu_{\mathrm{ez}}^2, \tag{A.5}$$

where η is the angle between $\hat{\alpha}$ and the x-axis, defined to be positive for rotation from the positive x-axis towards the positive y-axis. The formulas for C_1, C_2 and C_3 that express mixing of polarizations between the orthogonal directions due to the high numerical aperture have been derived elsewhere for use in the analysis of polarization microscopy measurements on bulk samples (Axelrod, 1989b):

$$C_1 = (1/8)\left[5 - 3\cos(\delta_{\mathrm{c}}) - \cos^2(\delta_{\mathrm{c}}) - \cos^3(\delta_{\mathrm{c}})\right]\left[1-\cos(\delta_{\mathrm{c}})\right]^{-1},$$

$$C_2 = (1/24)\left[1 - 3\cos(\delta_{\mathrm{c}}) + 3\cos^2(\delta_{\mathrm{c}}) - \cos^3(\delta_{\mathrm{c}})\right]\left[1-\cos(\delta_{\mathrm{c}})\right]^{-1},$$

$$C_3 = (1/6)\left[2 - 3\cos(\delta_{\mathrm{c}}) + \cos^3(\delta_{\mathrm{c}})\right]\left[1-\cos(\delta_{\mathrm{c}})\right]^{-1}. \tag{A.6}$$

Cantor, C.R., Schimmel, P.R., 1980. Biophysical Chemistry. Part II: Techniques for the Study of Biological Structure and Function. W.H. Freeman and Company, New York.
Carniglia, C.K., Mandel, L., Drexhage, K.H., 1972. Absorption and emission of evanescent photons. J. Opt. Soc. Am. 62, 479–486.
Chen, R.F., Bowman, R.L., 1965. Fluorescence polarization: measurement with ultraviolet-polarizing filters in a spectrophotofluorometer. Science 147, 729–732.
Clark, B.F.C., Thirup, S., Kjeldgaard, M., Nyborg, J., 1999. Structural information for explaining the molecular mechanism of protein biosynthesis. FEBS Lett. 452, 41–46.
Conibear, P.B., Kuhlman, P.A., Bagshaw, C.R., 1998. Measurement of ATPase activites of myosin at the level of tracks and single molecules. Adv. Exp. Med. Biol. 453, 15–26.
Corrie, J.E.T., Brandmeier, B.D., Ferguson, R.E., Trentham, D.R., Kendrick-Jones, J., Hopkins, S.C., van der Heide, U.A., Goldman, Y.E., Sabido-David, C., Dale, R.E., Criddle, S., Irving, M., 1999. Dynamic measurement of myosin light-chain-domain tilt and twist in muscle contraction. Nature 400, 425–430.
Corrie, J.E.T., Craik, J.S., Munasinghe, V.R.N., 1998. A homobifunctional rhodamine for labeling proteins with defined orientations of a fluorophore. Bioconjugate Chem. 9, 160–167.
Cubitt, A.B., Heim, R., Adams, S.R., Boyd, A.E., Gross, L.A., Tsien, R.Y., 1995. Understanding, improving and using green fluorescent proteins. TiBS 20, 448–455.
Dale, R.E., 1988. Some aspects of excited-state probe emission spectroscopy for structure and dynamics of model and biological membranes. In: Samori', B., Thulstrup, E.W. (Eds.), Polarized Spectroscopy of Ordered Systems. Kluwer Academic Publishers, Dordrecht, The Netherlands, pp. 491–567.
Dautet, H., Deschamps, P., Dion, B., MacGregor, A.D., MacSween, D., McIntyre, R.J., Trottier, C., Webb, P.P., 1993. Photon counting techniques with silicon avalanche photodiodes. Appl. Opt. 32, 3894–3900.
Dickson, R.M., Cubitt, A.B., Tsien, R.Y., Moerner, W.E., 1997. On/off blinking and switching behaviour of single molecules of green fluorescent protein. Nature 388, 355–358.
Dickson, R.M., Norris, D.J., Moerner, W.E., 1998. Simultaneous imaging of individual molecules aligned both parallel and perpendicular to the optic axis. Phys. Rev. Lett. 81, 5322–5325.
Dickson, R.M., Norris, D.J., Tzeng, Y.-L., Moerner, W.E., 1996. Three-dimensional imaging of single molecules solvated in pores of poly(acrylamide) gels. Science 274, 966–969.
Dominguez, R., Freyzon, Y., Trybus, K.M., Cohen, C., 1998. Crystal structure of a vertebrate smooth muscle myosin motor domain and its complex with the essential light chain: visualization of the pre-power stroke state. Cell 94, 559–571.
Doyle, D.A., Cabral, J.M., Pfuetzner, R.A., Kuo, A., Gulbis, J.M., Cohen, S.L., Chait, B.T., MacKinnon, R., 1998. The structure of the potassium channel: molecular basis of K^+ conduction and selectivity. Science 280, 69–77.
Dürig, U., Pohl, D.W., Rohner, F., 1986. Near-field optical-scanning microscopy. J. Appl. Phys. 59, 3318–3327.
Ehrhardt, A.G., Kang, H.C., Tuft, R.A., Fay, F.S., Ikebe, M., 1998. Labeling of calmodulin with a new bifunctional bodipy for improved fluorophore stability in anisotropy measurements. Biophys. J. 74, A380.
Empedocles, S.A., Neuhauser, R., Bawendi, M.G., 1999. Three-dimensional orientation measurements of symmetric single chromophores using polarization microscopy. Nature 399, 126–130.
Fass, D., Bogden, C.E., Berger, J.M., 1999. Quaternary changes in topoisomerase II may direct orthogonal movement of two DNA strands. Nat. Struct. Biol. 6, 322–326.
Forkey, J.N., Quinlan, M.E., Corrie, J.E.T., Goldman, Y.E., 1999. Single-molecule structural dynamics by fluorescence polarization microscopy. Biophys. J. 76, A20.
Funatsu, T., Harada, Y., Tokunaga, M., Saito, K., Yanagida, T., 1995. Imaging of single fluorescent molecules and individual ATP turnovers by single myosin molecules in aqueous solution. Nature 374, 555–559.
Goldman, Y.E., 1998. Wag the tail: structural dynamics of actomyosin. Cell 93, 1–4.
Griffin, B.A., Adams, S.R., Tsien, R.Y., 1998. Specific covalent labeling of recombinant protein molecules inside live cells. Science 281, 269–272.
Gutfreund, H., 1995. Kinetics for the Life Sciences. Cambridge University Press, New York.
Güttler, F., Croci, M., Renn, A., Wild, U.P., 1996. Single molecule polarization spectroscopy: pentacence in p-terphenyl. Chem. Phys. 211, 421–430.
Güttler, F., Sepiol, J., Plakhotnik, T., Mitterdorfer, A., Renn, A., Wild, U.P., 1993. Single molecule spectroscopy: fluorescence excitation spectra with polarized light. J. Lumin. 56, 29–38.

Ha, T., Enderle, Th., Chemla, D.S., Selvin, P.R., Weiss, S., 1996. Single molecule dynamics studied by polarization modulation. Phys. Rev. Lett. 77, 3979–3982.

Ha, T., Glass, J., Enderle, Th., Chemla, D.S., Weiss, S., 1998. Hindered rotational diffusion and rotational jumps of single molecules. Phys. Rev. Lett. 80, 2093–2096.

Ha, T., Laurence, T.A., Chemla, D.S., Weiss, S., 1999a. Polarization spectroscopy of single fluorescent molecules. J. Phys. Chem. B 103, 6839–6850.

Ha, T., Ting, A.Y., Liang, J., Caldwell, W.B., Deniz, A.A., Chemla, D.S., Schultz, P.G., Weiss, S., 1999b. Single-molecule fluorescence spectroscopy of enzyme conformational dynamics and cleavage mechanism. Proc. Natl. Acad. Sci. USA 96, 893–898.

Harms, G.S., Sonnleitner, M., Schütz, G.J., Gruber, H.J., Schmidt, Th., 1999. Single-molecule anisotropy imaging. Biophys. J. 77, 2864–2870.

Haugland, R.P., 1996. In: Spence, M.T.Z. (Ed.), Handbook of Fluorescent Probes and Research Chemicals, 6th Edition. Molecular Probes Inc, Eugene, OR, pp. 7–46.

Heim, R., Tsien, R.Y., 1996. Engineering green fluorescent protein for improved brightness, longer wavelengths, and fluorescence resonance energy transfer. Curr. Biol. 6, 178–182.

Hellen, E.H., Axelrod, D., 1987. Flourescence emission at dielectric and metal-film interfaces. J. Opt. Soc. Am. B. 4, 337–350.

Hirschfeld, T., 1976. Optical microscopic observation of single small molecules. Appl. Opt. 15, 2965–2966.

Hopkins, S.C., Sabido-David, C., Corrie, J.E.T., Irving, M., Goldman, Y.E., 1998. Fluorescence polarization transients from rhodamine isomers on the myosin regulatory light chain in skeletal muscle fibers. Biophys. J. 74, 3093–3110.

Houdusse, A., Kalabokis, V.N., Himmel, D., Szent-Gyorgyi, A.G., Cohen, C., 1999. Atomic structure of scallop myosin subfragment S1 complexed with MgADP: a novel conformation of the myosin head. Cell 97, 459–470.

Howard, J., Hunt, A.J., Baek, S., 1993. Assay of microtubule movement driven by single kinesin molecules. Methods Cell Biol. 39, 137–147.

Hubbell, W.L., Mchaourab, H.S., Altenbach, C., Lietzow, M.A., 1996. Watching proteins move using site-directed spin labeling. Structure 4, 779–783.

Iler, R.K., 1979. The Chemistry of Silica. Solubility, Polymerization, Colloid, and Surface Properties, and Biochemistry. Wiley, New York.

Inoué, S., Spring, K.R., 1997. Video Microscopy. The Fundamentals 2nd Edition.. Plenum Press, New York.

Jenkins, F.A., White, H.E., 1976. Fundamentals of Optics 4th Edition. McGraw-Hill, New York.

Johnson, I., 1998. Fluorescent probes for living cells. Histochem. J. 30, 123–140.

Kinosita Jr., K., Itoh, H., Ishiwata, S., Hirano, K., Nishizaka, T., Hayakawa, T., 1991. Dual-view microscopy with a single camera: real-time imaging of molecular orientations and calcium. J. Cell Biol. 115, 67–73.

Lakowicz, J.R., 1999. Principles of Fluorescence Spectroscopy, 2nd Edition. Plenum Press, New York.

Loots, E., Isacoff, E.Y., 1998. Protein rearrangements underlying slow inactivation of the Shaker K+ channel. J. Gen. Physiol. 112, 377–389.

Lu, H.P., Xun, L., Xie, X.S., 1998. Single-molecule enzymatic dynamics. Science 282, 1877–1882.

Macklin, J.J., Trautman, J.K., Harris, T.D., Brus, L.E., 1996. Imaging and time-resolved spectroscopy of single molecules at an interface. Science 272, 255–258.

Mannuzzu, L.M., Moronne, M.M., Isacoff, E.Y., 1996. Direct physical measure of conformational rearrangement underlying potassium channel gating. Science 271, 213–216.

Maxfield, F.R., 1989. Fluorescent analogs of peptides and hormones. Methods Cell Biol. 29, 13–28.

Mchaourab, H.S., Lietzow, M.A., Hideg, K., Hubbell, W.L., 1996. Motion of spin-labeled side chains in T4 lysozyme. Correlation with protein structure and dynamics. Biochemistry 35, 7692–7704.

Mendelson, R., Wilson, M.G.A., 1987. Chapter 4: Fluorescence polarization studies of myosin and muscle cross-bridges. In: Baskin, R.J., Yeh, Y. (Eds.), Optical Studies of Muscle Cross-Bridges. CRC Press, Boca Raton, FL, pp. 67–98.

Moerner, W.E., Kador, L., 1989. Optical detection and spectroscopy of single molecules in a solid. Phys. Rev. Lett. 62, 2535–2538.

Moerner, W.E., Orrit, M., 1999. Illuminating single molecules in condensed matter. Science 283, 1670–1676.

Mollaaghababa, R., Steinhoff, H-J., Hubbell, W.L., Khorana, H.G., 2000. Time-resolved site-directed spin-labeling studies of bacteriorhodopsin: loop-specific conformational changes in M. Biochemistry 39, 1120–1127.

Morikawa, K., Yanagida, M., 1981. Visualization of individual DNA molecules in solution by light microscopy: DAPI staining method. J. Biochem. 89, 693–696.

Nie, S., Zare, R.N., 1997. Optical detection of single molecules. Annu. Rev. Biophys. Biomol. Struct. 26, 567–596.

Orrit, M., Bernard, J., 1990. Single pentacene molecules detected by fluorescence excitation in a *p*-terphenyl crystal. Phys. Rev. Lett. 65, 2716–2719.

Orrit, M., Bernard, J., Personov, R.I., 1993. High-resolution spectroscopy of organic molecules in solids: from fluorescence line narrowing and hole burning to single molecule spectroscopy. J. Phys. Chem. 97, 10256–10268.

Ostap, E.M., White, H.D., Thomas, D.D., 1993. Transient detection of spin-labeled myosin subfragment 1 conformational states during ATP hydrolysis. Biochemistry 32, 6712–6720.

Pawley, J.B., 1995. Handbook of Biological Confocal Microscopy, 2nd Edition. Plenum Press, New York.

Pervushin, K., Riek, R., Wider, G., Wüthrich, K., 1997. Attenuated T_2 relaxation by mutual cancellation of dipole-dipole coupling and chemical shift anisotrophy indicates an avenue to NMR structures of very large biological macromolecules in solution. Proc. Natl. Acad. Sci. USA 94, 12366–12371.

Pierce, D.W., Hom-Booher, N., Vale, R.D., 1997. Imaging individual green fluorescent proteins. Nature 388, 338.

Quinlan, M.E., Forkey, J.N., Corrie, J.E.T., Goldman, Y.E., 1999. Tilting of the light chain region in single myosin molecules using total internal reflection fluorescence polarization microscopy. Biophys. J. 76, A165.

Rayment, I., Holden, H.M., Whittaker, M., Yohn, C.B., Lorenz, M., Holmes, K.C., Milligan, R.A., 1993a. Structure of the actin-myosin complex and its implications for muscle contraction. Science 261, 58–65.

Rayment, I., Rypniewski, W.R., Schmidt-Bäse, K., Smith, R., Tomchick, D.R., Benning, M.M., Winkelmann, D.A., Wesenberg, G., Holden, H.M., 1993b. Three-dimensional structure of myosin subfragment-1: a molecular motor. Science 261, 50–58.

Rodnina, M.V., Savelsbergh, A., Wintermeyer, W., 1999. Dynamics of translation on the ribosome: molecular mechanics of translocation. FEMS Microbiol. Rev. 23, 317–333.

Ruiter, A.G.T., Veerman, J.A., Garcia-Parajo, M.F., van Hulst, N.F., 1997. Single molecule rotational and translational diffusion observed by near-field scanning optical microscopy. J. Phys. Chem. 101, 7318–7323.

Sabido-David, C., Hopkins, S.C., Saraswat, L.D., Lowey, S., Goldman, Y.E., Irving, M., 1998. Orientation changes of fluorescent probes at five sites on the myosin regulatory light chain during contraction of single skeletal muscle fibres. J. Mol. Biol. 279, 387–402.

Saito, K., Tokunaga, M., Iwane, A., Yanagida, T., 1997. Polarization measurements of single fluorophores attached to motor protein in aqueous solution. Biophys. J. 72, A179.

Sase, I., Miyata, H., Corrie, J.E.T., Craik, J.S., Kinosita Jr., K., 1995. Real time imaging of single fluorophores on moving actin with an epifluorescence microscope. Biophys. J. 69, 323–328.

Sase, I., Miyata, H., Ishiwata, S., Kinosita Jr., K., 1997. Axial rotation of sliding actin filaments revealed by single-fluorophore imaging. Proc. Natl. Acad. Sci. USA 94, 5646–5650.

Schlichting, I., Almo, S.C., Rapp, G., Wilson, K., Petratos, K., Lentfer, A., Wittinghofer, A., Kabsch, W., Pai, E.F., Petsko, G.A., Goody, R.S., 1990. Time-resolved X-ray crystallographic study of the conformational change in Ha-Ras p21 protein on GTP hydrolysis. Nature 345, 309–315.

Schmidt, Th., Schütz, G.J., Baumgartner, W., Gruber, H.J., Schindler, H., 1995. Characterization of photophysics and mobility of single molecules in fluid lipid membrane. J. Phys. Chem. 99, 17662–17668.

Scholey, J.M., 1993. Methods in Cell Biology Volume 39 Motility Assays for Motor Proteins. Academic Press, Inc., New York.

Schütz, G.J., Schindler, H., Schmidt, Th., 1997. Imaging single-molecule dichroism. Opt. Lett. 22, 651–653.

Shera, E.B., Seitzinger, N.K., Davis, L.M., Keller, R.A., Soper, S.A., 1990. Detection of single fluorescent molecules. Chem. Phys. Lett. 174, 553–557.

Soper, S.A., Shera, E.B., Martin, J.C., Jett, J.H., Hahn, J.H., Nutter, H.L., Keller, R.A., 1991. Single-molecule detection of rhodamine 6G in ethanolic solutions using continuous wave laser excitation. Anal. Chem. 63, 432–437.

Swaminathan, R., Hoang, C.P., Verkman, A.S., 1997. Photobleaching recovery and anisotropy decay of green fluorescent protein GFP-S65T in solution and cells: cytoplasmic viscosity probed by green fluorescent protein translational and rotational diffusion. Biophys. J. 72, 1900–1907.

Sytnik, A., Vladimirov, S., Jia, Y., Li, L., Cooperman, B.S., Hochstrasser, R.M., 1999. Peptidyl transferase center activity observed in single ribosomes. J. Mol. Biol. 285, 49–54.
Thompson, N.L., McConnell, H.M., Burghardt, T.P., 1984. Order in supported phospholipid monolayers detected by the dichroism of fluorescence excited with polarized evanescent illumination. Biophys. J. 46, 739–747.
Trautman, J.K., Macklin, J.J., 1996. Time-resolved spectroscopy of single molecules using near-field and far-field optics. Chem. Phys. 205, 221–229.
Trautman, J.K., Macklin, J.J., Brus, L.E., Betzig, E., 1994. Near-field spectroscopy of single molecules at room temperature. Nature 369, 40–42.
Tsien, R.Y., 1998. The green fluorescent protein. Annu. Rev. Biochem. 67, 509–544.
Unger, M., Kartalov, E., Chiu, C-S., Lester, H.A., Quake, S.R., 1999. Single-molecule fluorescence observed with mercury lamp illumination. BioTechniques 27, 1008–1014.
Vallee, R.B., 1998. Methods in Enzymology: Volume 298 Molecular Motors and the Cytoskeleton Part B. Academic Press, San Diego.
van der Heide, U.A., Orbons, B., Gerritsen, H.C., Levine, Y.K., 1992. The orientation of transition moments of dye molecules used in fluorescence studies of muscle systems. Eur. Biophys. J. 21, 263–272.
VanderMeulen, D.L., Nealon, D.G., Gratton, E., Jameson, D.M., 1990. Excitation wavelength dependent fluorescence anisotropy of eosin-myosin adducts. Biophys. Chem. 36, 177–184.
Waggoner, A., 1995. Covalent labeling of proteins and nucleic acids with fluorophores. Methods Enzymol. 246, 362–373.
Walker, J.E., 1994. The regulation of catalysis in ATP synthase. Curr. Opin. Struct. Biol. 4, 912–918.
Wand, A.J., Ehrhardt, M.R., Flynn, P.F., 1998. High-resolution NMR of encapsulated proteins dissolved in low-viscosity fluids. Proc. Natl. Acad. Sci. USA 95, 15299–15302.
Wang, Y-L., 1989. Fluorescent analog cytochemistry: tracing functional protein components in living cells. Methods Cell Biol. 29, 1–12.
Warshaw, D.M., Hayes, E., Gaffney, D., Lauzon, A-M., Wu, J., Kennedy, G., Trybus, K., Lowey, S., Berger, C., 1998. Myosin conformational states determined by single fluorophore polarization. Proc. Natl. Acad. Sci. USA 95, 8034–8039.
Weiss, S., 1999. Fluorescence spectroscopy of single biomolecules. Science 283, 1676–1683.
Xie, X.S., Dunn, R.C., 1994. Probing single molecule dynamics. Science 265, 361–364.
Xie, X.S., Trautman, J.K., 1998. Optical studies of single molecules at room temperature. Annu. Rev. Phys. Chem. 49, 441–480.
Yang, N., George Jr., A.L., Horn, R., 1996. Molecular basis of charge movement in voltage-gated sodium channels. Neuron 16, 113–122.
Yang, T.-T., Sinai, P., Green, G., Kitts, P.A., Chen, Y.-T., Lybarger, L., Chervenak, R., Patterson, G.H., Piston, D.W., Kain, S.R., 1998. Improved fluorescence and dual color detection with enhanced blue and green variants of the green fluorescent protein. J. Biol. Chem. 273, 8212–8216.
Yariv, A., 1975. Quantum Electronics, 2nd Edition. Wiley, New York.
Yasuda, R., Noji, H., Kinosita Jr., K., Yoshida, M., 1998. F_1-ATPase is a highly efficient molecular motor that rotates with discrete 120° steps. Cell 93, 1117–1124.

Biology at the Single Molecule Level
S.H. Leuba an J. Zlatanova (Eds)
Published by Elsevier Science Ltd.

Review

Single molecule force spectroscopy in biology using the atomic force microscope

Jordanka Zlatanova[a,*], Stuart M. Lindsay[b], Sanford H. Leuba[c]

[a] *Biochip Technology Center, Argonne National Laboratory, 9700 South Cass Avenue, Bldg. 202-A253, Argonne, IL 60439, USA*

[b] *Department of Physics and Astronomy, Arizona State University, Tempe, AZ 85287, USA*

[c] *Physical Molecular Biology, Laboratory of Receptor Biology and Gene Expression, National Cancer Institute, NIH, Building 41, Room B507, Bethesda, MD 20892, USA*

Abstract

The importance of forces in biology has been recognized for quite a while but only in the past decade have we acquired instrumentation and methodology to directly measure interactive forces at the level of single biological macromolecules and/or their complexes. This review focuses on force measurements performed with the atomic force microscope. A general introduction to the principle of action is followed by review of the types of interactions being studied, describing the main results and discussing the biological implications. Published by Elsevier Science Ltd.

Keywords: Atomic force microscopy; DNA stretching; Intermolecular interactions; Intramolecular interactions; Protein unfolding; Single molecule force spectroscopy

---

Contents

*Corresponding author. Present address: Chemical Engineering, Chemistry and Materials Science Department, Polytechnic University, Brooklyn, NY 11201, USA. Tel.: +1-718-260-3176; fax: +1-718-260-3125.

E-mail address: jzlatano@duke.poly.edu (J. Zlatanova).

1. Introduction

Understanding the forces that govern specific molecular interactions is a challenging task in molecular and structural biology. Such specific interactions result from multiple weak, non-covalent bonds formed between defined portions of the interacting molecular partners. Various techniques have been recently employed to directly probe such weak interactions, including surface force apparatus (Israelachvili, 1992), pipette suction (Evans et al., 1991), magnetic beads (Smith et al., 1992), flow chamber apparatus (Pierres et al., 1996), and optical traps and tweezers (Ashkin et al., 1987; Ashkin, 1997). The techniques using optical traps have been particularly popular in view of their high force sensitivity. The disadvantages of the optical trap-based apparatuses are in the limited range of samples amenable to analysis (biopolymers must be generally longer than $\sim 2\,\mu m$), and general unsuitability for applying forces greater than ~ 150 picoNewtons (pN).

The advent of the probe microscopes, and in particular the atomic force microscope (AFM) (Binnig et al., 1986) has opened new horizons in force measurements. This review will focus on the use of AFM to study inter- and intramolecular interaction forces in biological macromolecules. It will also briefly illustrate the use of the techniques to evaluate mechanical properties of biological samples, like elasticity and viscosity.

2. The AFM for force measurements

2.1. *Principle of action*

The AFM (Fig. 1A) uses a sharp tip mounted at the end of a flexible cantilever to probe a number of properties of the sample, including its topological features and its mechanical characteristics. Precise lateral and vertical displacement of the sample with respect to the probe is achieved by a computer-controlled piezoceramics stage holding the sample, or conversely, the cantilever holder. Forces acting between the surface and the probe cause deflection of the

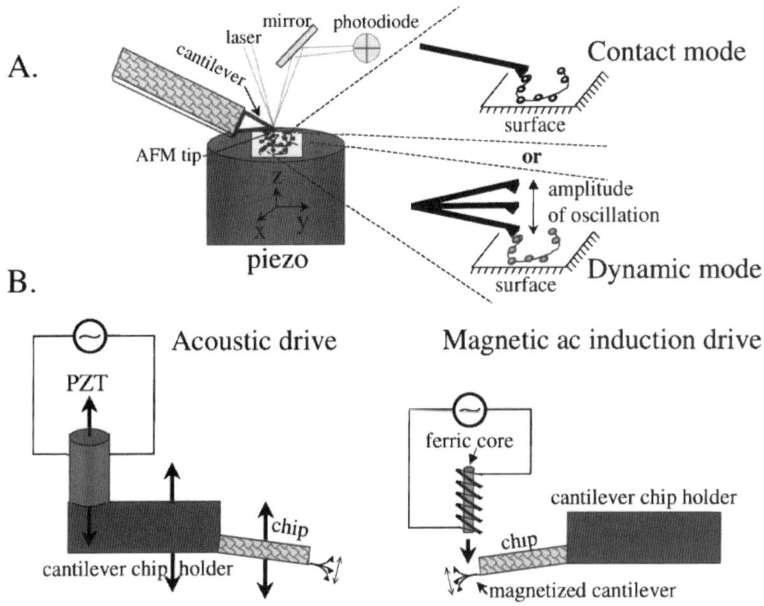

Fig. 1. Schematic of the principle of action of the atomic force microscope (A). The enlargements on the right-hand side illustrate the principle of the static mode (top) and the dynamic mode (bottom) of action. (B) Illustration of the acoustic and the magnetic-oscillation-induction dynamic modes of action.

cantilever that is registered by a laser beam reflected off the back of the cantilever. The cantilever deflections are used to create a topographic image of the sample when the probe is raster-scanned in the x–y direction, or to produce the so-called force curves, when the probe is moved in the z-direction, i.e., first moved downwards until it contacts the surface, and then upwards, till no interaction between probe and surface is felt. The resultant plot of cantilever deflection versus the separation between the probe and the sample is the force curve. The AFM combines high force sensitivity (down to a few pN) with high lateral resolution (often better than a nanometer, which is in the realm of molecular dimensions).

The high resolution of forces (for a detailed discussion see Sarid, 1991) results from the small spring constants of the cantilever (usually in the range of 0.5–0.01 N/m) to which the tip interacting with the sample is attached. The force sensitivity is limited by the stiffness of the cantilever and the laser power. The main noise contributions come from thermal excitation of the cantilever motion, the intensity distribution in the laser beam, and the bandwidth (the range of frequencies over which the signal is collected) (i.e., a slow pulling with limited bandwidth will give better signal-to-noise ratio, the noise being proportional to the square root of the bandwidth).

The high spatial resolution is achieved through the use of very sharp probes.

There are two different modes of operation of the AFM in its force recording regime (Fig. 1A). In the static (contact) mode, the tip approaches the surface, is pushed into the surface, and is then retracted by the piezo stage. In the dynamic mode, the cantilever is oscillated while it is moved to and from the surface. The measured parameter is the reduction in the amplitude of oscillation that is caused by the interaction of the tip and the surface.

There are two general methods to oscillate the cantilever, acoustical or magnetic (Fig. 1B). In acoustic, or 'tapping' mode (Zhong et al. (1993); for extension of this method to fluids see Hansma et al. (1994) and Putman et al. (1994)), the cantilever is oscillated above the surface near its resonance frequency. The second method uses a magnetic alternating current (ac) field to oscillate the cantilever that is coated with a thin magnetically susceptible film (for a recent detailed comparison of the two methods, see Lindsay, 2000).

A typical force versus displacement curve generated by an AFM is shown in Fig. 2. At the beginning, the probe is far away from the surface, so there is no interaction between the two (Fig. 2). As the probe–sample separation is reduced beyond a certain point, forces between atoms on the two surfaces begin to act, causing the flexible cantilever to bend toward the sample in the case of attractive forces (van der Waals and electrostatic), or away from the sample in the case of repulsive forces (electrostatic). At each distance, the cantilever bends until its elastic (restoring) force equals the probe–sample interaction force and the system is in equilibrium. The attractive forces can cause the probe to snap to the surface earlier (from a greater distance) than the expected time of contact in the absence of such forces. (Contact may be defined as the point when repulsion is first detectable.) The jump-to-contact observed in approach curves limits the range of data that may be obtained on the approach cycle and also the gentleness of the tip approach.

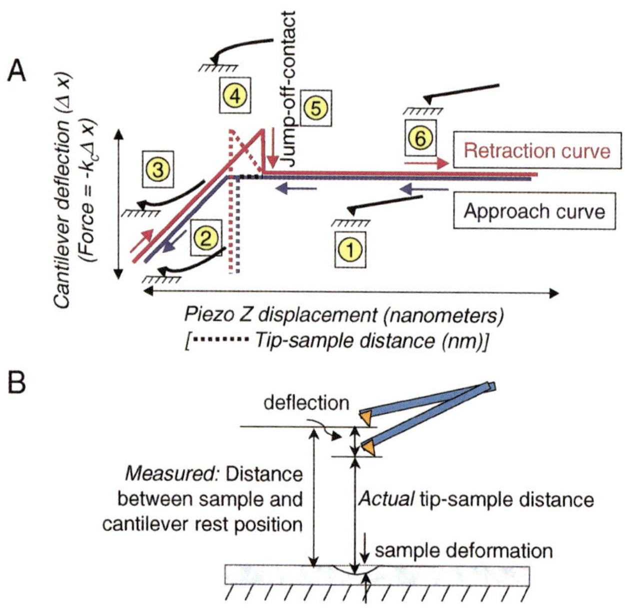

Fig. 2. Schematics of types of force curves and tip–sample interactions. (A) Typical contact force versus Z displacement curve generated by an AFM, and (B) schematic of the tip/surface interaction. In (A) the deflection of the cantilever is shown as a function of the piezo Z displacement. The numbers correspond to different states of the cantilever during the approach and retraction portions of the cycle: (1) AFM tip is not in contact with surface; (2) tip is being pushed into the surface, bending the cantilever; (3) tip is being withdrawn from the surface; (4) tips adheres to sample surface; (5) tip 'jumps-off-contact' from surface; (6) tip is not in contact with surface. The schematic in (B) of the tip–sample interaction is used to illustrate why there is a difference between the force versus piezo Z displacement curve [solid lines in (A)], and force versus tip–sample distance curve [dotted lines in (A)].

When the jump-to-contact is caused by electrostatic attraction, it can be minimized by operating in electrolytes that screen these interactions (Müller and Engel, 1997). Once in contact with the surface, the probe will experience an ever-increasing repulsive force (and the cantilever will bend away from the surface) as the electron orbitals of the atoms in the probe and the sample will begin to overlap. In this region of the force curve, referred to as the contact region, there may be elastic and/or plastic (reversible and/or irreversible) deformations of either or both the probe and the sample. These deformations may give additional information about the mechanic properties of the experimental sample, as will be discussed below.

As described by numerous authors (for recent reviews and references therein see Noy et al., 1997b; Heinz and Hoh, 1999a; Cappella and Dietler, 1999), there is a distinction between the force versus piezo Z displacement curve recorded by the AFM and the force versus tip–sample distance curve. We illustrate this distinction in Fig. 2 which shows that the actual tip–sample distance is the difference between the piezo Z displacement and the deflection of the cantilever (with soft samples, sample deformation needs to be taken into account too, see Section 5). At zero force, the points in both types of curves coincide; the maximum difference between the two plots is at the maximum force. When pulling long molecules like titin or chromatin fibers, the sample deformation at the surface and the cantilever deflection distance are negligible compared to the distance the tip travels up and down: thus, the difference between piezo Z displacement and tip/sample distance can be safely ignored.

Next, after a preset value of load is reached, the direction of motion is reversed and the probe moves away from the surface. During retraction of the probe, there may be manifestation of other forces: adhesion forces, created during contact and, in some specific cases, hydrophobic and solvation forces (Fig. 3). Of special interest to us is the adhesion force that is estimated from the deflection of the cantilever right before the jump-off-contact point. Jump-off-contact occurs when the effective elastic constant of the cantilever overcomes the adhesive interactions between probe and sample. Although the precise thermodynamic entity that is correlated to the adhesion force measured at jump-off-contact is still under discussion (see Moy et al., 1994a; Chilkoti et al., 1995), it is clear that this force can be used to compare interaction strength of different atomic arrangements within molecules or between molecules. The cantilever deflections occurring as a result of interatom interactions are converted into force using Hooke's law:

$$F = -k_c d,$$

where F is the force acting on the cantilever, k_c is the spring constant of the cantilever, and d is its deflection. There is considerable spread in the values of cantilever force constants and independent calibration is essential. Various authors have described different methods for calculating the spring constant. In one of these approaches, the spring constant is calculated from the cantilever geometry (Sader and White, 1993; Sader, 1995; Chen et al., 1994, 1995). This approach, however, assumes that the material properties of the thin films used in the manufacture of cantilevers are stable, reproducible, and known. Another set of approaches relies on measuring static deflection in response to a known force (Senden and Ducker, 1994; Torii et al., 1996; Gibson et al., 1996). Yet a third approach measures changes in dynamical properties of vibrating cantilevers (Hutter and Bechhoefer, 1993; Cleveland et al., 1993; Sader et al., 1995). We have found the static methods to be the most straightforward (Li et al., 1993).

Force	Description	Shape of Curve
From approach curve		
Attractive van der Waals force	Creates jump-to-contact peak; depends on force between tip and sample and the system geometry; Meniscus forces must be eliminated by working in a low humidity environment or in liquid	
Repulsive double-layer electrostatic force	Due to charging both sample and tip surfaces with like charges (adsorption of ions from solution and dissociation of functional groups on surfaces)	
Repulsive hydration force	Comes from repellant interactions between hydrated ions bound to tip and sample surfaces; apparent at high salt concentration	
Solvation force	When tip and sample are within a few molecular diameters from each other, oscillation between attraction and repulsion may be observed; caused by ordering of non-polar liquid molecules between the two liquid-solid interfaces; force follows density of solvent	
From retraction curve		
Adhesion	Due to van der Waals forces and indentation of sample by tip (which increases contact area); Meniscus force must be eliminated	
Hydrophobic force	Appear on imaging hydrophobic samples in water; Results in a gradual pull-off instead of a jump-off-contact	

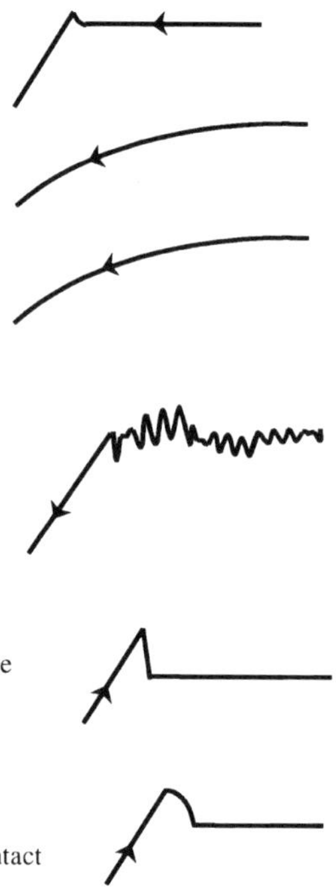

Fig. 3. Types of interaction forces felt by the cantilever during the approach and retraction portion of the force curve.

AFM force curves are rich in information (for reviews see Burnham et al., 1993; Cappella et al., 1997; Burnham and Kulik, 1998; Vinckier and Semenza, 1998; Heinz and Hoh, 1999a; Cappella and Dietler, 1999). They have allowed, among other things, measurements of the strength of individual hydrogen bonds (Hoh et al., 1992), and of different kinds of covalent bonds (Grandbois et al., 1999). Fig. 3 summarizes the types of forces sensed by the cantilever, and illustrates the effect of each type of interaction on the appearance of the force curve. In practice, force curves are rather complex since different forces may affect the behavior of the cantilever at any given value of probe/sample separation. This complexity requires specific care in the interpretation, the correctness of which depends on the results of sets of control experiments.

Although the dynamic mode presents several important advantages over the contact mode (Lindsay, 2000), the interpretation of contact force curves is rather straightforward (see below), whereas curves recorded in the dynamic mode are more difficult to interpret in terms of forces. It has been shown that the contact force curve is reproduced by integration of the oscillation amplitude curves when the forces are conservative, i.e., the process is reversible (Liu et al., 1999). In the event that the amplitude damping is purely a consequence of an elastic interaction (and dissipative processes do not contribute to it), the stiffness of the molecule is given by

$$S(z) = -k\left(\frac{A_0}{A(z)} - 1\right), \tag{1}$$

where A_0 is the undamped amplitude (typically $\sim$5 nm), $A(z)$ is the amplitude at a distance z, and k, the spring constant of the cantilever. The force can be obtained from Eq. (1) by integration:

$$F(z) = -k \int \left(\frac{A_0}{A(z)} - 1 \right) \mathrm{d}z + C. \tag{2}$$

Comparison of contact and magnetic ac curves may be extremely useful in the interpretation of the data, because, in the event that they reflect reversible, purely elastic processes, the two are simply related (Liu et al., 1999). Thus, the pair of curves recorded simultaneously serves to distinguish elastic from dissipative processes.

For biological applications tips are often modified to carry specific chemical groups or molecules, which allows assessing the weak van der Waals, electrostatic and hydration forces acting between the molecules on the tip and those on the surface. It is well established that a thin wetting film of liquid (water) covers both the probe and the sample when the AFM is operated in air (Grigg et al., 1992; Burnham et al., 1993; Burnham and Kulik, 1998). This thin film of water creates capillary forces that may have a dramatic effect on the force curves. The overall magnitude of the capillary force can be large enough to obscure the weak van der Waals forces. In view of this, experiments aimed at quantitation of interaction forces should be performed in liquids, in vacuum, or under dry nitrogen. Aqueous solution operation is desirable with biological molecules because it preserves their native structure.

The major parameter affecting bond rupture is the applied loading rate (the change of force with time — $\mathrm{d}F/\mathrm{d}t$). Loading rate is also given as the product $k_c v$ of the tip velocity, v, and the spring constant, k_c, of the cantilever (Evans and Ritchie, 1997). The probability of thermal rupturing of the bond increases as $\exp(-[E_B - Fx_B]/k_B T)$, where E_B is the potential energy barrier of bond disruption, F is the applied force in the bond direction, x_B is the stretched distance in this direction, k_B is the Boltzman constant, and T is the temperature. This description of loading rate applies if the loading force is generated by the cantilever alone. However, when the force is transmitted by a compliant polymer, the loading rate at some site (the bond to be ruptured) within that polymer also depends on the compliance and the extension of the polymer. Evans and Ritchie (1999) have illustrated this point on the example of the unfolding of a titin domain as a function of the pulling speed (Fig. 4), based on the original data from Rief et al. (1997a) and Kellermayer et al. (1997). The upper set of points (falling on the line in this log–log plot, Fig. 4) is for titin molecules of 500 nm in length. The lower cluster of data points is for longer molecules (several microns). The longer molecules generate a force that increases more slowly with time for the same overall pulling speed. When the internal loading rate is taken into account, the data fall on a universal line. Thus, comparison of bond-rupture or unfolding force data from various experiments requires control of both the external loading rate and the sample size (Evans and Ritchie, 1999).

2.2. General requirements for sample preparation and deposition

Studying biological macromolecules with the AFM puts several requirements to the way the sample is prepared and attached to the surface. Extreme care should be taken in using solutions of highest possible purity, since contamination may affect both imaging and force measurements. In addition, special attention should be paid to strongly attaching the molecules to the surface. Good

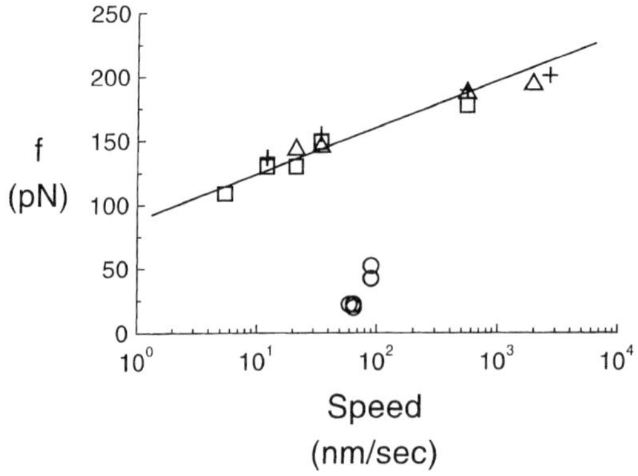

Fig. 4. Relationship of measured force of unfolding a titin domain as a function of pulling speed (Fig. 2 from Evans and Ritchie (1999), with permission of Biophysical Journal).

attachment is a must in this technique since the scanning probe may shift a weakly bound molecule around, thus precluding stable imaging or force registration. While the attachment should be firm enough to avoid undesired motions, it should still allow certain freedom of the molecule to change its conformation during biological activity or interaction with its molecular partners. Too firm an attachment along considerable portion of the molecular surface may lead to undesirable denaturing effects and should be avoided. The above requirements have resulted in considerable effort to properly attach molecules to the surface, both via non-specific adsorption and by covalent bonding. Recent useful descriptions of available techniques can be found in Tendler et al. (1996), Müller et al. (1997), Vinckier and Semenza (1998), Wagner (1998), Siedlecki and Marchant (1998), and Engel et al. (1999).

3. Intermolecular interactions

As mentioned above, it is possible to use the capability of AFM to sense interaction forces between atoms at the end of the probe and in the specimen to study the weak, non-covalent, usually short-range forces involved in molecular recognition reactions. In order to be able to do so, the partners in the molecular recognition reaction have to be immobilized onto the surface and the probe. Probes containing certain chemical functionalities can be used for mapping the spatial arrangement of chemical groups on a surface, in an approach termed *chemical force microscopy* (Frisbie et al. (1994); for a review see Noy et al. (1997b)). In this pioneering work from C. Lieber's laboratory it was demonstrated that probes functionalized with $-CH_3$ or –COOH groups can specifically interact with similar groups on the surface, with the spatial pattern of interaction reproducing the spatial distribution of functional groups on the surface (Frisbie et al., 1994). A significant technological advance in chemical force microscopy was achieved by introducing covalently functionalized nanotubes as AFM probes (Wong et al., 1998). Nanotubes offer several important advantages over conventional AFM tips: they have high aspect ratio allowing probing of deep crevices, small effective radius increasing the lateral resolution of topographical and force imaging, and are very flexible, avoiding the common problem of tip breakage. Functionalized nanotubes allowed chemically sensitive imaging of patterned samples, and more importantly in the context of this review, measuring interaction forces between biotin/streptavidin (Wong et al., 1998; Table 1).

Measuring intermolecular interactions between partners in molecular recognition reactions utilizes an analogous approach. It immobilizes each of the partners on either the probe or the surface, and then takes the probe through approach/retraction cycles. The magnitude of the cantilever deflection at the jump-off-contact peak is taken to reflect the rupture force needed to break the molecular interactions holding the two partners in close contact; breaking of the interaction bonds restores the cantilever to its neutral non-contact, non-deflected position.

There are several groups of intermolecular interaction forces experimentally measured with the AFM (Tables 1 and 2). Typical force curves recorded for each type of molecular pair are presented in Fig. 5, to facilitate perception of the characteristic features in each case and to allow comparisons. It must be noted that the specific intermolecular interactions are registered against the background probe/surface interactions, which may be as high as, or even higher, than the biological interaction of interest (Stuart and Hlady (1995); for further possible artifacts, see Stuart and Hlady (1999)). This requires careful choice of specificity controls. These generally include the use of non-functionalized probes or surfaces, blocking the interaction between the immobilized molecular partners with free ligands in the medium, or changing the pH or the salt concentration in the medium (for detailed information on specificity controls used in specific applications, see Tables 1 and 2).

3.1. Receptor/ligand interactions

A lot of experimental effort has gone into studying receptor/ligand interactions, on the example of the small ligand biotin interacting with the closely related proteins streptavidin or avidin. The biotin/streptavidin ligand/receptor pair has been used as a model system because of its unusually high affinity and the availability of structural and thermodynamic data. The interpretation of the force data has been greatly facilitated by the use of site-directed mutants of the receptor proteins, as well as of structural derivatives of biotin itself. As seen from Table 1, measurements from different laboratories, performed under a wide variety of experimental setups and conditions, agree pretty well, reinforcing the credibility of the AFM approach for force measurements.

The biotin/streptavidin system has been also utilized to study the dependence of the measured interaction forces on the loading rate. Evans and Ritchie (1997) (see also above) have argued, on the basis of theoretical considerations concerning bond rupture under application of force, that bond strength progresses through three dynamic regimes of loading rate: a slow-loading, a fast-loading, and an ultrafast-loading regime. In each of these regimes, the dependence of the bond strength on the loading rate is different. Thus, in order to expose the energy landscape that governs bond strength, molecular adhesion forces must be examined over an enormous span of time scales. A convincing experimental proof of these theoretical considerations was obtained in experimental measurements of biotin/avidin interaction forces (Merkel et al., 1999; Table 1).

It is worth noting that a different conclusion has been recently reached by Schwesinger et al. (2000) who studied the loading-rate-dependent unbinding forces for different series of ligand–receptor pairs. It was reported that for nine different fluorescein–anti-fluorescein antibody pairs

Table 1
Interaction forces between ligand and receptor pairs

Molecular partners	Experimental setup	Specificity controls	Average forces (pN)	Reference
Avidin/biotin	Biotinylated bovine serum albumin (BSA) adsorbed on tip, with subsequent avidin adsorption; surface is biotinylated	Free avidin, free biotin, free BSA; biotin-coated tips	15,000–2000[a]	Moy et al. (1994b)
Avidin/biotin Streptavidin/biotin	Biotinylated BSA adsorbed on tip, with subsequent avidin or streptavidin adsorption; ligand (biotin or derivatives) immobilized on surface	Immobilized biotin derivatives; low or high pH	~160 ~260	Moy et al. (1994a)
Avidin/biotin or derivatives	Biotinylated BSA adsorbed on tip, then incubated with avidin; biotinylated agarose beads as substrate	Free avidin or biotin in medium	~160	Florin et al. (1994)
Biotin/streptavidin	Biotinylated BSA adsorbed on glass microspheres (glued to cantilevers) and mica surface; surface further incubated with streptavidin	BBSA surface (no streptavidin); streptavidin surface blocked with biotin	~340	Lee et al. (1994b)
Biotin/streptavidin site-directed mutants	Biotinylated BSA adsorbed on tip and mica surface; surface further incubated with wild-type or mutant streptavidin	Free biotin in medium	Between 100 and 450 for mutants	Chilkoti et al. (1995)
Biotin/antibiotin antibody (Ab)	Biotinylated BSA covalently bound to tip via linker; Ab covalently bound to surface via linker; also reverse configuration	Non-biotinylated BSA on tip; biotin and streptavidin in medium; non-specific Ab on surface; low or high pH	~110	Dammer et al. (1996)
Biotin/streptavidin	Biotin covalently bound to nanotube tips; streptavidin linked to surface by biotin groups	Free biotin in medium; unmodified nanotube tips	~200	Wong et al. (1998)
Fluorescein/ antifluorescein single-chain Ab	Fluorescein covalently bound via linker to tip; Ab attached to gold surface by engineered C-terminal cysteine	Free fluorescein	~50	Ros et al. (1998)
Biotin/streptavidin or avidin	Biotin covalently attached via linkers to glass micro-beads and surfaces; avidin was further adsorbed to both, such that free biotin groups were still available for infrequent bond formation	Linkers terminated in chemical groups inactive in attaching biotin; free biotin	Between 5 and 170, depending on loading rate	Merkel et al. (1999)

[a] Number of interacting molecular pairs estimated to be ~100.

Table 2
Interaction forces between various pairs of protein

Molecular partners	Experimental setup	Specificity controls	Average forces (pN)	Reference
Antigen/antibody (Ag/Ab)				
IgG-Ab/protein G	Biotinylated BSA adsorbed on tip, with subsequent avidin adsorption; further functionalization of above tip with biotinylated IgG-Ab; surface is protein G-covered	Free protein G	3000–4000[a]	Moy et al. (1994b)
Fluorescein/antifluorescyl Ab	Fluorescein covalently bound to a silica bead glued to cantilever; Ab covalently bound to surface	Non-specific Ab on surface; non-derivatized tip;[b] no protein on surface	~200[c]	Stuart and Hlady (1995)
Anti-human serum albumin/human serum albumin	Covalently bound Ab to tip and Ag to surface via long flexible spacers	Free serum albumin in medium	~250	Hinterdorfer et al. (1996)
Ferritin/antiferritin Ab	Both Ag and Ab covalently attached to tip and surface, respectively	Non-functionalized tip and tips at different stages of chemical treatment used for immobilizaiton; non-specific Ab on surface	~50	Allen et al. (1997)
Anti-intercellular adhesion molecule-1 (ICAM-1)/ICAM-1	Covalently bound Ab to tip and Ag to surface via long flexible spacers		~100	Willemsen et al. (1998)
Other protein/protein interactions				
Actin/actin in actin filaments			Not given	Moy et al. (1994a)
Cell adhesion proteoglycans	Proteoglycans covalently attached to organic monolayers on tip and surface through the protein moiety	Surfaces covered with monolayers; Ca^{2+} or Mg^{2+} in various concentrations; monoclonal Ab to a carbohydrate epitope and a non-specific Ab	Up to 400; polyvalent binding with steps of 40	Dammer et al. (1995)
Insulin/insulin	Monomers convalently attached to both tips and surface via specific residues in ways to either favor or disfavor dimer formation	Free insulin; anti-insulin Ab	1300[d]	Yip et al. (1998)
Two protein substrates citrate synthase or β-lactamase/*E. coli* chaperonin GroEL	Protein substrates covalently bound to tip; GroEL adsorbed to mica with the substrate binding site on top	Native or denatured substrates; presence or absence of ATP; bare mica or non-functionalized hydrophilic or hydrophobic tips	~240 and ~420 for lactamase, and citrate synthase, respectively	Vinckier et al. (1998)

Table 2 (*continued*)

Molecular partners	Experimental setup	Specificity controls	Average forces (pN)	Reference
Recombinant P-selectin/P-selectin glycoprotein ligand-1 (PSGL-1)	Biotinylated P-selectin attached to avidin-coated coverslips; biotinylated PSGL-1 attached to avidin-coated tips	Buffer supplemented with EDTA, instead of Ca^{2+}; non-glycosylated or glycosylated PSGL-1	~160	Fritz et al. (1998)
Myelin basic protein/lipid bilayers	Lipid adsorbed to surface and further incubated with myelin; tip coated with myelin	Bare mica surface or surface with adsorbed lipid; low or high ionic strength	~140[e]	Mueller et al. (1999)

[a] Number of interacting molecular pairs not determined.
[b] Large discontinuities in the force curves were observed in the first two controls and attributed to non-specific discrete interactions between protein and the AFM spherical bead tip.
[c] Measured within 2 s of initial contact; extended contact leads to increased forces.
[d] Multiple unbinding events and elongation of monomer prior to dissociation.
[e] No elastic stretching in myelin when on lipid, in contrast to when on mica.

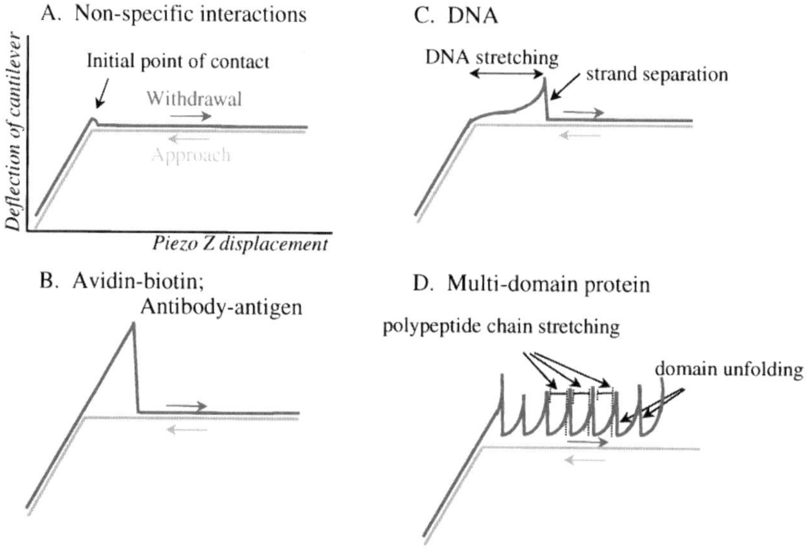

Fig. 5. Illustration of the major types of force curves obtained as a result of different types of interactions between biological macromolecules. (A) A control curve recording the non-specific interactions between an AFM tip and a hard surface. (B) Curve reflecting specific 'adhesion' interactions between pairs of molecular partners immobilized on the tip and surface, respectively. Such curves have been recorded in the case of biotin–avidin (or streptavidin) interactions, and in the case of antigen–antibody interactions (see Tables 1 and 2). (C) Force curve obtained upon stretching of DNA. Note the different shape of the curve in the portion preceding the jump-off-contact. This shape can be fitted by the worm-like chain model and results from the elastic stretching of the DNA from B- to S-conformation (see text). The jump-off-contact reflects the melting of the DNA duplex. (D) Typical force curves obtained on stretching of multi-domain proteins (for schematic interpretation of such multi-peak curves see Fig. 6). Similar multi-peak curves with somewhat less regular appearance are obtained upon stretching of chromatin (see text).

the measured unbinding forces correlated well with the respective thermal off-rates in solution; the dependence of the force on the logarithm of the loading rates was linear over six orders of magnitude of the latter parameter. The authors attribute the difference between their results and those from Evans' laboratory (see above) to certain atypical structural features of the avidin/biotin pair.

3.2. Protein/protein interactions

The second populated class of intermolecular force measurements involved protein/protein interactions, including the highly specific antigen/antibody interactions (Table 2). Again, the variety of experimental setups was wide, and the molecular pairs studied were very diverse with respect to their structure and their physiological functions. Once again, since different loading regimes have been used by different investigators, it is not straightforward to directly compare the measured interaction forces.

3.3. Interactions between complementary strands of DNA

The interactions between the two strands of a double-helical DNA molecule may also be classified as intermolecular interactions, and should, as such, be discussed in this section. In a pioneering work, Lee et al. (1994a) measured the interaction forces between complementary 20-base strands covalently immobilized to a silica probe and surface. The sequence of the oligonucleotides was designed so that they did not contain self-complementary regions, and base pairing was restricted to the formation of duplexes of discrete lengths: 20, 16, 12, 8 and 4 bp (only the first three of these were expected to be thermodynamically stable). Non-complementary oligonucleotides served as specificity controls. Measurements of interaction force gave ~70 pN per base pair.

In a different experimental approach, Boland and Ratner (1995) used self-assembled purines and pyrimidines on planar gold surfaces and gold-coated AFM tips. Specific hydrogen-bonding interactions between the molecules on the tip and surface could only be measured when complementary base pairing could take place. A rupture force of 54 pN was reported for a single AT pair, close to the value reported by Lee et al. (1994a). These values are much higher than the value of 10–15 pN per base pair reported form the unzipping experiments of Essevaz-Roulet et al. (1997) performed using bendable micro-needles as force sensors. A possible explanation for this discrepancy could lie in the significant non-specific interactions between the tip and the surface in the AFM setup, as discussed by Samori (1998).

More recently, Noy et al. (1997a) have used self-assembled monolayers for immobilizing complementary 14-mers to gold-coated tip and surface. With increasing applied force, the final separation of the complementary strands proceeded after DNA stretched to a stable form of approximately twice the length of the B-form. This structural transition of double-stranded DNA from the B-form to the so-called S- (stretched) form was first identified in long DNA molecules in experiments using optical fibers (Cluzel et al., 1996) and optical tweezers (Smith et al., 1996). It is worth noting that the structural transition to the extended, S-form DNA in the AFM experiments was observed at a force of 120 ± 50 pN, close to the value of 140 pN estimated from modeling studies (Lebrun and Lavery, 1996), and much higher than the force measured in both the optical tweezers and optical fibers experiments (~70 pN). The adhesion forces between the complementary oligonucleotides were distributed around 460 ± 180 pN.

A similar sequence of structural transitions — stretching followed by melting — has been recently reported for a restriction fragment of λ-DNA stretched with the AFM (Rief et al., 1999a). The force for the stretching transition was found to be 65 pN, and that for the melting ~150 pN. Stretching poly(dG–dC) and poly(dA–dT) allowed the expected sequence dependence of melting to be directly demonstrated, with base pair unbinding force for G–C of ~20 pN, and that for A–T of ~9 pN. Evidently, more work is needed to resolve some apparent inconsistencies among the results obtained in different laboratories, and between experimental measurements and theory.

3.4. Stretching chromatin

Finally, the AFM has been used to stretch chromatin fibers, with the goal of determining the forces needed to break the bonds between the DNA and the histone octamers in the nucleosomal particles. Such studies are of particular importance since processes like transcription, replication, and repair that use DNA as a template must, in one way or another, move the histones out of the way of the corresponding enzymatic machineries. Stretching of native chromatin fibers isolated from chicken erythrocyte nuclei (Leuba et al., 1999) produced complex multi-peak patterns, reminiscent of those obtained on stretching multi-domain proteins (see below). Moreover, the presence or absence of linker histones in the fiber affected the magnitude of the peaks (Leuba et al., 1999). In order to understand the complex force curves obtained on native chromatin fibers, we switched to a defined system: chromatin fibers of relatively regular structure obtained by reconstituting a tandemly repeated nucleosome positioning DNA sequence with core histones (Leuba et al., 2000). Stretching such arrays of 12 nucleosomes produced force curves containing one or more peaks, with peak-to-peak distances much too small to arise from unraveling of the nucleosomal DNA from around the histone octamer. Careful analysis of the data led to the conclusion that, under the experimental conditions used, nucleosomes did not unravel even at forces exceeding 300 pN. The most likely explanation of these results is that the linker DNA (the DNA connecting nucleosomes) stretches under the applied tension, followed by detachment of the individual nucleosome from the surface. Further experiments are in progress to study chromatin behavior under applied tension.

It should be noted that chromatin fibers have been recently stretched using optical tweezers. Cui and Bustamante (2000) pulled on long stretches of chicken erythrocyte chromatin fibers, whereas Bennick et al. (submitted) directly reconstituted single chromatin fibers on λ-DNA molecules already attached between two polystyrene beads in the liquid cell of the apparatus. Both groups observed irreversible changes in fiber structure when the stretching force exceeded 20 pN. The data resolution in the Bennink et al. experiments was high enough to allow measurements of individual opening events that were attributed to the unraveling of individual nucleosomes in the chromatin fiber. It is highly desirable to compare the forces measured in the optical tweezers experiments and those obtained in the AFM; such a comparison may become possible after slight modifications of the experimental setup presently used in our AFM experiments.

4. Intramolecular structural transitions in polysaccharides, DNA, and multi-domain proteins

The AFM has proven particularly useful in studying intramolecular interactions. We have already mentioned the stretching of the DNA double helix that precedes its melting. A major research effort has focused on the unfolding of multi-domain protein molecules or of individual protein domains. The interested reader may greatly benefit from the several recent enlightening reviews covering this area (Fisher et al., 1999a, b; Carrion-Vazquez et al., 2000). We will limit ourselves to schematically illustrating the principle of these experiments (Fig. 6), and to summarizing the available literature in a table format (Table 3).

At small extensions (relative to the chain length), polymers generate a restoring force that is mainly entropic in origin. If force is applied to a polymer chain, an opposing force is created as a result of the reduction in entropy. The behavior of polymers under mechanical stress has been described by the worm-like chain model of elasticity (Flory, 1989), with the persistence length and the contour length comprising the adjustable parameters of the model. A particularly well-studied example of an entropic elastic polymer is DNA stretched at forces up to 65 pN (Bustamante et al., 1994; Marko and Siggia, 1995). The behavior of DNA, however, deviates from entropic elasticity at relatively high forces: application of forces above a certain threshold level to DNA leads to conformational changes beyond simple straightening of the chain and results in extensions beyond the contour length. As mentioned before, DNA undergoes a stretching B- to S-transition above 65 pN (Cluzel et al., 1996; Smith et al., 1996; Noy et al., 1997a; Rief et al., 1999a).

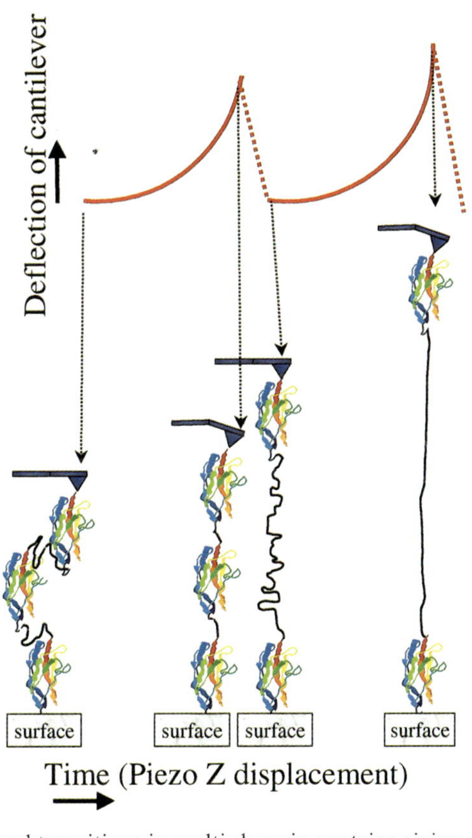

Fig. 6. Schematic showing the structural transitions in multi-domain proteins giving rise to the multi-peak force curves, as those represented in Fig. 5D.

Table 3
Stretching of multi-domain proteins

Native protein, recombinant fragment, or polyhomodomain	Structure of the protein	Extension per sub-unit/domain (ΔL_c)	Average force (pN)	Reference
α_2-macroglobulin	Four identical subunits, each of 1451 aa, containing 11 intra- and 2 inter-subunit SS bonds	150 nm for each of four identical sub-units; such an extension is possible without breaking the SS bonds	>250	Mitsui et al. (1996)
Titin (native protein or recombinant fragments containing 8 or 4 immunoglobulin (Ig-like domains)	Ig-like and fibronectin (Fn)-III-like tandemly repeated domains	25–28 nm	150–300	Rief et al. (1997a)
Titin (recombinant fragments from different parts of the molecule containing 6–8 Ig- or Fn-III-like domains)	Ig-like and Fn-III like tandemly repeated domains	28.2–31.5 nm	180–240[a]	Rief et al. (1998)
Tenascin (native protein or recombinant fragment containing 15 Fn-III-like domains)	SS-linked hexamers, each made of tandemly repeated epidermal growth factor-like and Fn-III-like domains. Fn-III domain is a seven-stranded β-barrel, similar to that of Ig-like domains	28.5 nm	~140	Oberhauser et al. (1998)
Twelve identical repeats of the Ig-like domain 27 of titin [$(I27)_{12}$]	β-barrel consisting of seven β-strands; H-bonds perpendicular to the direction of applied force			Carrion-Vazquez et al. (1999) and Marszalek et al. (1999)
Wild type		24.8 nm (75 amino acids)	~200–210	
Mutants containing clusters of five glycines inserted				
In the folded region		1.9 nm (5 amino acids) increase over wild type		
Outside the folded region		No increase		
Twelve identical repeats of the Ig-like			~265[b]	Marszalek et al. (1999)

Table 3 (*continued*)

Native protein, recombinant fragment, or polyhomodomain	Structure of the protein	Extension per sub-unit/domain (ΔL_c)	Average force (pN)	Reference
domain 28 of titin [$(I28)_{12}$]				
Nine identical repeats of the C2A domain of synaptotagmin I[$(C2A)_9$]	β-barrel with N- and C-terminal ends pointing in the same direction; H-bonds parallel to the direction of applied force	38 nm (106 amino acids)	~60	Carrion-Vazquez et al. (2000)
Spectrin (native protein or recombinant fragment) containing repeats 13–18; α-actinin (four spectrin-like domains α1–α4)	Spectrin contains homologous repeats; each repeat forms a triple-helical, antiparallel coiled-coil	31.7 nm	~25–35	Rief et al. (1999b)
Four identical repeats of a rat calmodulin domain CaM_4	Seven α-helices in dumbbell shape; homogeneous distribution of intrahelix H-bonds	The protein extends in one step; 225 nm	~600	Carrion-Vazquez et al. (2000)

[a] Although structurally highly homologous, the Fn-III-like domains of titin and tenascin unfold at different forces: 180–200 pN for titin, and only ~110 pN for tenascin.

[b] Each complete unfolding of a domain is preceded by a fast initial extension of ~7 Å which reflects the existence of an unfolding intermediate.

A well-expressed deviation from entropic elasticity in the high-force stretching regime has also been observed for certain polysaccharides (Rief et al., 1997b; Marszalek et al., 1998). This deviation was attributed to either twisting and bending of bond angles (Rief et al., 1997b), or to chair–boat transitions of the glucopyranose ring of the stretched polysaccharide (Marszalek et al., 1998).

Stretching of multi-domain proteins presents an even more complicated case, where the force–extension curves are strings of successive enthalpic and entropic portions, reflecting the unfolding of individual domains in the multi-domain polypeptide chain, followed by stretching of the unfolded domain (Fig. 6). As such proteins are elongated as a result of the initial application of force, they undergo a typical entropic stretching at the beginning. At a certain force, however, one of the folded domains unfolds, adding significant length to the chain and relaxing the stress on the cantilever, which returns to its non-deflected state. The denatured portion of the polypeptide chain can now undergo entropic stretching, behaving like a typical polymer chain. Further extension creates forces high enough to unfold a second domain, which is then stretched entropically, etc. The unfolding and stretching of each individual domain creates an individual peak in the force curve, leading to the characteristic saw-tooth pattern, illustrated in Figs. 5 and 6. Table 3 enumerates the reported studies performed on different protein substrates. Considerable progress in interpreting the force curves has been achieved through the use of recombinant proteins consisting of identical, tandemly repeated folded domains (Table 3; see Fisher et al., 1999a, b; Carrion-Vazquez et al., 2000 and references cited therein), and of steered molecular dynamic simulations (Lu et al., 1998; Marszalek et al., 1999). An alternative way to interpret peaks in multi-peak force curves has been to image the material before and after pulling. For example, in the case of pulling well-defined two-dimensional bacterial surface layers, it has been possible to identify by subsequent imaging which bacterial pore domains have been unzipped from the close spatial association with neighboring subunits (Müller et al., 1999). These authors interpret the multiple, well-spaced peaks in a force curve as discrete extensions of the portions of the polypeptide chains that connect individual subunits. Using the same technique, Oesterhelt et al. (2000) were able to image purple membranes before and after extraction of individual bacteriorhodopsin molecules from the membranes. The force–extension curves were recorded after the initial imaging and interpreted in terms of sequential extraction and unfolding of individual α-helices. Some of the seven transmembrane α-helices constituting bacteriorhodopsin unfolded pair-wise, whereas others unfolded one by one; moreover, details in the structure of specific portions of the force curve allowed resolution of complex extraction and unfolding paths for some individual helices.

The data obtained thus far (Table 3) already allow some interesting insights into the stability of different domains. Folded β-barrel domains, such as the immunoglobulin-like domains of the giant muscle protein titin and the fibronectin-III-like domains of the extracellular matrix protein tenascin, unfold under high force, between 200 and 260 pN. Such a force is needed to simultaneously break several hydrogen bonds in the β-barrel that are situated perpendicularly to the direction of applied force. Another β-barrel, present in the C2A domain of the membrane protein synaptotagmin I, unfolds at lower forces; in this domain the hydrogen bonds stabilizing the structure are parallel to the direction of applied force, and are broken sequentially, requiring much lower force, ~60 pN (Carrion-Vazquez et al., 2000). The triple-helical, coiled-coils in the folded domains of spectrin, a component of the membrane-associated skeleton in erythrocytes, unravel at even lower forces, between 25 and 35 pN (Rief et al., 1999b). In contrast to the β-barrel structures, the tertiary structure of the spectrin domain is not stabilized by hydrogen bonds (hydrogen bonds stabilize the α-helices themselves but not the bundle as a whole); the hydrophobic interactions maintaining the integrity of the bundle structure are much weaker than the hydrogen bonds, and are thus ruptured at much lower force. Finally, proteins like calmodulin, although containing repeated folded domains, do not produce peaks in the force curves but behave as random elastic polypeptide chains under force (Carrion-Vazquez et al., 2000). Their "all α" domains that are stabilized by intrahelical hydrogen bonds do not resist unfolding, and the protein yields to force in one step. More recently it has been demonstrated with an artificial homopolymer of repeating α-helical T4 lysozymes molecules that it is possible to observe small individual peaks per protein monomer (Yang et al., 2000).

5. Measuring the viscoelastic properties of biological structures and macromolecules

It is clear that whenever the effective stiffnesses of the cantilever and the biological sample on the surface are comparable, and the probe is pushed into the sample, the sample undergoes measurable indentation of the local surface at the point of contact of the tip. When the stress

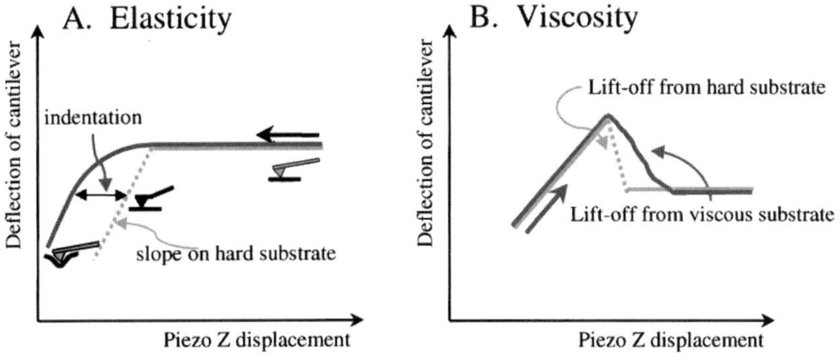

Fig. 7. Appearance of force curves when the sample indents under the tip (A) and when the jump-off-contact is slowed down due to viscosity of the sample (B).

(deformation force) and the strain (the amount of deformation) are linearly related, the deformation of the material is elastic, and the material will regain its original form upon relaxation. The depth of indentation can be used to measure local elasticity (Radmacher et al. (1992, 1994a, b) and Fritzsche and Henderson (1997); for reviews see Vinckier and Semenza (1998) and Heinz and Hoh (1999a)) in terms of Young's elastic modulus (the mechanical resistance of a material while elongating or compressing) (Fig. 7A). A detailed treatise of how AFM indentation data can be used to measure micromechanical properties of soft biological samples has been recently published (Costa and Yin, 1999).

The capability of AFM to provide information on the elastic properties of biological structures has been used to study different types of differentiated cells and organelles (for further references see Vinckier and Semenza, 1998; Heinz and Hoh, 1999a). The elastic modulus of thin films of gelatin was estimated to be 0.02×10^6 Pa (Pascal or N/m^2) (Domke and Radmacher, 1998), that of human chromosomes $0.05–0.1 \times 10^6$ Pa (Ikai et al., 1997), of microtubules 3×10^6 Pa (Vinckier et al., 1996), and that of lyzozyme 500×10^6 Pa (Radmacher et al., 1994b). The elastic modulus of DNA was shown to change with conformation of the molecule: B-DNA was characterized by a modulus of 290×10^6 Pa, whereas the modulus of the stretched (S-form) was estimated to be 2000×10^6 Pa, i.e., there was a 7-fold increase in the stiffness of the DNA double helix upon B- to S-transition (Noy et al., 1997a). For comparison, estimates from optical tweezers experiments gave a value of 350×10^6 Pa for B-form DNA (Smith et al., 1996).

The behavior of soft biological samples shows another type of deviation from that of a hard surface, this time on the retraction curve. Whereas the lift-off occurs quickly on hard surfaces, it may be considerably slowed down in the case of soft biological samples (Fig. 7B). The lift-off speed may be used to estimate the viscosity of the sample (Radmacher et al., 1994b).

There is little doubt that the use of AFM for elasticity and viscosity measurements will be broadened in the future.

6. Measuring interactions between cells

Recently, force–distance curves have been also utilized to identify cell partners that interact specifically in certain biological reactions. By functionalizing AFM tips with whole cells of a given type and studying their interaction with monolayers of other cell types, it was possible to identify the cell type in the uterine epithelium that interacts specifically with cells in the embryo during implantation (Thie et al., 1998). The technique introduced in this work offers novel approaches to the study of cell–cell interactions that are essential in many of the biological processes taking place in multicellular organisms.

7. Force maps

Laterally resolved force curves can be recorded during raster scanning in the *x–y* direction (Martin et al., 1987; Radmacher et al., 1994a, b; Baselt and Baldeschwieler, 1994). Individual curves can be assembled into a three-dimensional force volume (reviewed in Heinz and Hoh, 1999a). An alternative approach to producing spatially resolved force measurements is to create isoforce images across the sample, by assigning each point of the surface a separation distance at which a certain force is measured (Heinz and Hoh, 1999a). Collecting series of isoforce images at different forces can, in principle, reconstruct the same force volume as the one created by collecting individual laterally resolved force curves. The wealth of information in the force volume can be used to produce sample/surface maps reflecting different properties of the surface: adhesion, viscosity, elasticity, electrostatic interactions, etc. (e.g., Ludwig et al., 1997; Willemsen et al., 1998; Heinz and Hoh, 1999b).

A much faster approach to the point by point force curve measurement mentioned above has been to use the binding of a tethered antibody-modified tip to the substrate of the antibody to generate an effective force volume image (Raab et al., 1999). In this method, the image is distorted by enlarged widths and increased heights due to stretching of the antibody tether when the antibody binds the target antigen. This method permits location of the antigens, but quantitative force measurements require that a conventional force curve be acquired once the target is found in the image.

8. Concluding remarks

It is becoming increasingly clear that AFM can be used in a variety of different applications to study biologically relevant inter- and intramolecular interactions. The field of force spectroscopy is in an exponential phase of development: each year we witness not only numerous new examples of applications, but also significant developments in instrumentation. There is little doubt that the AFM, combined with other techniques, such as the laser optical tweezers, will further promote our comprehension of the complexity and intricacy of molecular interactions in a way unperceivable until recently.

Acknowledgements

We thank Dr. M. Karymov for discussion. S.H.L. is a National Cancer Institute Scholar.

References

Allen, S., Chen, X., Davies, J., Davies, M.C., Dawkes, A.C., Edwards, J.C., Roberts, C.J., Sefton, J., Tendler, S.J.B., Williams, P.M., 1997. Detection of antigen–antibody binding events with the atomic force microscope. Biochemistry 36, 7457–7463.

Ashkin, A., 1997. Optical trapping and manipulation of neutral particles using lasers. Proc. Natl. Acad. Sci. USA 94, 4853–4860.

Ashkin, A., Dziedzic, J.M., Yamane, T., 1987. Optical trapping and manipulation of single cells using infrared laser beams. Nature 330, 769–771.

Baselt, D.R., Baldeschwieler, J.D., 1994. Imaging spectroscopy with the atomic force microscope. J. Appl. Phys. 76, 33–38.

Bennick, U.L., Leuba, S.N., Leno, G.H., Zlatanova, J., de Grooth, B.G., Greve, J., Unfolding individual nucleosomes by stretching single chromatin fibres with optical tweezers. Submitted.

Binnig, G., Quate, C.F., Rohrer, C., 1986. Atomic force microscope. Phys. Rev. Lett. 56, 930–933.

Boland, T., Ratner, B.D., 1995. Direct measurement of hydrogen bonding in DNA nucleotide bases by atomic force microscopy. Proc. Natl. Acad. Sci. USA 92, 5297–5301.

Burnham, N.A., Colton, R.J., Pollock, H.M., 1993. Interpretation of force curves in force microscopy. Nanotechnology 4, 64–80.

Burnham, N.A., Kulik, A.J., 1998. Surface forces and adhesion. In: Bhushan, B. (Ed.), Handbook of Micro/Nanotribology. CRC Press, Boca Raton, FL.

Bustamante, C., Marko, J.F., Siggia, E.D., Smith, S., 1994. Entropic elasticity of λ-phage DNA. Science 265, 1599–1600.

Cappella, B., Baschieri, P., Frediani, C., Miccoli, P., Ascoli, C., 1997. Force–distance curves by AFM. IEEE Engng. Med. Biol. 16, 58–65.

Cappella, B., Dietler, G., 1999. Force–distance curves by atomic force microscopy. Surf. Sci. Rep. 34, 1–104.

Carrion-Vazquez, M., Marszalek, P.E., Oberhauser, A.F., Fernandez, J.M., 1999. Atomic force microscopy captures length phenotypes in single proteins. Proc. Natl. Acad. Sci. USA 96, 11288–11292.

Carrion-Vazquez, M., Oberhauser, A.F., Fisher, T.E., Marszalek, P.E., Li, H., Fernandez, J.M., 2000. The mechanical topology of protein studied by AFM. Prog. Biophys. Mol. Biol., 74, 63–91.

Chen, G.Y., Warmack, R.J., Huang, A., Thundat, T., 1995. Harmonic response of near-contact scanning force microscopy. J. Appl. Phys. 78, 1465–1469.

Chen, G.Y., Warmack, R.J., Thundat, T., Allison, D.P., Huang, A., 1994. Resonance response of scanning force microscopy cantilevers. Rev. Sci. Instr. 65, 2532–2537.

Chilkoti, A., Boland, T., Ratner, B.D., Stayton, P.S., 1995. The relationship between ligand-binding thermodynamics and protein–ligand interaction forces measured by atomic force microscopy. Biophys. J. 69, 2125–2130.

Cleveland, J.P., Manne, S., Bocek, D., Hansma, P.K., 1993. A nondestructive method for determining the spring constant of cantilevers for scanning force microscopy. Rev. Sci. Instr. 64, 403–405.

Cluzel, P., Lebrun, A., Heller, C., Lavery, R., Viovy, J.-L., Chatenay, D., Caron, F., 1996. DNA: an extensible molecule. Science 271, 792–794.

Costa, K.D., Yin, F.C.P., 1999. Analysis of indentation: implications for measuring mechanical properties with atomic force microscopy. J. Biomech. Engng. 121, 462–471.

Cui, Y., Bustamante, C., 2000. Pulling a single chromatin fiber reveals the forces that maintain its higher-order structure. Proc. Natl. Acad. Sci. USA 97, 127–132.

Dammer, U., Hegner, M., Anselmetti, D., Wagner, P., Dreier, M., Huber, W., Guntherodt, H.-J., 1996. Specific antigen/antibody interactions measured by force microscopy. Biophys. J. 70, 2437–2441.

Dammer, U., Popescu, O., Wagner, P., Anselmetti, D., Guntherodt, H.-J., Misevic, G.N., 1995. Binding strength between cell adhesion proteoglycans measured by atomic force microscopy. Science 267, 1173–1175.

Domke, J., Radmacher, M., 1998. Measuring the elastic properties of thin polymer films with the atomic force microscope. Langmuir 14, 3320–3325.

Engel, A., Lyubchenko, Y., Müller, D., 1999. Atomic force microscopy: a powerful tool to observe biomolecules at work. Trends Cell Biol. 9, 77–80.

Essevaz-Roulet, B., Bockelmann, U., Heslot, F., 1997. Mechanical separation of the complementary strands of DNA. Proc. Natl. Acad. Sci. USA 94, 11935–11940.

Evans, E., Berk, D., Leung, A., 1991. Detachment of agglutinin-bounded red blood cells. I. Forces to rupture molecular-point attachments. Biophys. J. 59, 838–848.

Evans, E., Ritchie, K., 1997. Dynamic strength of molecular adhesion bonds. Biophys. J. 72, 1541–1555.

Evans, E., Ritchie, K., 1999. Strength of a weak bond connecting flexible polymer chains. Biophys. J. 76, 2439–2447.

Fisher, T.E., Marszalek, P.E., Oberhauser, A.F., Carrion-Vazquez, M., Fernandez, J.M., 1999a. The micro-mechanics of single molecules studied with atomic force microscopy. J. Physiol. 520, 5–14.

Fisher, T.E., Oberhauser, A.F., Carrion-Vazquez, M., Marszalek, P.E., Fernandez, J.M., 1999b. The study of protein mechanics with the atomic force microscope. Trends Biochem. Sci. 24, 379–384.

Florin, E.L., Moy, V.T., Gaub, H.E., 1994. Adhesion forces between individual ligand–receptor pairs. Science 264, 415–417.

Flory, P.J., 1989. Statistical Mechanics of Chain Molecules. Hanser, Munich.

Frisbie, C.D., Rozsnyai, L.F., Noy, A., Wrighton, M.S., Lieber, C.M., 1994. Functional group imaging by chemical force microscopy. Science 265, 2071–2974.

Fritz, J., Katopodis, A.G., Kolbinger, F., Anselmetti, D., 1998. Force-mediated kinetics of single P-selectin/ligand complexes observed by atomic force microscopy. Proc. Natl. Acad. Sci. USA 95, 12283–12288.

Fritzsche, W., Henderson, E., 1997. Mapping elasticity of rehydrated metaphase chromosomes by scanning force microscopy. Ultramicroscopy 69, 191–200.

Gibson, C.T., Watson, G.S., Myhra, S., 1996. Determination of the spring constants of probes for force microscopy/spectroscopy. Nanotechnology 7, 259–262.

Grandbois, M., Beyer, M., Rief, M., Clausen-Schaumann, H., Gaub, H.E., 1999. How strong is a covalent bond? Science 283, 1727–1730.

Grigg, D.A., Russell, P.E., Griffith, J.E., 1992. Tip–sample forces in scanning probe microscopy in air and vaccum. J. Vac. Sci. Technol. A 10, 680–683.

Hansma, P.K., Cleveland, J.P., Radmacher, M., Walters, D.A., Hillner, P.E., Bezanilla, M., Fritz, M., Vie, D., Hansma, H.G., Prater, C.B., Massie, J., Fukunaga, L., Gurley, J., Elings, V., 1994. Tapping mode atomic force microscopy in liquids. Appl. Phys. Lett. 64, 1738–1740.

Heinz, W.F., Hoh, J.H., 1999a. Spatially resolved force spectroscopy of biological surfaces using the atomic force microscope. Trends Biotechnol. 17, 143–150.

Heinz, W.F., Hoh, J.H., 1999b. Relative surface charge density mapping with the atomic force microscope. Biophys. J. 76, 528–538.

Hinterdorfer, P., Baumgartner, W., Gruber, H.J., Schilcher, K., Schindler, H., 1996. Detection and localization of individual antibody–antigen recognition events by atomic force microscopy. Proc. Natl. Acad. Sci. USA 93, 3477–3481.

Hoh, J.H., Cleveland, J.P., Prater, C.B., Revel, J.-P., Hansma, P.K., 1992. Quantized adhesion detected with the atomic force microscope. J. Am. Chem. Soc. 114, 4917–4918.

Hutter, J.L., Bechhoefer, J., 1993. Calibration of atomic-force microscope tips. Rev. Sci. Instr. 64, 1868–1873.

Ikai, A., Mitsui, K., Tokuoka, H., Xu, X.M., 1997. Mechanical measurements of a single protein molecule and human chromosomes by the atomic force microscope. Mater. Sci. Engng. C 4, 233–240.

Israelachvili, J.N., 1992. Intermolecular and Surface Forces. Academic Press, New York.

Kellermayer, M.S., Smith, S.B., Granzier, H.L., Bustamante, C., 1997. Folding–unfolding transitions in single titin molecules characterized with laser tweezers. Science 276, 1112–1116.

Lebrun, A., Lavery, R., 1996. Modeling extreme stretching of DNA. Nucleic Acids Res. 24, 2260–2267.

Lee, G.U., Chrisey, L.A., Colton, R.J., 1994a. Direct measurement of the forces between complementary strands of DNA. Science 266, 771–773.

Lee, G.U., Kidwell, D.A., Colton, R.J., 1994b. Sensing discrete streptavidin–biotin interactions with atomic force microscopy. Langmuir 10, 354–357.

Leuba, S.H., Karymov, M.A., Liu, Y.Z., Lindsay, S.M., Zlatanova, J., 1999. Mechanically stretching single chromatin fibers. Gene Ther. Mol. Biol. 4, 297–301.

Leuba, S.H., Zlatanova, J., Karymov, M.A., Bash, R., Liu, Y.-Z., Lohr, D., Harrington, R.E., Lindsay, S.U., 2000. The mechanical properties of single chromatin fibers under tension. Single Mol. 1, 185–193.

Li, Y.Q., Tao, N.J., Pan, J., Garcia, A.A., Lindsay, S.M., 1993. Direct measurement of interaction forces between colloidal particles using the scanning force microscope. Langmuir 9, 637–641.

Lindsay, S.M., 2000. The scanning probe microscope in biology. In: Bonnell, D. (Ed.), Scanning Tunneling Microscopy and Related Techniques. Wiley, New York, in press.

Liu, Y.Z., Leuba, S.H., Lindsay, S.M., 1999. Relationship between stiffness and force in single molecule pulling experiments. Langmuir 15, 8547–8548.

Lu, H., Isralewitz, B., Krammer, A., Vogel, V., Schulten, K., 1998. Unfolding of titin immunoglobulin domains by steered molecular dynamics simulation. Biophys. J. 75, 662–671.

Ludwig, M., Dettmann, W., Gaub, H.E., 1997. Atomic force microscope imaging contrast based on molecular recognition. Biophys. J. 72, 445–448.

Marko, J.F., Siggia, E.D., 1995. Stretching DNA. Macromolecules 28, 8759–8770.

Marszalek, P.E., Lu, H., Li, H., Carrion-Vazquez, M., Oberhauser, A.F., Schulten, K., Fernandez, J.M., 1999. Mechanical unfolding intermediates in titin modules. Nature 402, 100–103.

Marszalek, P.E., Oberhauser, A.F., Pang, Y.P., Fernandez, J.M., 1998. Polysaccharide elasticity governed by chair–boat transitions of the glucopyranose ring. Nature 396, 661–664.

Martin, Y., Williams, C.C., Wickramasinghe, H.K., 1987. Atomic force microscope-force mapping and profiling on a sub 100 D scale. J. Appl. Phys. 61, 4723–4729.

Merkel, R., Nassoy, P., Leung, A., Ritchie, K., Evans, E., 1999. Energy landscapes of receptor–ligand bonds explored with dynamic force spectroscopy. Nature 397, 60–63.

Mitsui, K., Hara, M., Ikai, A., 1996. Mechanical unfolding of □$_2$-macroglobulin molecules with atomic force microscope. FEBS Lett. 385, 29–33.

Moy, V.T., Florin, E.-L., Gaub, H.E., 1994a. Intermolecular forces and energies between ligands and receptors. Science 266, 257–259.

Moy, V.T., Florin, E.-L., Gaub, H.E., 1994b. Adhesive forces between ligand and receptor measured by AFM. Colloid Surf. A 93, 343–348.

Mueller, H., Butt, H.-J., Bamberg, E., 1999. Force measurements on myelin basic protein adsorbed to mica and lipid bilayer surfaces done with the atomic force microscope. Biophys. J. 76, 1072–1079.

Müller, D.J., Baumeister, W., Engel, A., 1999. Controlled unzipping of a bacterial surface layer with atomic force microscopy. Proc. Natl. Acad. Sci. USA 96, 13170–13174.

Müller, D.J., Engel, A., 1997. The height of biomolecules measured with the atomic force microscope depends on electrostatic interactions. Biophys. J. 73, 1633–1644.

Müller, D.J., Engel, A., Amrein, M., 1997. Preparation techniques for the observation of native biological systems with the atomic force microscope. Biosens. Bioelect. 12, 867–877.

Noy, A., Vezenov, D.V., Kayyem, J.F., Meade, T.J., Lieber, C.M., 1997a. Stretching and breaking duplex DNA by chemical force microscopy. Chem. Biol. 4, 519–527.

Noy, A., Vezenov, D.V., Lieber, C.M., 1997b. Chemical force microscopy. Annu. Rev. Mater. Sci. 27, 381–421.

Oberhauser, A.F., Marszalek, P.E., Erickson, H.P., Fernandez, J.M., 1998. The molecular elasticity of the extracellular matrix protein tenascin. Nature 393, 181–185.

Oesterhelt, F., Oesterhelt, D., Pfeiffer, M., Engel, A., Gaub, H.E., Müller, D.J., 2000. Unfolding pathways of individual bacteriorhodopsins. Science 288, 143–146.

Pierres, A., Benoliel, A.-M., Bongrand, P., 1996. Measuring bonds between surface-associated molecules. J. Immunol. Methods 196, 105–120.

Putman, C.A., van der Werf, K.O., de Grooth, B.G., van Hulst, N.F., Greve, J., 1994. Tapping mode atomic force microscopy in liquids. Appl. Phys. Lett. 64, 2454–2456.

Raab, A., Han, W., Badt, D., Smith-Gill, S.J., Lindsay, S.M., Schindler, H., Hinterdorfer, P., 1999. Antibody recognition imaging by force microscopy. Nat. Biotechnol. 17, 901–905.

Radmacher, M., Cleveland, J.P., Fritz, M., Hansma, H.G., Hansma, P.K., 1994a. Mapping interaction forces with the atomic force microscope. Biophys. J. 66, 2159–2165.

Radmacher, M., Fritz, M., Cleveland, J.P., Walters, D.A., Hansma, P.K., 1994b. Imaging adhesion forces and elasticity of lysozyme adsorbed on mica with the atomic force microscope. Langmuir 10, 3809–3814.

Radmacher, M., Tillmann, R.W., Fritz, M., Gaub, H.E., 1992. From molecules to cells: imaging soft samples with the Atomic Force Microscope. Science 257, 1900–1905.

Rief, M., Clausen-Schaumann, H., Gaub, H.E., 1999a. Sequence-dependent mechanics of single DNA molecules. Nat. Struct. Biol. 6, 346–349.

Rief, M., Gautel, M., Oesterhelt, F., Fernandez, J.M., Gaub, H.E., 1997a. Reversible unfolding of individual titin immunoglobulin domains by AFM. Science 276, 1109–1112.
Rief, M., Gautel, M., Schemmel, A., Gaub, H.E., 1998. The mechanical stability of immunoglobulin and fibronectin III domains in the muscle protein titin measured by atomic force microscopy. Biophys. J. 75, 3008–3014.
Rief, M., Oesterhelt, F., Heymann, B., Gaub, H.E., 1997b. Single molecule force spectroscopy on polysaccharides by atomic force microscopy. Science 276, 1295–1297.
Rief, M., Pascual, J., Saraste, M., Gaub, H.E., 1999b. Single molecule force spectroscopy of spectrin repeats: low unfolding forces in helix bundles. J. Mol. Biol. 286, 553–561.
Ros, R., Schwesinger, F., Anselmetti, D., Kubon, M., Schafer, R., Pluckthun, A., Tiefenauer, L., 1998. Antigen binding forces of individually addressed single-chain Fv antibody molecules. Proc. Natl. Acad. Sci. USA 95, 7402–7405.
Sader, J.E., 1995. Parallel beam approximation for v-shaped atomic force microscope cantilevers. Rev. Sci. Instr. 66, 4583–4587.
Sader, J.E., Larson, I., Mulvaney, P., White, L.R., 1995. Method for the calibration of atomic force microscope cantilevers. Rev. Sci. Instr. 66, 3789–3798.
Sader, J.E., White, L., 1993. Theoretical analysis of the static deflection of plates for atomic force microscope applicaitons. J. Appl. Phys. 74, 1–9.
Samori, B., 1998. Stretching, tearing, and dissecting single molecules of DNA. Angew. Chem. Int. Ed. 37, 2198–2200.
Sarid, D., 1991. Scanning Force Microscopy. Oxford University Press, New York.
Schwesinger, F., Ros, R., Strunz, T., Anselmetti, D., Güntherodt, H.-J., Honegger, A., Jermutus, L., Tiefenauer, L., Plückthun, A., 2000. Unbinding forces of single antibody–antigen complexes correlate with their thermal dissociation rates. Proc. Natl. Acad. Sci. USA, 197, 9972–9977.
Senden, T.J., Ducker, W.A., 1994. Experimental determination of spring constants in atomic force microscopy. Langmuir 10, 1003–1004.
Siedlecki, C.A., Marchant, R.E., 1998. Atomic force microscopy for characterization of the biomaterial interface. Biomaterials 19, 441–454.
Smith, S.B., Cui, Y., Bustamante, C., 1996. Overstretching B-DNA: the elastic response of individual double-stranded and single-stranded DNA molecules. Science 271, 795–799.
Smith, S.B., Finzi, L., Bustamante, C., 1992. Direct mechanical measurements of the elasticity of single DNA molecules by using magnetic beads. Science 258, 1122–1126.
Stuart, J.K., Hlady, V., 1995. Effects of discrete protein–surface interactions in scanning force microscopy adhesion force measurements. Langmuir 11, 1368–1374.
Stuart, J.K., Hlady, V., 1999. Reflection interference contrast microscopy combined with scanning force microscopy verifies the nature of protein–ligand interaction force measurements. Biophys. J. 76, 500–508.
Tendler, S.J.B., Davies, M.C., Roberts, C.J., 1996. Molecules under the microscope. J. Pharm. Pharmacol. 48, 2–8.
Thie, M., Röspel, R., Dettmann, W., Benoit, M., Ludwig, M., Gaub, H.E., Denker, H.-W., 1998. Interactions between trophoblast and uterine epithelium: monitoring of adhesive forces. Hum. Reprod. 13, 3211–3219.
Torii, A., Sasaki, M., Hane, K., Okuma, S., 1996. A method for determining the spring constant of cantilevers for atomic force microscopy. Meas. Sci. Technol. 7, 179–184.
Vinckier, A., Dumortier, C., Engelborghs, Y., Hellemans, L., 1996. Dynamic and mechanistic study immobilized microtubules with atomic force microscopy. J. Vac. Sci. Technol. B. 14, 1427–1431.
Vinckier, A., Gervasoni, P., Zaugg, F., Ziegler, U., Lindner, P., Groscurth, P., Plückthun, A., Semenza, G., 1998. Atomic force microscopy detects changes in the interaction forces between GroEL and substrate proteins. Biophys. J. 74, 3256–3263.
Vinckier, A., Semenza, G., 1998. Measuring elasticity of biological materials by atomic force microscopy. FEBS Lett. 430, 12–16.
Wagner, P., 1998. Immobilization strategies for biological scanning probe microscopy. FEBS Lett. 430, 112–115.
Willemsen, O.H., Snel, M.M.E., van der Werf, K.O., de Grooth, B.G., Greve, J., Hinterdorfer, P., Gruber, H.J., Schindler, H., van Kooyk, Y., Figdor, C.G., 1998. Simultaneous height and adhesion imaging of antibody–antigen interactions by atomic force microscopy. Biophys. J. 75, 2220–2228.
Wong, S.S., Joselevich, E., Woolley, A.T., Cheung, C.L., Lieber, C.M., 1998. Covalently functionalized nanotubes as nanometre-sized probes in chemistry and biology. Nature 394, 52–55.

Yang, G., Cecconi, C., Baase, W.A., Vetter, I.R., Breyer, W.A., Haack, J.A., Matthews, B.W., Dahlquist, F.W., Bustamante, C., 2000. Solid-state synthesis and mechanical unfolding of polymers of T4 lysozyme. Proc. Natl. Acad. Sci. USA 97, 139–144.

Yip, C.M., Yip, C.C., Ward, M.D., 1998. Direct force measurements of insulin monomer–monomer interactions. Biochemistry 37, 5439–5449.

Zhong, Q., Inniss, D., Kjoller, K., Elings, V.B., 1993. Fractured polymer/silica fiber surface studied by tapping mode atomic force microscopy. Surf. Sci. Lett. 290, L688–L692.

Biology at the Single Molecule Level
S.H. Leuba an J. Zlatanova (Eds)

Review

Mechanical design of proteins studied by single-molecule force spectroscopy and protein engineering

Mariano Carrion-Vazquez*, Andres F. Oberhauser, Thomas E. Fisher, Piotr E. Marszalek, Hongbin Li, Julio M. Fernandez

Department of Physiology and Biophysics, Mayo Foundation, 1-159 Medical Sciences Building, Rochester, MN 55905, USA

Abstract

Mechanical unfolding and refolding may regulate the molecular elasticity of modular proteins with mechanical functions. The development of the atomic force microscopy (AFM) has recently enabled the dynamic measurement of these processes at the single-molecule level. Protein engineering techniques allow the construction of homomeric polyproteins for the precise analysis of the mechanical unfolding of single domains. α-Helical domains are mechanically compliant, whereas β-sandwich domains, particularly those that resist unfolding with backbone hydrogen bonds between strands perpendicular to the applied force, are more stable and appear frequently in proteins subject to mechanical forces. The mechanical stability of a domain seems to be determined by its hydrogen bonding pattern and is correlated with its kinetic stability rather than its thermodynamic stability. Force spectroscopy using AFM promises to elucidate the dynamic mechanical properties of a wide variety of proteins at the single molecule level and provide an important complement to other structural and dynamic techniques (e.g., X-ray crystallography, NMR spectroscopy, patch-clamp). © 2000 Elsevier Science Ltd. All rights reserved.

Contents

*Corresponding author. Tel.: +1-507-284-9442; fax: +1-507-284-0521.
E-mail address: carrion.mariano@mayo.edu (M. Carrion-Vazquez).

---

1. Introduction and overview

The importance of mechanical forces at the molecular level of biological systems has become increasingly apparent during the last few years (Bensimon, 1996; Evans and Ritchie, 1997; Erickson, 1994, 1997; Hynes, 1999; Chicurel et al., 1998; Ohashi et al., 1999; Shaub, 1999; Bustamente et al., 2000). Since many proteins are exposed to mechanical tension, a variety of topological solutions to use or resist force may have arisen through evolution. An example of a protein under mechanical tension is the muscle protein titin, which is responsible of the passive elasticity of muscle. Proteins from the extracellular matrix (ECM) and the cytoskeleton also operate under tension. Evidence from in vivo studies suggests that tension on the ECM protein fibronectin triggers domain unfolding exposing cryptic binding sites that are important for the ECM assembly (Ohashi et al., 1999). The physical properties underlying the mechanical functions of proteins are largely unknown, but one structural characteristic common to many proteins exposed to mechanical tension is that they contain multiple individually folded domains.

Atomic force microscopy (AFM) has been applied recently to the analysis, at the single-molecule level, of the mechanical properties of modular proteins with mechanical functions from muscle, ECM, and cytoskeleton. This has made it possible to examine the elastic properties of proteins composed of a variety of heterogeneous tandem modules, such as the immunoglobulin (Ig) and fibronectin type III (FN-III) domains of titin (Rief et al., 1997a, 1998), the FN-III domains of tenascin (Oberhauser et al., 1998), and the triple helical coiled-coils of spectrin (Rief et al., 1999).

The combination of protein engineering techniques with single-molecule AFM has recently allowed the study of the mechanical properties of single modules (Carrion-Vazquez et al., 1999a,b;

Yang et al., 2000). A prerequisite for these studies was the construction of identical tandem repeats of the domain under study. Recombinant DNA technology has made homomeric polyproteins available, which has opened the door to the analysis of the mechanical properties of individual modules. These techniques have built upon information derived from NMR spectroscopy and dynamic molecular simulations to identify the determinants of mechanical stability in an immunoglobulin domain from human cardiac titin. This combined approach allows us to examine the determinants of stability in different types of protein folds and to compare their mechanical properties.

Several reviews have recently appeared covering the emerging field of single-protein force spectroscopy by AFM (Engel et al., 1999; Ludwig et al., 1999; Fisher et al., 1999a,b; Fisher et al., 2000a,b; Strick et al., 2000; Janshoff et al., 2000). This review will outline the recent combination of AFM and molecular biology to study the molecular determinants of the mechanical stability of proteins. In particular, it will discuss the application of this approach to compare the mechanical stability of proteins that are α-helical to those that are β-barreled and to examine the mechanical properties of proteins that lack known mechanical function.

2. The mechanical unfolding of single proteins by AFM

In contrast to classical imaging instruments such as light and electron microscopes, scanning probe microscopes use a stylus to scan a sample surface and monitor a variety of interactions between the stylus tip and the sample. Since its invention in 1986 (Binnig et al., 1986), the AFM has been the most widely used scanning probe microscope for biological applications. The AFM is a remarkably simple instrument that works like a miniature phonograph and is capable of imaging single macromolecules with submolecular resolution (Bustamante et al., 1997; Czajkowsky and Shao, 1998). The "force spectroscopy" or "force-measuring" mode of the AFM was later developed to measure nanomechanical properties of materials (Burnham and Colton, 1989) and has allowed the measurement of the binding forces of complementary DNA strands (Lee et al., 1994) and receptor–ligand systems (Florin et al., 1994), as well as the forces that maintain the conformation of sugar rings in a polysaccharide (Rief et al., 1997b; Marszalek et al., 1998, 1999a) and folded domains in a modular protein (Rief et al., 1997a). This technique is capable of measuring the forces required to unfold protein domains with piconewton (pN) sensitivity and the length changes with Ångstrom resolution. In this configuration, a single protein is suspended between a sharp tip of silicon nitride, mounted on a cantilever, and a substrate mounted on a high precision piezoelectric positioner to allow control of its vertical movement. In a typical experiment (see Fig. 1), the tip of the cantilever is brought into contact with a layer of protein attached to the substrate, then the piezoelectric positioner retracts. When a portion of a single protein molecule is picked up at random by the tip through adsorption, the retraction of the positioner stretches the suspended segment of the protein. The first source of resistance to extension of the protein are entropic forces (like any other polymer, proteins tend to coil up to maximize the disorder of their constituent segments). Extension of the molecule reduces its entropy producing a restoring force that results in bending of the cantilever. This deflection changes the angle of reflection of a laser beam bounced off the surface of the cantilever and is measured as a change in output from a photodetector. Using the spring constant of the cantilever, the relationship between deflection and force can be calibrated and the force calculated. The entropic elasticity of the protein can be formally described by the worm-like chain (WLC) model of polymer elasticity (Fig. 1). The additional extension of the protein may cause unraveling of the protein fold increasing the effective length of the polymer between the tip and the substrate and dropping the force acting on the cantilever to near zero.

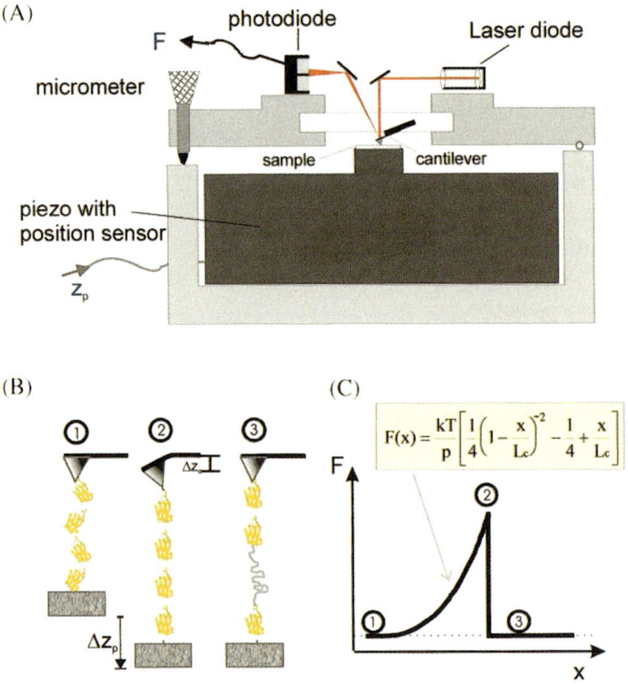

Fig. 1. Using the atomic force microscope to measure the molecular elasticity of single proteins. (A) A schematic of our custom-built AFM. The AFM uses a Digital Instruments AFM detector head mounted on top of a single-axis piezoelectric positioner from Physik Instrumente. The positioner has a capacitative sensor with a Z-axis resolution of 1 Å. Data acquisition and the voltage control of the piezoelectric positioner are achieved using a PC-mounted data acquisition board controlled by custom-made LabView (National Instruments) software. The force is measured by the deflection of the cantilever and the extension can be calculated from the position of the piezoelectric actuator (z_p). (B) The unfolding of a protein domain by an external force. When axial stress is applied to a folded domain the protein will unravel. As the distance between substrate and cantilever increases (from states 1 to 2) the protein elongates and the reduction of its entropy generates a restoring force that bends the cantilever. When a domain unfolds (state 3) the contour length of the protein increases, returning the force on the cantilever to near zero. Further extension again results in force on the cantilever (state 1 again). (C) The entropic elasticity of proteins can be described by the worm-like chain (WLC) model of polymer elasticity (inset). This equation predicts the entropic restoring force (F) generated upon extension (x) of a protein in terms of its persistence length (p) and its contour length (L_C) (Rief et al., 1997a; Marko and Siggia, 1995). In general, polymer elasticity results from the tendency of a chain to coil in order to maximize its conformational freedom (entropy) under the drive of thermal fluctuations. Extending a relaxed polymer chain generates an opposing force that is predicted from the reduction in the entropy of the polymer. This effect is known as entropic elasticity and is the basis of the elasticity of proteins. The saw-tooth pattern of peaks on the force-extension relationship corresponds to sequential unraveling of individual domains of a modular protein like the one shown here. The numbers correspond to the stages marked in 1(B).

Until recently, the only way to measure the stability of a protein was to change its physical (e.g., with heat or high pressure) or chemical environment (e.g., with guanidinium chloride or urea, acidic or alkaline pH) and monitor the loss of protein conformation, using spectroscopic techniques from which folding free energies could be obtained. However, the series of conformational changes that a protein undergoes in response to mechanical stress, as may occur in vivo, may be very different. AFM force spectroscopy offers several advantages over thermal or chemical denaturation. First of all, AFM allows the measurement of folding reactions at the single-molecule level and therefore can uncover rare unfolding events that are not observed by using traditional bulk methods, in which measured parameters are averaged over populations of many molecules. It acts directly on the protein molecule, leaving its environment untouched. It can increase the rate of unfolding exponentially, reducing considerably the experimental time. It allows a direct measurement of the mechanical stability of the fold. The process of mechanical unfolding is therefore defined as a function of extension (here the reaction coordinate is well defined, i.e. length), which in the case of proteins exposed to stress in vivo, may be a more physiological perturbation. In contrast to other structural techniques (i.e. X-ray crystallography), force spectroscopy is dynamic and it works in real time. Tiny amounts of protein are needed (typically a few μg), it is not limited to a range of molecular weights, and it works in aqueous solution, so crystal formation is not required.

3. The use of engineered polyproteins in force spectroscopy

3.1. The advantage of polyproteins for AFM studies

When a cantilever tip is pressed against a layer of protein and then withdrawn, many protein molecules attached to the substrate will interact with the tip and thus transmit a force that will bend the cantilever. Only rarely will the entire properly folded molecule of interest be suspended between substrate and tip. More commonly, the interaction will involve only a fragment of the protein, which may be in a denatured form. The consequences of such "nonspecific" interactions are illustrated in Fig. 2. The early part of the force–extension curve, which reflects the time during which the cantilever tip is close to the substrate, is typically characterized by large fluctuations in force. This poses two difficulties to the interpretation of AFM force–extension curves. First, the forces involved in extending short molecules may be masked by the "noise" caused by such nonspecific interactions. Second, since almost all recordings show force peaks reflecting such interactions between the cantilever tip and the adsorbed protein layer, a means is needed to differentiate true unfolding events from spurious interactions.

The problem of nonspecific interactions was surmounted in the pioneer AFM studies by the use of multidomain proteins. Thus, the presence of multiple Ig and FN-III domains in titin and tenascin enabled the detection of domain unfolding (Rief et al., 1997a; Oberhauser et al., 1998). However, modular proteins are constituted by heterogeneous populations of modules (differing in size, structure, sequence and stability) and it is not possible to identify the force peaks corresponding to the unfolding of specific domains. The construction of homomeric polyproteins (i.e., tandem repeats of a single domain) now allows the study of the mechanical properties of

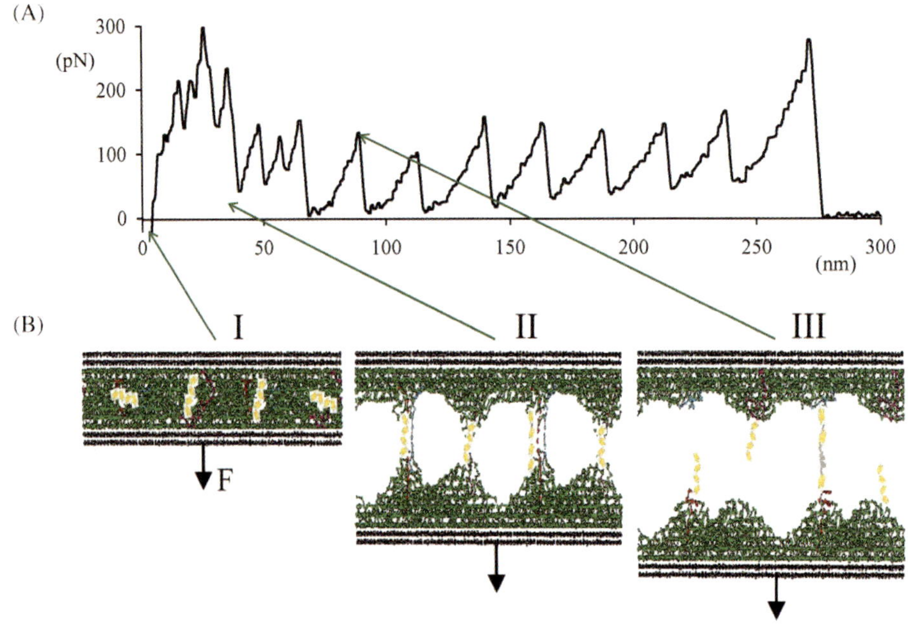

Fig. 2. (A) Force–extension recording of a single polyprotein measured with AFM techniques. A typical individual protein domain of 100 amino acids would have a folded length of 1.5 to 6 nm but an unfolded length of about 40 nm (length of an amino acid: 0.40 nm). In AFM, the first 30 to 75 nm are typically obscured by nonspecific interactions between the cantilever tip and the polyprotein layer. (B) A hypothetical series of events as an AFM tip is pressed against a monolayer of polymers and then withdrawn (modified from Baljon and Robbins, 1996). Colors represent different conformations of the same polyprotein, yellow being the native fold. Panel I shows the compression of the protein layer as the tip is pressed against it. As the substrate is withdrawn (panel II) some of the molecules are adsorbed to the tip. The bonds between the two layers are gradually broken until a single protein forms the only connection between the two surfaces. On further extension, the independent domains of the protein begin to unfold (panel III).

specific domains (Carrion-Vazquez et al., 1999a). Unlike native modular proteins, polyproteins have a perfect repetitive structure that results in a unique periodical saw-tooth pattern in force-extension recordings. This "fingerprint" clearly demonstrates that domains from a single molecule are being unfolded and enables the measurement of the force required to unfold those domains as well as their size.

Studies with the I27 module of human cardiac titin (see below) have shown several more advantages of polyproteins in force spectroscopy. As we shall see in Section 5.1.1, polyproteins produce a linear increase of the spatial resolution which is positively correlated with the increasing number of modules in the polyprotein (i.e., there is an increase of the signal-to-noise ratio). The use of this "amplification" by repetition of protein modules allows amino acid resolution to be easily obtained in force–extension recordings and enables accurate module sizing of both the folded and unfolded domains (Carrion-Vazquez et al., 1999b). This amplification by polyproteins has also uncovered subtle folding intermediates in the mechanical unfolding of Ig domains from titin (Marszalek et al., 1999b).

3.2. Periodic patterns of force peaks measured by AFM represent true unfolding events

Proteins tend to denature to some extent when adsorbed onto a surface (Buijs et al., 1996; Hlady and Buijs, 1998). Therefore, it is important to demonstrate that the patterns observed with AFM correspond to true unfolding events rather than to the detachment of denatured proteins adsorbed onto a surface. With the use of polyproteins in force spectroscopy, a body of experimental evidence has accumulated that refutes protein desorption as an explanation for the force–extension curves observed. The following is a list of these arguments (many of which are based on observations described in Section 5.1.1):

(1) The rate of module unfolding of I27 measured by AFM ($3.3 \times 10^{-4}\,s^{-1}$) is comparable to that obtained from chemical denaturation experiments ($4.9 \times 10^{-4}\,s^{-1}$) (Carrion-Vazquez et al., 1999a). Similar results are obtained using poly-I28, which has also comparable mechanical ($2.8 \times 10^{-5}\,s^{-1}$) and chemical ($1.6 \times 10^{-5}\,s^{-1}$) unfolding rates (Li et al., 2000). These similarities would be difficult to explain using a disorption model and provide strong evidence that saw-tooth patterns represent true protein unfolding events.

(2) When a polyprotein is stretched by AFM, the interval between peaks is highly reproducible and corresponds closely to that expected based on NMR and molecular dynamics simulations. In the case of the I27 module from human cardiac titin (see Section 5.1.1 for details) the folded core encompasses about 72 amino acids (there is uncertainty about its boundaries) and the observed distance between peaks (28.1 ± 0.17 nm) corresponds to 70 amino acids, considering that, as we shall see in Section 5.1.1.5., the length of each amino acid residue is 0.40 nm ($70 \times 0.40\,\text{nm} \approx 28.1\,\text{nm}$). Since the I27 domain has a total of 89 amino acids, these results imply that the "spacer sequence" between the folded cores of the polyprotein is fully extended prior to the unraveling events. This is inconsistent with a desorption hypothesis, which would predict equal intervals corresponding to the full folded length of each domain or an alternating pattern of two different intervals corresponding to two elements in series ("spacers" and "folded core"). Furthermore, when insertions of 5 glycine residues where made inside the folded core the spacing between peaks increased by the expected distance; similar insertions made outside of this region did not change the intervals (Carrion-Vazquez et al., 1999b). These experiments confirm that the spacer regions are extended prior to unfolding.

(3) Force–extension recordings of a $(I27–I28)_4$ heterodimeric polyprotein (constructed out of 4 repeats of a I27–I28 dimer) show saw-tooth patterns with force peaks clustered into two classes, the first 4 peaks being ~200 pN and the remaining 4 peaks being ~300 pN (pulling speed: 0.6 nm/ms). The force of unfolding of I27 and I28 homomeric polyproteins suggests that the weaker peaks represent unfolding of I27 domains while the stronger ones represent unfolding of I28 modules. If the force peaks were due to sequential desorption of the modules an alternating pattern of force peaks would be expected for this construct, according with its linear arrangement (Fig. 3; Li et al., 2000).

(4) When titin modules are stretched, the force peaks of the force–extension curves tend to be ordered from weakest to strongest (Rief et al., 1997a). This suggests that in the protein modules are unfolding in an order according to their mechanical stability rather than desorbing in a linear fashion from the point of pulling.

(5) Additional evidence that favors the unfolding model comes from the fact that the forces (pulling speed: 0.6 nm/ms) and the spacing between peaks are distant and different for different

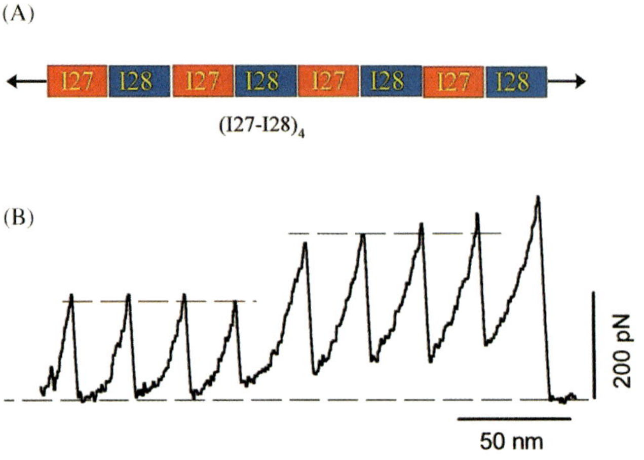

Fig. 3. The extension of a heterodimeric polyprotein rules out desorption as an explanation for the force spectroscopy observations. (A) Schematic diagram of the (I27–I28)$_4$ polyprotein. (B) The force–extension curve of a full length (I27–I28)$_4$ polyprotein shows two levels of unfolding forces (pulling speed: 0.6 nm/ms) with the less stable domains unfolding first (at about 200 pN, as expected for the I27 domain) followed by the more stable domains (at about 300 pN, close to that expected for I28). The unfolding of domains according to their mechanical stability, irrespective of their order in the heteropolyprotein, is inconsistent with an alternative hypothesis in which the force peaks represent detachment of modules adsorbed to the substrate. Note that the last peak corresponds to the extension and detachment of the fully unfolded protein from the AFM tip.

folds, as we will show in this review. Whereas I27 domains unfold at ~200 pN with spacing of 28.1 nm, the corresponding values for the C2A domain are ~60 pN and 38 nm and <20 pN and unknown for calmodulin.

Taken together, these observations effectively rule out the possibility that the AFM observations are an artifact of desorption from the substrate.

3.3. Does the pulling geometry affect the AFM measurements?

In our discussion up to now we have assumed an orthogonal geometry of pulling, i.e. that the polyprotein is being pulled at a 90° angle, with the two attachment points of the protein (to the cantilever tip and to the substrate) in a vertical line. A more realistic picture is that the polyprotein would be pulled at a slight angle. Fig. 4A illustrates this point. Prior to extension, the distance between the two ends of a folded domain will be 4.4 nm (as estimated by NMR; Improta et al., 1996). The angle between the fully extended domain (28.1 nm) and the substrate will therefore be 77° (see Fig. 4B). This angle would result in an error in the estimate of protein length of less than 1%, which represents the worst-case scenario. In a multidomain protein the maximum distance between the attachment points will be given by 4.4 nm multiplied by the number of modules; the proportional increase and therefore the angle of extension would remain the same. Such geometrical errors in the alignment of the attachment points are therefore negligible for our measurements.

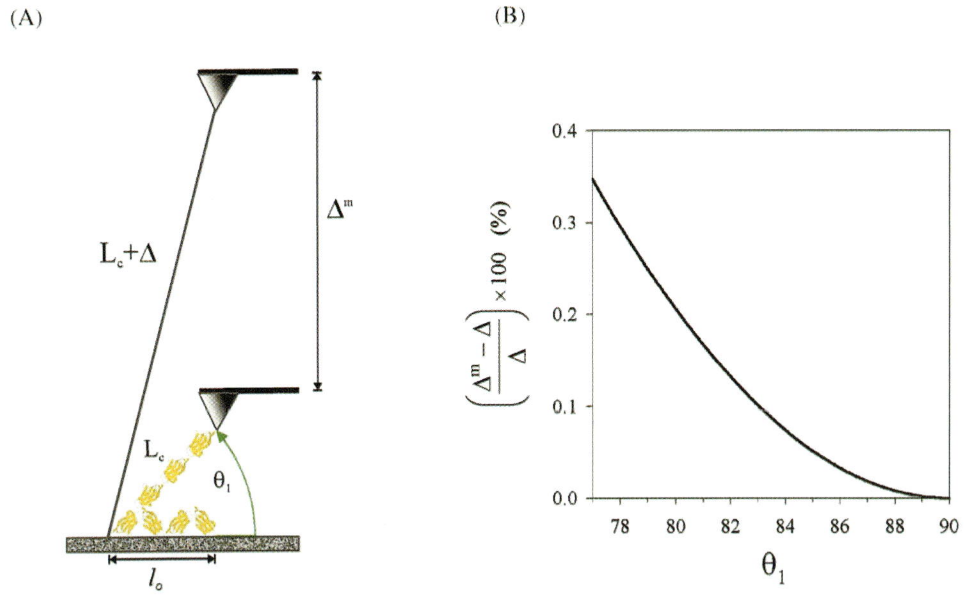

Fig. 4. Geometrical errors in the measurements of the contour length of a polyprotein. (A) A relaxed polyprotein with a length l_0 is picked up by the AFM tip and stretched by force to a folded contour length L_c. The stretched polyprotein is pulled at an angle θ_1. Module unfolding extends the protein to a contour length $L_c+\Delta$. The measured increase, Δ^m, is a function of θ_1. **(B)** We calculate this error as a function of θ_1 for a typical polyprotein composed of 10 identical I27 domains with a random end-to-end length of $l_0 = 8.6$ nm and a contour length of $L_c = 38$ nm and $\Delta = 280$ nm. Under these conditions, the maximal error in the measurement of Δ occurs at a $\theta_1 = 77°$ and is less than 1%. Molecules stretched at higher θ_1 values have a lower error. When $\theta_1 = 90°$ the theoretical error is zero (see text). Modified from Carrion-Vazquez et al. (1999b).

4. Constructing polyproteins

Most of the reported methods to synthesize tandem repeats of a DNA sequence are laborious and have therefore been used only rarely (Graham and Maio, 1992). The method we use routinely is based in a single-step cloning of Ava I concatemers (Hartley and Gregori, 1981) into a custom-made expression vector (Carrion-Vazquez et al., 1999a).

Our method relies on the concatemerization of DNA monomers, flanked by restriction enzyme sticky ends, using in vitro self-ligation. The restriction enzyme Ava I is used since its restriction site is asymmetrical, allowing directional assembly of the monomers (i.e., head to tail). Monomers of the expressed polyprotein will therefore be connected by a Leu-Gly linker, which corresponds to the translation of codons of the Ava I site. This technique was originally reported by Hartley and Gregori (1981) and considered as one of the most reliable methods to make tandem repeats of a DNA sequence (Graham and Maio, 1992). Fig. 5A shows a ladder of ligation products separated by analytical electrophoresis on an agarose gel.

pET vectors are considered to be the most powerful system developed for cloning and expression of recombinant proteins in *E. coli* (Studier et al., 1990). A new pET vector tailored to our specific needs was constructed ("pET Ava I"). This consists of a short His-tag on the N-terminus (for the affinity purification of the protein), an enterokinase site (for the eventual cleavage

(A) (B)

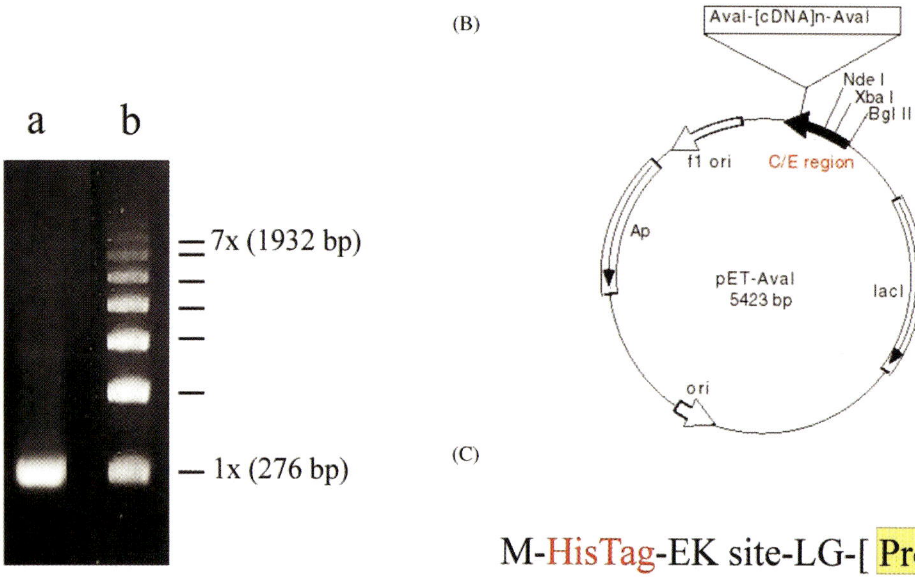

(C)

M-HisTag-EK site-LG-[Protein-LG]$_n$-CC

Fig. 5. Engineering polyproteins for AFM. (A) Directional DNA concatemerization of monomers is done by self-ligation of the sticky ends of the nonpalindromic CTCGGG *Ava*I restriction site. This figure shows an agarose gel stained with ethidium bromide showing the I27 monomer (lane a) and a ladder of concatemers with various numbers of I27 monomers (lane b). (B) Map of our custom-made expression vector designed for single-step cloning of the Ava I concatemers. Its cloning/expression region includes an N-terminal His-tag (for affinity purification of the polyprotein), an enterokinase site (for easy cleavage of the His-tag out of the fusion protein), the CTCGGG AvaI cloning site, two cysteine residues at the C-terminus (for covalent attachment of the polyprotein to gold-coated coverslips) and a stop codon. (C) Structure of the repetitive protein generated by this method.

of the His-tag), a unique CTCGGG Ava I cloning site (for cloning of the Ava I multimers), and two cysteine residues on the C-terminus (for the covalent attachment of the polyprotein to gold coated coverslips) (Fig. 5B). The resulting polyproteins have a short His-tag and an enterokinase site on the N-terminus, two cysteine residues on the C-terminus, and a Leu-Gly linker between monomers (Fig. 5C). There is evidence that such His-tags remain unstructured in solution and do not affect the properties of the flanking regions (Politou et al., 1994a,b, 1996; Improta et al., 1996).

The first step in this method is the synthesis by PCR of the monomer carrying CTCGGG Ava I sites on both ends. After Ava I digestion, the sticky-ended monomer is gel purified and then self-ligated at high concentration. These concatemers are then ligated into "pET Ava I" and then transformed into a recombinase-defective *E. coli* cloning strain to prevent the deletion of tandem repeats by recombination. Clones bearing high-order multimers are selected and transformed into a recombinase-defective expression strain to express the polyprotein. Purification is usually done specifically by affinity chromatography using Ni^{2+}-columns, which bind the His-tagged polyproteins.

4.1. Construction of Ava I monomers

The first step in our method is the synthesis of PCR mutagenic primers bearing the appropriate sequence modifications to include the Ava I nonpalindromic sites at both ends. Since the fidelity of copy is critical in protein expression, we use for this amplification Pfu DNA polymerase (Stratagene), which has the lowest error rate available due to its proofreading activity. We have used three different PCR templates in our studies: titin template was a cDNA insert carrying 8 Ig domains (I27–I34) of human cardiac titin in a pET plasmid [I27–I34-CC/pET8c-(His)], obtained from Dr. M. Gautel (Rief et al., 1997a). The calmodulin template was a cDNA encoding rat calmodulin in a pET 14b plasmid (pETCM clone), obtained from Dr. K. Titani (Hayashi et al., 1998). The C2A template was a cDNA clone from rat synaptotagmin I (pCMV65-5 clone), obtained from Dr. T.H. Südhof (Perin et al., 1990). For I27 and calmodulin an extra Gly codon was added upstream of the Leu-Gly linker encoded by the Ava I site.

The PCR products were subcloned and sequenced following standard techniques (Sambrook et al., 1989).

4.2. *Synthesis of concatemers*

The subcloned monomers are digested with Ava I, gel purified and self-ligated overnight at 16°C. Ligation products are then separated by analytical electrophoresis in an agarose gel and stained with ethidium bromide.

To prevent self-ligation of the concatemers, which would result in circular concatemers unable to ligate into a vector, we start with a high concentration of monomers.

4.3. *Cloning of concatemers into pET Ava I*

The Ava I concatemers are subcloned into a dephosphorylated pET Ava I expression vector (Fig. 5B) by bulk ligation. The ligation products are transformed into SURE 2 (rec B^-, rec J^-) supercompetent cells (Stratagene) and the colonies analyzed directly by DNA purification and restriction digestion or by PCR colony screening using Platinum Taq DNA polymerase (Gibco-BRL). This is the most critical and tedious step since bulk ligation is used. We typically screen 50–300 clones to get a concatemer of eight or more monomers.

4.4. *Expression and purification of polyproteins*

Clones carrying the appropriate multimeric insert are transformed into BLR(DE3) competent cells (rec A^-) for expression. Expression is then induced in an exponential culture in the presence of 1 mM IPTG. Cells are lysed by sonication and the lysate is centrifuged to collect the soluble fraction, which frequently contains most of the recombinant protein. Polyproteins are then purified from this fraction using Ni^{2+}-affinity chromatography (Novagene). Calmodulin was also purified by phenyl-sepharose hydrophobic chromatography (Pharmacia) according to Gopalakrishna and Anderson (1982). Proteins were analyzed by SDS-PAGE, concentrated and dialyzed into the final buffer by ultrafiltration using Centriplus and Microcon (Amicon-Millipore). The final buffer was either PBS/5 mM DTT or 0.1 M imidazole pH 6.0 for poly-I27 and poly-C2A, and 0.1 M Tris pH 7.5 for calmodulin.

5. Protein folds have different mechanical stability

In this section we will discuss the mechanical properties of three different folds. Two of them are considered to be "all beta" structures (Murzin et al., 1995; Bateman et al., 1999) and have a so-called β-sandwich topology. One of them is from a typical mechanical protein (an immunoglobulin domain from titin) and the other from a protein involved in secretion (the C2A domain from synaptotagmin I). The third case is calmodulin, a ubiquitous regulatory protein considered to be an "all alpha" protein.

5.1. Titin immunoglobulin domains present high mechanical stability

The immunoglobulin (Ig) domain is one of the most abundant protein modules in nature, being present in a large variety of proteins including antibodies, cell adhesion molecules, cell surface receptors, enzymes, chaperones and muscle proteins (Bork et al., 1994: Gerstein and Levitt, 1997). The Ig superfamily consists of several topologies; the "I" set ("intermediate") is considered to have the ancestral topology, based on its hybrid structure (i.e., it combines features from all the other topological classes) and on phylogenetic comparisons (Chothia and Jones, 1997). The I set, which includes the majority of Ig domains occurring in cell adhesion proteins, surface receptors and muscle proteins (Harpaz and Chothia, 1994), was the first topology to be analyzed in detail by AFM in the muscle protein titin (Rief et al., 1997a).

Titin is a gigantic modular protein (~3000 kDa) responsible for the passive elastic properties of muscle (reviewed in Trinick and Tskhovrebova, 1999). Although the molecular determinants of titin elasticity are still a matter of debate (reviewed in Horowits, 1999; Trinick and Tskhovrebova, 1999; Linke, 2000), recent results show that unfolding of just a few Ig modules could fully account for the viscoelastic properties of skeletal myofibers (Minajeva et al., submitted). Furthermore, there is some evidence that the unfolding/refolding of Ig modules could occur in the physiological range of force (Linke et al., 1996). AFM studies have shown that a group of I-type Ig modules from human cardiac titin (I27–I34; "I" refers here to the sarcomeric I band, not to the aforementioned "intermediate" topology) displays a hierarchical ordering of mechanical stability from low (150 pN) to high (300 pN) unfolding forces (Rief et al., 1997a). AFM could not, however, establish the correspondence between force peaks and modules to identify the individual stability of specific modules. As we have already discussed, the solution to this problem came from molecular biology.

5.1.1. The I27 module

The I27 module of human cardiac titin was originally chosen for AFM studies because it is very stable thermodynamically (melting point, 73°C; Politou et al., 1995), its tertiary structure had been already established by NMR spectroscopy (Improta et al., 1996), and its stretching had been simulated using a technique known as steered molecular dynamics (SMD; Lu et al., 1998).

The I27 module, which is 89 amino acids long, has a typical Ig I topology composed of seven β-strands (strands A–G), which fold into two face-to-face β-sheets through backbone hydrogen bonds and hydrophobic core interactions (Fig. 6C). With the exception of the parallel pair A′G, all adjacent β-strands in both sheets are antiparallel ("parallel" and "antiparallel" refer to the orientations of two apposed strands with respect to their N- and C-termini; two strands that are

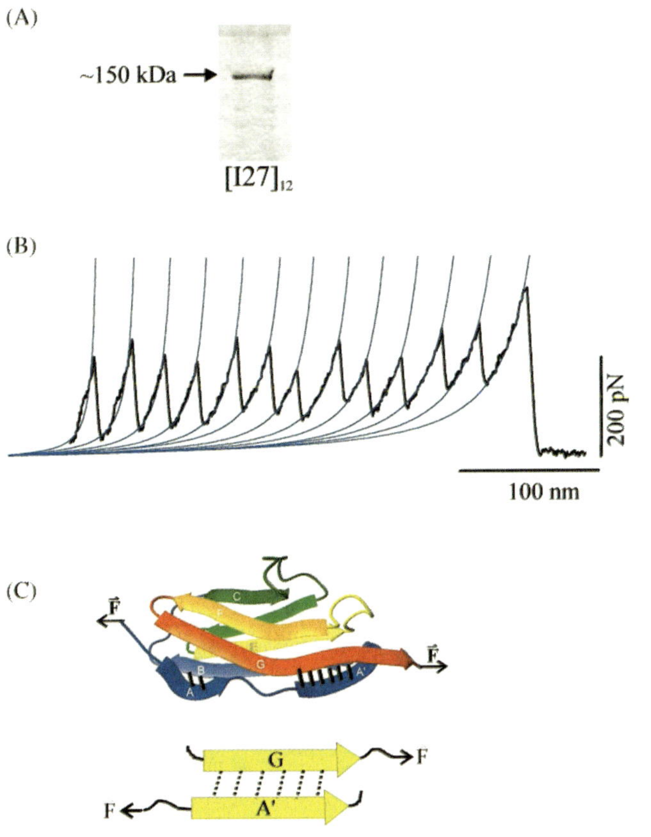

Fig. 6. Mechanical properties of single human cardiac titin immunoglobulin domains. (A) Coomassie blue staining of the purified $I27_{12}$ protein separated using SDS-PAGE. The measured molecular weight of the polyprotein (~150,000) is in good agreement with the predicted molecular weight of a $I27_{12}$ concatemer including a His-tag-EK region, 13 GLG linkers, and 2 terminal cysteine residues. (B) Stretching $I27_{12}$ with AFM (at a pulling speed of 0.6 nm/m s) resulted in a force-extension curve with peaks that vary randomly in amplitude about a value of 204 ± 26 pN ($n = 266$). The last peak represents the final extension of the unfolded protein prior to detachment from the AFM tip. Note the random distribution of the unfolding forces for each peak and the deviation from entropic behaviour ("the hump") evident in the early peaks. Fits of the WLC model to the force-extension curves of $I27_{12}$ give a persistence length of $p = 0.39 \pm 0.07$ nm ($n = 10$) and a variable contour length of $L_c = 25$ to 496 nm (blue lines). Consecutive peaks were fitted by the WLC with the same persistence length and a contour length increment of $\Delta L_c = 28.1 \pm 0.17$ nm ($n = 16$). The persistence length is a measure of the distance over which the polymer retains memory of a direction. This persistence length is of the size of a single amino acid (0.40 nm). (C) The mechanical topology of I27. Top. 3-D structure of titin I27 domain (from SCOP database; Murzin et al., 1995) showing the "shear pattern" model for the H-bond breakage in the I27 domain, where the H-bonds break simultaneously. Seven β-strands (each of them shown as a ribbon arrow of a different color) are folded into two β-sheets, one comprising strands A,B,D,E and the other including strands A′,C,F,G. The dashed lines depict the topology of the critical H-bonds between β-strands under an applied force. Bottom. Detail of the hydrogen bond patch between A′ and G strands are shown as discontinuous lines. The direction and sense of the applied force is indicated by black arrows.

oriented in the same direction are parallel while strands oriented in opposite direction are antiparallel). SMD simulations identified the interactions likely to underlie the mechanical stability of I27 (Lu et al., 1998). Simulations of the application of axial force to the terminal ends of this module showed that a critical set of six hydrogen bonds, linking strands A′ and G, provided the point of maximal resistance to unfolding. Since the hydrogen bonds in this patch are perpendicular to the direction of the applied force (a "shear" topology), unfolding requires the simultaneous rupture of this cluster of hydrogen bonds (see Fig. 6C, bottom). Upon the rupture of this patch of hydrogen bonds, the rest of the folded polypeptide unravels with little resistance. These theoretical predictions fit well with experimental observations (Carrion-Vazquez et al., 1999a,b) as is described below. A similar hydrogen-bonding topology has been proposed, based on X-ray crystallography, for the first Ig domain of the adhesion protein ICAM-1 (Casasnovas et al., 1998).

5.1.1.1. Polyproteins demonstrate the stochastic nature of mechanical unfolding. To analyze the mechanical properties of the I27 domain by AFM, we constructed a $I27_{12}$ polyprotein, as discussed in Section 4 (Fig. 6A). In this construct, an additional glycine residue was added to the linker (GLG instead of LG) to more closely mimic the GGG linkers used in previous studies (Improta et al., 1998). Electron micrographs of rotary shadowed $I27_{12}$ show rod-like structures ~58 nm in length (H. Erickson, pers. comm.), which corresponds well to the expected length based on the NMR spectroscopic measurements of single I27 modules (each of which has a length of 4.4 nm; Improta et al., 1998).

Random segments of the $I27_{12}$ polyprotein were picked up by adsorption to the AFM tip. The resulting force–extension curves showed a saw-tooth pattern with as many as 12 equally spaced force peaks. The fitting of consecutive unfolding peaks to the WLC model showed that each unfolding event added 28.1 ± 0.17 nm to the protein length. The unfolding force varied randomly around a mean value of ~200 pN (Fig. 6B). A histogram of force peaks (unfolding force) revealed an asymmetrical distribution of events with a maximum at about ~200 pN (Carrion-Vazquez et al., 1999a), which is in contrast to the pattern of rising forces observed with the I27–I34 titin recombinant fragment, where peaks ranged from 150 pN to 300 pN (Rief et al., 1997a). These experiments demonstrate that domain unfolding does not occur above a "threshold force" but rather is a probabilistic event, which may therefore be observed at any force, albeit with different probability (Carrion-Vazquez et al., 1999a). The force at which an unfolding event occurs reflects the mechanical stability of a particular folded module under the specific experimental conditions used. The unfolding force, and therefore the mechanical stability, depends on the rate of extension, the faster the pulling speed, the higher the force required to cause the unfolding of the domain.

5.1.1.2. Polyproteins spontaneously refold upon relaxation. The domains of modular proteins like titin have been proposed to undergo many unfolding/refolding cycles during their physiological activity (Erickson, 1994). The refolding process can be reproduced in vitro by AFM in both native modular proteins (Rief et al., 1997a; Oberhauser et al., 1998) and polyproteins (Carrion-Vazquez et al., 1999a). Once the polyprotein has been stretched and its domains unraveled, it can be relaxed by returning the substrate to the original position. The unfolded domains recover their fold spontaneously in a time-dependent fashion with the proportion of recovered modules depending on the relaxation time between extensions. This procedure can be repeated several times, allowing the measurement of the refolding rate of individual protein domains.

5.1.1.3. Is the fidelity of refolding decreased in polyproteins?. Since native modular proteins are not strictly repetitive, it could be argued that the repetitive nature of polyproteins could favor abnormal intermodule refolding. AFM refolding experiments with poly-I27 have identified missing peaks ("skips") in their force–extension curves (Oberhauser et al., 1999). The interval between peaks in a skip corresponds to the size of two I27 domains plus the length of the linker between domains. This suggests that skips represent misfolding events in which the A′ strand of one domain binds with the G strand of the adjacent domain creating a larger fold with a stability similar to that of the native I27. However, the frequency of skips in poly-I27 (~2%) was found to be of the same order as that of the FN-III domains of tenascin (~4%) indicating that the frequency of misfolding is not higher in polyproteins than in heterogeneous modular proteins.

5.1.1.4. Polyproteins allow an accurate analysis of the unfolding kinetics. The unfolding of a domain under an applied force is a probabilistic event that can be modeled as a two-state process with a rate that depends exponentially on the axial force and the unfolding distance along which the force is applied. This description gives a probability of observing an unfolding event as $P = N\alpha\Delta t$, where N is the number of modules, Δt the pulling interval, and α is the unfolding rate constant. The unfolding rate constant at a given force is given by $\alpha(F) = \alpha_0 e^{F\Delta x/kT}$. For a given α_0 (unfolding rate at zero force), if the unfolding distance (Δx, the extension required to disrupt the fold) is small, the force of unfolding will be high, while if the forces that maintain the fold are distributed over a greater distance the unfolding force will be lower.

One can model the probability of unfolding versus the applied force using Monte Carlo techniques to estimate the unfolding distance and the unfolding rate at zero force (Rief et al., 1997; Oberhauser et al., 1998; Kellermayer et al., 1997). The best fit for the histogram of unfolding forces for I27 was obtained using an unfolding rate at zero force of $3.3 \times 10^{-4}\,s^{-1}$ and an unfolding distance of 0.25 nm (Carrion-Vazquez et al., 1999a).

The short unfolding distance measured for I27 (roughly the size of a molecule of H_2O) is consistent with the SMD simulations (Lu et al., 1998), which predict that unfolding occurs as a single event with low force after the rupture of the hydrogen bonds of the A′G patch. Furthermore, the interval between peaks (28.1 ± 0.17 nm) observed using AFM suggests an elongation of 70 amino acids, considering that the length of an amino acid residue is 0.40 nm (see Section 5.1.1.5.) ($70 \times 0.40\,\text{nm} \approx 28.1\,\text{nm}$). This agrees well with the length of the "force hidden" region predicted by the SMD simulation (about 72 amino acids; there is uncertainty in the boundaries of this region).

5.1.1.5. Amplification by polyproteins allows the mechanical dissection of a fold with amino acid resolution. To demonstrate the validity of this structural model, and to measure the spatial resolution of the AFM, a mutant polyprotein was constructed in which a cluster of 5 glycine residues was inserted into the hairpin loop joining the F and G β-strands of I27, which is within the predicted "force-hidden" region. The interval between force peaks was lengthened, on average, by 2.00 nm per domain. This value can be used to estimate the length of an amino acid residue (2.00 nm/5 = 0.40 nm). The length 0.40 ± 0.06 nm obtained this way is somewhat bigger than that determined from the geometry of the polypeptide chain (0.36 nm; Oesterhelt et al., 2000). The difference is potentially due to the stretching of amino acid residues by the external force, and partially it originates from applying the WLC model to determine the contour length of the protein. Since this caliper, 0.40 nm/amino acid residue, already includes all uncertainties of our methodology, it can be used confidently to count the number of amino acids.

The corresponding control polyproteins in which the glycine cluster was inserted before or after the A′G patch (i.e., outside the force-hidden fold) did not show an increase in the unfolding interval, presumably because these regions are already pre-stretched before unfolding takes place (Fig. 7; Carrion-Vazquez et al., 1999b).

These experiments reveal the remarkable resolution (i.e., to the amino acid level) attainable in AFM by using polyproteins to amplify linearly the mechanical features of a single module.

5.1.1.6. Amplification by polyproteins allows the study of unfolding intermediates. In addition to a major force barrier to unfolding, SMD simulations (Lu et al., 1998) predict that a weaker barrier, the AB patch, which consists of two H-bonds in a shear topology (see Fig. 6C), should be disrupted by force-induced extension prior to the rupture of the A′G patch. Linear amplification by polyproteins allowed this phenomenon to be observed and analyzed. The first unfolding peaks in Fig. 6B show clear deviations from the entropic behavior predicted by the WLC model of polymer elasticity. These "humps" of low force (~100 pN) are likely to represent simultaneous rupture of the AB hydrogen bonds in each of the domains. The presence of a smaller hump in subsequent unfolding events suggests that the remaining domains refold very quickly to their native conformation before the next unfolding event occurs. This hypothesis was confirmed by site-directed mutagenesis with the construction of a mutant polyprotein in which a proline substitution disrupting the AB patch abolished the hump (Marszalek et al., 1999b).

The molecular determinants of titin elasticity are poorly understood (Trinick and Tskhovrebova, 1999; Horowits, 1999). In cardiac titin, three regions have been proposed to be responsible for the global elastic properties of the molecule: the Ig domains and the so-called PEVK and N2B regions (Granzier et al., 1997; Linke et al., 1999). The discovery of this unfolding intermediate in Ig modules of titin uncovers a new elastic component in this protein that occurs at lower forces prior to unfolding and would result in an extension similar in size to that generated by the elongation of the PEVK region in cardiac titin (Marszalek et al., 1999b).

5.1.1.7. Hydrogen bonding pattern and the mechanical properties of I27. As discussed above, SMD simulations (Lu et al., 1998) predict that there are two patches of hydrogen bonds that are relevant for the mechanical properties of I27 (Fig. 6C, top). As the fold is stretched, the AB patch unfolds first, at low force (~100 pN), causing deformation of the first peaks (the "hump"). At higher forces (~200 pN) the A′G patch is disrupted and the hydrophobic core unfolds with little resistance. The topology of these hydrogen bond patches appears to be critical to understanding the mechanical properties of I27. The SMD studies predict that the simultaneous breakage of the A′G cluster of H-bonds causes the high-energy barrier opposing the forced unfolding of the I27 domain (Fig. 6C, bottom). This simultaneous H-bond rupture would require higher force than a sequential pattern of breakage, in which the H-bonds break one after the other (see below) (Lu and Schulten, 1999). The AB patch, which has only two H-bonds in a shear topology, unfolds

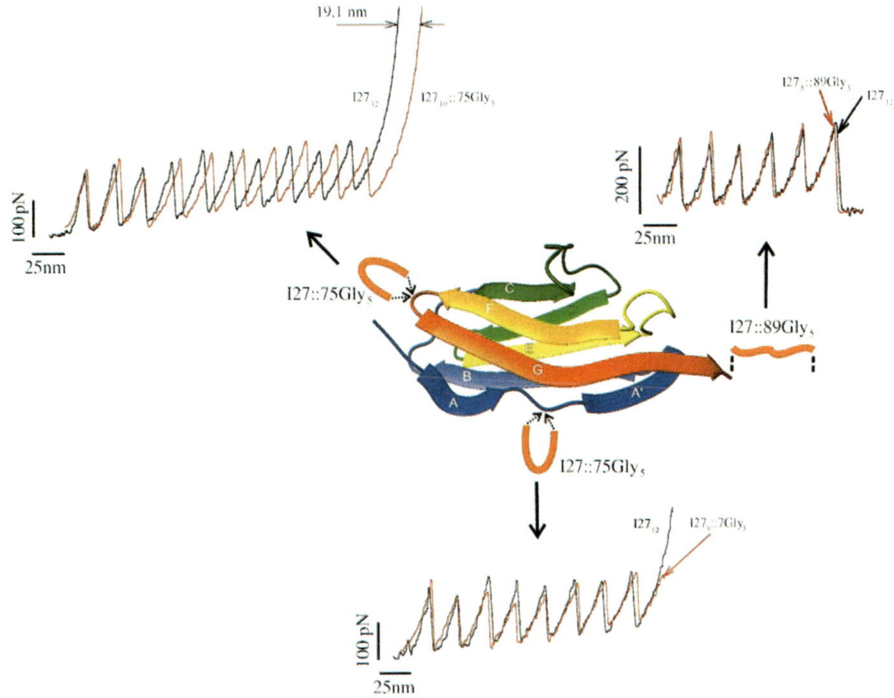

Fig. 7. Direct observation of a length phenotype in an immunoglobulin module. (Center) Structure of the I27 immunoglobulin module and location of the glycine insertion sites. The N- and the C-termini of I27 are parallel and are in close apposition over the A′G strands. A mechanical linkage between the A′ and G strands is thought to provide continuity of force between the N- and C-termini of the folded module. Rupture of this linkage exposes the amino acids that are hidden in the fold, extending the contour length of the protein. We constructed polyproteins based on the wild type (I27) and mutant forms of the I27 module with a Gly_5 insert in the FG hairpin loop at position 75 (I27::75Gly_5), in the N-terminal region at position 7 (I27::7Gly_5), and after the C-terminus at position 89 (I27::89Gly_5). (Top left) Insertion of 5 glycine residues into the FG loop of the I27 module increases the contour length of the unfolded module by ~2 nm (5 × 0.40 nm). A comparison of the force–extension curves of a wild type I27 polyprotein ($I27_{12}$; black trace) and a mutant polyprotein, $I27_{10}$::75Gly_5 (red trace), shows that the mutant polyprotein extends more with each unfolding event. After 10 consecutive unfolding events, this particular mutant polyprotein is 19.1 nm longer than the wild type polyprotein (arrows), corresponding to a difference of 1.91 nm/module. (Top right and bottom) The saw-tooth pattern in the force-extension curves obtained from polyproteins engineered with 5 glycine insertions placed either in position 89 ($I27_5$::89Gly_5; red trace in top right traces) or in position 7 ($I27_9$::7Gly_5; red trace in bottom traces) superimposed on the saw-tooth pattern obtained from the wild type polyprotein ($I27_{12}$; black trace).

with less force that the A′G patch, which has six H-bonds with the same mechanical topology (Fig. 6C).

Proline mutagenesis has been used to test the hypothesis that backbone hydrogen bonds are responsible for the two mechanical barriers present during the unfolding of the I27 module. As mentioned, a proline substitution in the A strand, to prevent the interaction with the B strand, abolished the formation of the low force barrier (Marszalek et al., 1999b). When valines at positions 11, 13 and 15 of the A′ strand were substituted by prolines, to prevent the formation of hydrogen bonds with the G strand, the force required to unfold the domain was decreased, as expected. However, a mutation of tyrosine to proline in position 9 actually increased the mechanical stability of the domain (Li et al., in press). This result indicates that other non-covalent interactions may also contribute to the mechanical resistance of the fold.

5.1.2. Mechanical properties of other titin modules

Ig-like domains of the Ig and FN-III types are 7-stranded β-sandwich structures with the N- and C-terminal strands parallel to each other and pointing in opposite directions. These domains seem to have evolved to withstand forces when connected in series and, because of the shear topology of their critical hydrogen bonds, these domains offer a high resistance to mechanical stretching (Rief et al., 1997a, 1998; Oberhauser et al., 1998; Carrion-Vazquez et al., 1999a). Interestingly, Ig domains that occur in muscle proteins, cell adhesion proteins, and surface receptors usually belong to the "I" class (Harpaz and Chothia, 1994) and are likely to have a similar mechanical topology.

By engineering polyproteins it has been possible to analyze in detail the molecular determinants of the elastic properties of one of those modules, the I27. What can be expected from the study of the mechanical properties of other modules? First of all we need to consider that homologous titin modules from distant species (orthologs) are much more similar than analogous repeats within the same species (paralogs) (Witt et al., 1998). This observation indicates that the specific amino acid sequence of each of the titin modules has been subject to strong selective pressure through evolution and therefore may be relevant for function. Since the main function of titin seems to be mechanical, the primary structure of each of its modules may be critical for determining its mechanical properties. Hence each of titin Ig modules seems to contribute a specific mechanical stability to the passive elastic properties of titin and may not be merely equivalent repetitions. AFM experiments using polyproteins from a contiguous module, the I28, show that it unfolds at higher forces ($\sim$260 pN) and presents a more stable "hump" ($\sim$150 pN) (Marszalek et al., 1999b). Future structural information about I28 could shed light into the molecular origin of these differential properties. Further research will also be necessary in order to understand how these properties can be related to the global elastic properties of titin. The first clues towards the understanding of this combinatorial logic come from AFM experiments using a heterodimeric polyprotein I27–I28 (Fig. 3). The mechanical stability of the I28 in this chimera is increased to $\sim$300 pN. Therefore, it seems that the topological connectedness of modules is a critical factor for their mechanical stability (Li et al., 2000).

5.2. The mechanical stability of other protein folds

Do Ig-like domains respond to stretching in a quantitatively different fashion than do domains that are not normally exposed to mechanical strain? To further understand the principles of the mechanical design of proteins a comparison was made with proteins that do not have a known mechanical function. Using the polyprotein approach, we investigated the C2A domain from synaptotagmin I and calmodulin.

5.2.1. The C2 domain: a fold with low mechanical stability

The C2 domain is a conserved "all beta" module present in more than 40 different proteins, many of them involved in membrane interactions and signal transduction (e.g. synaptotagmins, rabphilins, protein kinase C isozymes, phospholipase A_2, perforins). The first C2 domain of synaptotagmin I (C2A) is believed to be the calcium sensor that initiates membrane fusion during neurotransmitter release (Sudhof and Rizo, 1996; Shao et al., 1996). The 3D structure of C2A was found to be a β-sandwich composed of 127 amino acids arranged into eight antiparallel strands with the N- and C-terminal strands pointing in the same direction (Fig. 8C, left; Sutton et al., 1995). In contrast to the I27 domain, in which the A′G patch of H-bonds have a "shear" topology, those in the AG patch of the C2A domain are in a "zipper" configuration (i.e., parallel to the direction of the applied force; Fig. 8C, right). SMD simulations (Lu and Schulten, 1999) predicted very low forces of unfolding for this domain. In AFM experiments the force–extension curve of C2A polyproteins showed a saw-tooth pattern with force peaks of ~60 pN separated by ~38 nm (Fig. 8B). The measured size of the folded core of the domain agrees well with that

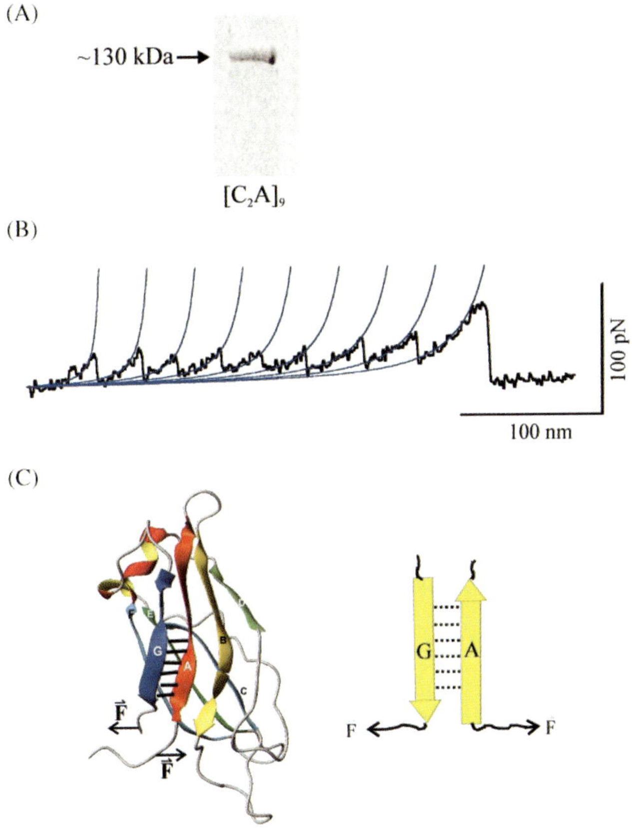

Fig. 8. Mechanical properties of a single I C2A domain. (A) Coomassie blue staining of the purified $C2A_9$ protein separated using SDS-PAGE. (B) Force–extension relationship for the C2A polyprotein measured with AFM (at a pulling speed of 0.6 nm/m). The saw-tooth pattern show peaks with an unfolding force of ~60 pN. Consecutive peaks were fitted by the WLC with the same persistence length (p = 0.4 nm) and a contour length increment of $\Delta L_c = 38$ nm. (C) Left. Schematic diagram of the topology of the β-sandwich structure of C2A domain (from SCOP database; Murzin et al., 1995). Right. The "zipper" model for the H-bond breakage in C2A domains in which each bond breaks sequentially. The hydrogen bonds between A and G strands are shown as black lines. The direction and sense of the applied force is indicated by the black arrows.

predicted based on the number of amino acids between the end of the A strand and the beginning of the G strand ($106 \times 0.4 = 42$ nm).

Circular permutants (e.g., topological mutant proteins that maintain the amino acid sequence with different N- and C-termini) of the C2A domain could be used to test the hypothesis of the lower mechanical resistance of the "β-zipper" versus the "β-shear" H-bonds in a β-barrel.

5.2.2. Calmodulin: a compliant protein structure

Calmodulin is a highly conserved antiparallel "all α" protein that acts as a primary calcium-dependent regulator of many intracellular processes. The three-dimensional structure of calmodulin (148 residues) is dumbbell shaped and consists of 7 α-helixes distributed in a helical central region capped by two globular regions, each containing two helix-loop-helix EF-hand motifs that are responsible for Ca^{2+}-binding (Fig. 9C, left; Babu et al., 1985, 1988). In contrast to the β-sandwich topology, where there is a nonhomogeneous distribution of the inter-strand H-bonds (in either shear or zipper configuration), the α-helix has a homogeneous distribution of intrahelix H-bonds. SMD simulations predict that stretching of the structure would cause sequential breaking of the intra-helix hydrogen bonds, leading to the elongation and separation of α-helices, a process that would require little force. The predicted unfolding pathway does not require breakage of clustered H-bonds (K. Schulten, personal communication) as was predicted by SMD for cytochrome c6, another all α-helical protein (Lu and Schulten, 1999). Therefore no force peaks are expected upon stretching of calmodulin.

To test these predictions a rat polycalmodulin (CaM_4) was constructed (Fig. 9A). CaM_4 was purified either by Ni^{2+}-affinity chromatography or by phenyl-shepharose affinity chromatography, with identical results. Only CaM bound to Ca^{2+} in its native conformation binds to phenyl-sepharose; CaM in the presence of low concentration of Ca^{2+} does not bind (Gopalakrishna and Anderson, 1982). The stretching of CaM_4 did not yield any force peaks, indicating that the unfolding forces must be below our AFM noise level ($\sim$20 pN). Only very compliant structures of the approximate size of the tetrameric polyprotein were found ($4 \times 148 \times 0.40 = 237$ nm) confirming SMD predictions (Fig. 9B). Both SMD and AFM therefore agree that calmodulin behaves like a random polypeptide chain under force. Unlike the I27 domain, whose tertiary fold has been designed to withstand force, calmodulin does not show mechanical barriers to prevent its unfolding. Paradoxically, one of the most striking characteristics of calmodulin is its thermodynamic stability, particularly in the presence of Ca^{2+}. The protein may be exposed to 95°C or to a 9 M urea solution with retention of biological activity (Wallace et al., 1980). The Ca^{2+}-free form of CaM has a melting temperature of $\sim$55°C, while the Ca^{2+}-bound form denatures only at temperatures exceeding 90°C (Brzeska et al., 1983). We have not detected, however, any difference in mechanical stability in the presence or absence of Ca^{2+}.

The only other α-helical structure that has been stretched by AFM is spectrin, a cytoskeletal protein. However, this protein forms bundles of α-helices, rather than single α-helices. The force–extension relationship of this protein showed very low force peaks (25–35 pN; Rief et al., 1999; Fisher et al., 1999b). This range of forces is intermediate between the compliance of a single α-helix and the low force range of a zipper β-sandwich.

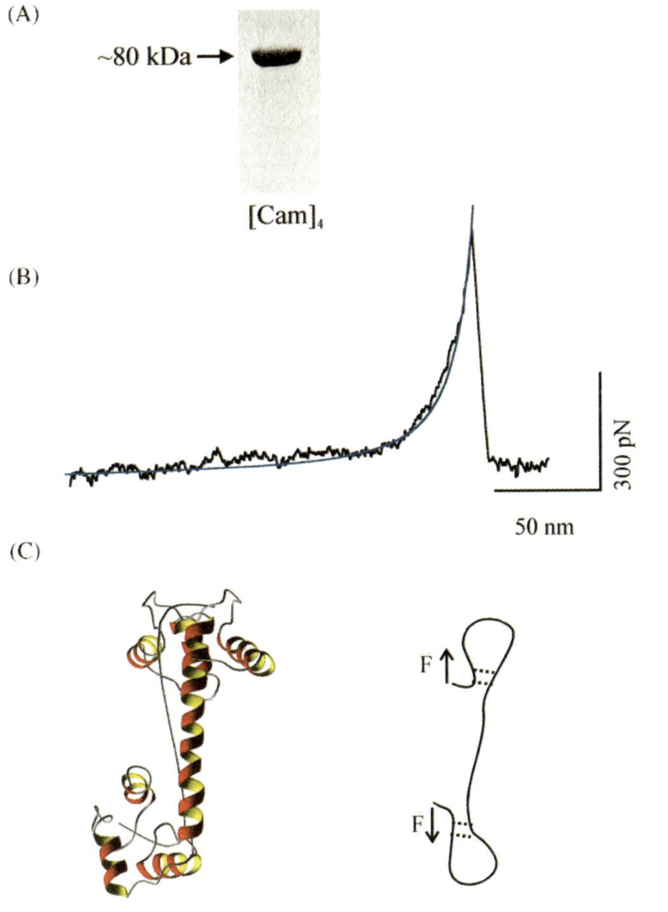

Fig. 9. The force-extension relationship for poly-calmodulin. (A) Coomassie blue staining of the purified CaM_4 protein separated using SDS-PAGE. (B) Stretching single calmodulin polyproteins (at a pulling speed of 0.6 nm/m s) give force–extension curves with no evident force-peaks. The force curve is well described by the WLC model (continuous lines) using a contour length of 212 nm (194 ± 15 nm, $n = 26$) and a persistence length of 0.32 nm. This contour length is similar to the expected contour length of a protein 580 amino acids in length (each CaM domain is 148 aa; 148aa × 4 monomers × 0.40 nm/residue = 237 nm). (C) Left. 3-D Structure of rat CaM showing its α-helical structure (from SCOP database; Murzin et al., 1995). CaM is made of seven antiparallel α-strands and two short antiparallel β-sheet hairpins with a "zipper" topology (not shown) that provide a structural link between the two Ca^{2+}- motifs (EF hands) of each globular region. Right. Putative mechanical topology of a calmodulin domain. The two small β-hairpins of calmodulin (see Fig. 8C, right) are in a "zipper' configuration, and therefore should offer little resistance to mechanical unfolding.

6. The topological determinants of mechanical stability of proteins

Proteins are designed to perform specific biological functions and typically maintain a stable native conformation. In contrast, mechanical modules like the I27 of titin may be designed to undergo continuous unfolding/refolding cycles (Erickson, 1994). The topological design of this module includes two patches of H-bonds that act as mechanical barriers. One of these barriers, the AB patch, offers low resistance (~100 pN), while the A′G patch has a higher mechanical resistance

(~200 pN), which seems to protect the hydrophobic core from unfolding. It appears that proteins that experience mechanical stress (e.g., titin or ECM proteins) are composed of modules, like the I27 or FN-III, that are designed to withstand force while working in continuous unfolding/refolding cycles. They have strong backbone H-bonds patches that act as mechanical barriers to unfolding, with the β-shear H-bond topology being a stronger mechanical constraint than the zipper topology. An extreme case of a mechanical barrier to unfolding would be the disulphide bond, a covalent bond present in most Ig domains and other modules from antibodies, ECM and other extracellular proteins. The disulphide bond seems to have evolved to prevent the unfolding of the hydrophobic core over the physiological range of forces. Although the physiological role of a disulphide bond in a module is not clear, it may protect the mechanically isolated region (e.g., the hydrophobic core) from unfolding when the function of the protected region (e.g., ligand binding, dimerization) is incompatible with its mechanical unfolding.

In contrast, the proteins with no mechanical function that we have studied have very weak mechanical barriers to unfolding, as in the case of calmodulin. Between these two extremes there seems to be a spectrum of intermediate topologies, from the low resistance of the α-helical bundles of spectrin (25–35 pN) to the intermediate β-zippered sandwiches of the C2 domain (60 pN). Recently, an "alpha and beta" protein (Murzin et al., 1995), the bacteriophage T4 lysozyme, has been reported to exhibit unfolding forces of ~60 pN (Yang et al., 2000). Interestingly, this value is intermediate between the stability of the C2 domain and that of calmodulin, considering that the pulling speed in these experiments was about twice that used in our experiments (for a discussion on the dependence of the force with the pulling speed see Carrion-Vazquez et al., 1999a).

What are the physicochemical parameters that could help us to predict the mechanical stability of a module? The stability of a module, measured by equilibrium denaturation, may not be related to its probability of unfolding at a given force. We have found that mechanical stability does not necessarily correlate with thermodynamic stability. For example, the kinetic stability (during mechanical unfolding) of I27 and I28 is about 22.2 and 23.7 kcal/mol (Li et al., 2000), while their thermodynamic stability is about 7.6 and 3.0 kcal/mol, respectively (the melting temperatures of these modules are 72.6 and 54.1°C, respectively; Politou et al., 1995, 1996). A lack of correlation between thermal and mechanical stability is also observed for the C2 domain, CaM (Table 1) and spectrin repeats, which present a melting temperature rather similar to that of I28 (53°C; DeSilva et al., 1997). Thermodynamic and kinetic stability does not always correlate (Jackson, 1998; Plaxco et al., 1998). For example, the 10th FN-III domain from fibronectin is 4 kcal/mol more stable than the third FN-III domain from tenascin, yet it unfolds two times faster (Clarke et al., 1999). Folding kinetic experiments provide detailed information that allows a reconstruction of the energy diagram for unfolding. In the case of I27 (and I28) both the height of the unfolding energy barrier and the position of the transition state are similar for chemical and mechanical unfolding (Carrion-Vazquez et al., 1999a; Li et al., 2000), which indicates that kinetic stability (i.e., the height of the unfolding energy barrier) may be a good predictor of mechanical stability. However, the chemical unfolding experiments can only define a "solvent exposure" reaction coordinate whereas AFM experiments define the reaction coordinate as length of the protein during extension (Carrion-Vazquez et al., 1999a).

SMD simulations predict very well the qualitative behavior of a protein domain under force but they predict unfolding forces that are ten times larger than those observed by AFM

Table 1
Mechanical stability of single domains and proteins compared with their thermodynamic stability[a]

Domain	Conformational topology	Mechanical topology	Melting Temp. T_m (°C)	Thermodynamic stability $\Delta G^{(H_2O)}$ (kcal/mol)	Unfolding force at 0.6 nm/m s (pN)	Kinetic stability $\Delta G^{F=0}$ (kcal/mol)
CaM	α-Helix	Zipper	55°C[b]		<20[c]	ND
			>90°C (+Ca^{2+})[b]	—		
C2A	β-Sandwich	Zipper	55°C[d]		~60[c]	ND
			74°C (+Ca^{2+})[d]	—		
127	β-Sandwich	Shear	72.6°C[e]	~7.6[e,f]	~200[g]	~22.2[g]
128	β-Sandwich	Shear	54.1°C[e]	~3.0[e,f]	~260[h,f]	~23.7[f]

[a] Major discrepancies are highlighted in bold. Note that thermal and chemical stability is not a good predictor of mechanical stability, which can be defined, as a first approximation, by the force necessary to unfold a domain (this unfolding force depends on the pulling speed).

[b] Data compiled from Brzeska et al. (1983).

[c] Data compiled from this study.

[d] Data compiled from Shao et al. (1996).

[e] Data compiled from Politou et al. (1995)

[f] Data compiled from Li et al. (2000).

[g] Data compiled from Carrion-Vazquez et al. (1999a).

[h] Data compiled from Marszalek et al. (1999).

(Lu et al., 1998). SMD simulations cover only a very brief length of time (~1 ns) and therefore cannot fully incorporate contributions of the thermal motion (entropy) to the unfolding event. Furthermore, due to the brief simulation times involved, SMD cannot simulate the folding reaction.

We are just starting to understand the characteristics that define an elastic domain. For a given mechanical stability, the force of unfolding will be high if the disruption of the fold requires little extension, and will be lower if the forces that maintain the fold are distributed over a longer unfolding distance. The force required to unfold a domain is therefore highly dependent on the topology of the H-bonds within the fold. The spatial orientation (relative to the force vector), number, location and strength of these bonds appears to determine not only the mechanical stability, but also the dependence of the rates of unfolding and refolding on the applied force.

7. Conclusions

Force spectroscopy has been used to study the mechanical properties of heterogeneous populations of modules in modular proteins with known mechanical roles. The construction of homomeric polyproteins, using recombinant DNA technology, has recently allowed the study of the mechanical properties of a single module of the muscle protein titin (i.e., I27). We have applied this technology to proteins and domains with no known mechanical function, the C2A domain of synaptotagmin I and calmodulin. In contrast to the mechanical stability of I27, which requires forces of about 200 pN for unfolding, the C2 domain was found to be less stable (~60 pN) while calmodulin was mechanically compliant, with no force peaks present (<20 pN), at similar extension rates (0.6 nm/ms). These results indicate that the interactions that resist mechanical unfolding in an α-helix (calmodulin) are much weaker than those maintaining the structure of a zipper H-bonded β-sandwich (C2A domain), which is in turn less stable than a shear H-bonded β-sandwich topology (I27). We also have compared the mechanical stability of these structures with their thermodynamic and kinetic stability and conclude that the mechanical stability of a protein fold is not directly correlated with its thermodynamic stability but with its kinetic stability, which seems to be determined by its topology, the spatial pattern of its backbone hydrogen-bonding.

7.1. A niche for single-molecule force spectroscopy in the analysis of protein structure and dynamics

As an analytical tool, the primary attribute of AFM is that it provides mechanical information from individual molecules, not vast ensembles. This is the case for structural techniques, whether the molecules are ordered as in X-ray crystallography or dispersed as in NMR spectroscopy, and for techniques for monitoring protein folding, such as stopped-flow fluorimetry. AFM provides both structural (mechanical topology) and functional (mechanical properties) information from single protein domains in aqueous solution and in real time. Therefore, AFM promises, in theory, the possibility of directly probing the so-called folding energy landscape (Brockwell et al., 2000).

In spite of its recent progress, force spectroscopy is a developing technique still in its infancy. There is ample room for technical improvements in both the AFM setup (instrumentation, recording techniques, analysis software, etc.) and in protein engineering (polyprotein construction, conditions to favor native folding, protein attachment and orientation, etc.). An ongoing development with future potential is the combination of force spectroscopy with fluorescence dynamics at the single-molecule level (Brockwell et al., 2000).

7.2. *Perspectives*

Current research in molecular biology is highly engaged in compiling a "horizontal", rather descriptive, catalog of protein sequences with their biological roles, but the "vertical" investigation of the physicochemical mechanisms of how proteins perform their functions (i.e., what eventually would allow us to understand biological systems at a deeper level) is in danger of falling behind. In particular, we expect that the analysis of the mechanical properties of protein folds will uncover in the near future a variety of architectures and structural elements and will set the foundations for a "mechanical biochemistry" of proteins. We are likely to encounter a variety of mechanical topologies in proteins inside and outside the cell to prevent unwanted unfolding, or to provide elasticity.

Mechanical unfolding in modular proteins has been proposed to modulate the kinetics of the interactions with their receptors or ligands (Oberhauser et al., 1998). Thus, mechanical design may be crucial not only for known mechanical proteins but for the regulation of the lifetime of protein-ligand interactions and for the function of the tensed network that, via cell-adhesion receptors and ECM proteins, supports mechanochemical sensing and transduction (Chicurel et al., 1998). Mechanical unfolding of protein modules may also be responsible for the high tensile strength and high toughness of natural adhesives (Smith et al., 1999). Research in these areas promises a broad spectrum of applications ranging from industry (e.g., development of glues with new properties) to medicine (e.g., high-resistance organic materials for orthopedics or dentistry).

Furthermore, mechanical unfolding in the cell may not be limited to proteins with a mechanical function. Most proteins in the cell seem to be mechanically unfolded at some point or another during their lifespan. Several cell compartments (i.e., endoplasmic reticulum, mitocondrion, chloroplast and proteasome) possess protein translocases that appear to use mechanical forces to unfold proteins during their normal turnover, typically prior to their assisted folding by chaperones or to their degradation (Rothman and Kornberg, 1986; Glick, 1995; Pfanner and Meijer, 1995; Baumeister et al., 1997; Shtilerman et al., 1999). The "nanomachinery" involved in this process is still poorly understood and, although the pulling geometry in vivo may be different, progress in force spectroscopy may benefit also our understanding of these biological unfoldases and vice versa.

The mechanical design behind the structure of proteins with a mechanical function probably represents precise evolutionary solutions to a variety of mechanical challenges that organisms have encountered through their evolution. The combination of force spectroscopy and molecular biology has now opened the way to the study of the mechanical properties of individual protein folds at the level of single molecules with nanoscopic resolution.

Acknowledgements

We are grateful to Drs. Mathias Gautel, Thomas H. Südhof and Koiti Titani for their generous gifts of the cDNA clones referred to in the text, and to Carmenlu Badilla-Fernandez and

William J. Greenleaf for excellent technical assistance. This work was funded by R01 grants from the National Institutes of Health to JMF, AFO and PEM, and by a National Science Foundation grant to PEM, JMF and AFO.

References

Babu, Y.S., Sack, J.S., Greenhough, T.J., Bugg, C.E., Means, A.R., Cook, W.J., 1985. Three-dimensional structure of calmodulin. Nature 315, 37–40.

Babu, Y.S., T.J., Bugg, C.E., Cook, W. J., 1988. Structure of calmodulin refined at 2.2 Å resolution. J. Mol. Biol. 204, 191–204.

Baljon, A.R.C., Robbins, M.O., 1996. Energy disipation during rupture of adhesive bonds. Science 271, 482–484.

Bateman, A., Birney, E., Durbin, R., Eddy, S.R., Finn, R.D., Sonnhammer, E.L., 1999. Pfam 3.1: 1313 multiple alignments match the majority of proteins. Nucleic Acid Res. 27, 260–262.

Baumeister, W., Cejka, Z., Kania, M., Seemüller, E., 1997. The proteasome: a macromolecular assembly designated to confine proteolysis to a nanocompartment. Biol. Chem. 378, 121–130.

Bensimon, D., 1996. Force: a new structural control parameter? Curr. Biol. 4, 885–889.

Binnig, G., Quate, C.F., Gerber, C., 1986. Atomic force microscope. Phys. Rev. Lett. 56, 930.

Bork, P., Holm, L., Sander, C., 1994. The immunoglobulin fold. Structural classification, sequence patterns and common core. J. Mol. Biol. 242, 309–320.

Brockwell, D.J., Smith, D.A., Radford, S.E., 2000. Protein folding mechanisms: new methods and emerging ideas. Curr. Opin. Str. Biol. 10, 16–25.

Brzeska, H., Venyaminov, S.V., Gravarek, Z., Drabikowski, W., 1983. Comparative studies on thermostability of calmodulin, skeletal muscle troponin C and their triptic fragments. FEBS Lett. 153, 169–173.

Buijs, J., Norde, W., Lichtenbelt, J.W.T., 1996. Changes in the secondary structure of adsorbed IgG and F(ab′)(2) studied by FTIR spectroscopy. Langmuir 12, 1605–1613.

Burnham, N.A., Colton, R.J., 1989. Measuring the nanomechanical properties and surface forces of materials using an atomic force microscope. J. Vac. Sci. Technol. A7, 2906–2913.

Bustamante, C., Rivetti, C., Keller, D.J., 1997. Scanning force microscopy under aqueous solutions. Curr. Opin. Str. Biol. 7, 709–716.

Bustamante, C., Smith, S.B., Liphardt, J., Smith, D., 2000. Single-molecule studies of DNA mechanics. Curr. Opin. Str. Biol. 10, 279–285.

Carrion-Vazquez, M., Oberhauser, A.F., Fowler, S.B., Marszalek, P.E., Broedel, S.E., Clarke, J., Fernandez, J.M., 1999a. Mechanical and chemical unfolding of a single protein: A comparison. Proc. Natl. Acad. Sci. USA 96, 3694–3699.

Carrion-Vazquez, M., Marszalek, P.E., Oberhauser, A.F., Fernandez, J.M., 1999b. Atomic force microscopy captures length phenotypes in single proteins. Proc. Natl. Acad. Sci. USA. 96, 11288–11292.

Casasnovas, J.M., Stehle, T., Liu, J.H., Wang, J.H., Springer, T.A., 1998. A dimeric crystal structure for the N-terminal two domains of intercellular adhesion molecule-1. Proc. Nat. Acad. Sci. USA 95, 4134–4139.

Chicurel, M.E., Chen, C.S., Ingber, D.E., 1998. Cellular control lies in the balance of forces. Curr. Opin. Cell. Biol. 10, 232–239.

Chothia, C., Jones, E.Y., 1997. The molecular structure of cell adhesion molecules. Annu. Rev. Biochem. 66, 823–862.

Clarke, J., Cota, E., Fowler, S.B., Hamill, S.J., 1999. Folding studies of immunoglobulin-like beta-sandwich proteins suggest that they share a common folding pathway. Str. Fold. Des. 7, 1145–1153.

Czajkowsky, D.M., Shao, Z., 1998. Submolecular resolution of single macromolecules with atomic force microscopy. FEBS Lett. 430, 51–54.

DeSilva, T.M., Harper, S.L., Kotula, L., Hensley, P., Curtis, P.J., Otvos Jr., L., Speicher, D.W., 1997. Physical properties of a single-motif erythrocyte spectrin peptide: a highly stable independently folding unit. Biochemistry 36, 3991–3997.

Engel, A., Gaub, H.E., Muller, D.J., 1999. Atomic force microscopy: a forceful way with single molecules. Curr. Biol. 9, R133–R136.

Erickson, H.P., 1994. Reversible unfolding of fibronectin type III and immunoglobulin domains provides the structural basis for stretch and elasticity of titin and fibronectin. Proc. Natl. Acad. Sci. USA 91, 10114–10118.
Erickson, H.P., 1997. Stretching single protein molecules: titin is a weird spring. Science 276, 1090–1092.
Evans, E., Ritchie, K., 1997. Dynamic strength of molecular adhesion bonds. Biophys. J. 72, 1541–1555.
Fisher, T.E., Marszalek, P.E, Oberhauser, A.F., Carrion-Vazquez, M., Fernandez, J.M., 1999a. The micro-mechanics of single molecules studied with atomic force microscopy. J. Physiol. -London 520, 5–14.
Fisher, T.E., Oberhauser, A.F., Carrion-Vazquez, M., Marszalek, P.E., Fernandez, J.M., 1999b. The study of protein mechanics with the atomic force microscope. Trends Biochem. Sci. 24, 379–384.
Fisher, T.E., Marszalek, P.E., Fernandez, J.M., 2000a. Stretching single molecules into novel conformations using the atomic force microscope. Nat. Struct. Biol. 7, 719–724.
Fisher, T.E., Carrion-Vazquez, M., Oberhauser, A.F., Li, H., Marszalek, P.E., Fernandez, J.M., 2000b. Single molecule force spectroscopy of modular proteins in the nervous system. Neuron 27, 435–446.
Florin, E.L., Moy, V.T., Gaub, H.E., 1994. Adhesive forces between individual ligand-receptor pairs. Science 264, 415–417.
Gerstein, M., Levitt, M., 1997. A structural census of the current population of protein sequences. Proc. Natl. Acad. Sci. USA 94, 11911–11916.
Glick, B.S., 1995. Can Hsp70 proteins act as force-generating motors?. Cell 80, 11–14.
Gopalakrishna, R., Anderson, W.B., 1982. Ca^{2+}-induced hydrophobic site on calmodulin: application for purification of calmodulin by phenyl-sepharose affinity chromatography. Biochem. Biophys. Res. Commun. 104, 830–836.
Graham, G.J., Maio, J.J., 1992. A rapid and reliable method to create tandem arrays of short DNA sequences. BioTechniques 13, 780–789.
Granzier, H., Kellermayer, M., Helmes, M., Trombitas, K., 1997. Titin elasticity and mechanism of passive force development in rat cardiac myocytes probed by thin-filament extraction. Biophys. J. 73, 2043–2053.
Harpaz, Y., Chothia, C., 1994. Many of the immunoglobulin superfamily domains in cell adhesion molecules and surface receptors belong to a new structural set which is close to that containing variable domains. J. Mol. Biol. 238, 528–539.
Hartley, J.L., Gregori, T.J., 1981. Cloning multiple copies of a DNA segment. Gene 13, 347–353.
Hayashi, N., Matsubara, M., Takasaki, A., Titani, K., Taniguchi, H., 1998. An expression system of rat calmodulin using T7 phage promoter in Escherichia coli. Prot. Express. Purif. 12, 25–28.
Hlady, V., Buijs, J., 1998. Local and global optical spectroscopic probes of adsorbed proteins. In: Malmsten, M. (Ed.), Biopolymers at Interfaces. Dekker, New York, pp. 181–220.
Horowits, R., 1999. The physiological role of titin in striated muscle. Rev. Physiol. Biochem. Pharmacol. 138, 57–96.
Hynes, R.O., 1999. The dynamic dialogue between cells and matrices: implications of fibronectin's elasticity. Proc. Natl. Acad. Sci. USA 96, 2588–2590.
Improta, S., Politou, A.S., Pastore, A., 1996. Immunoglobulin-like modules from titin I-band: extensible components of muscle elasticity. Structure 4, 323–337.
Improta, S., Krueger, J.K., Gautel, M., Atkinson, R.A., Lefevre, J.F., Moulton, S., Trewhella, J., Pastore, A., 1998. The assembly of immunoglobulin-like modules in titin: implications for muscle elasticity. J. Mol. Biol. 284, 761–777.
Jackson, S.E., 1998. How do small single-domain proteins fold? Fold. Des. 3, R81–R91.
Janshoff, A., Neitzert, M., Oberdörfer, Y., Fuchs, H., 2000. Force spectroscopy of molecular systems — single molecule spectroscopy of polymers and biomolecules. Augew. Chem. Int. Ed. 39, 3212–3237.
Kellermayer, M.S., Smith, S.B., Granzier, H.L., Bustamante, C., 1997. Folding-unfolding transitions in single titin molecules characterized with laser tweezers. Science 276, 1112–1116.
Lee, G.U., Chrisey, L.A., Colton, R.J., 1994. Direct measurement of the forces between complementary strands of DNA. Science 266, 771–773.
Li, H., Oberhauser, A.F., Fowler, S.B., Clarke, J., Fernandez, J.M., 2000. Atomic force microscopy reveals the mechanical design of a modular protein. Proc. Natl. Acad. Sci. USA 97, 6527–6531.
Li, H., Carrion-Vazquez, M., Oberhauser, A.F., Marszalek, P.E., Fernandez, J.M., in press. Point mutations alter the mechanical stability of immunoglobulin modules. Nature Str. Biol.
Linke, W.A., 2000. Stretching molecular springs: elasticity of titin filaments in vertebrate striated muscle. Histol. Histopathol. 15, 799–811.

Linke, W.A., Ivemayer, M., Olivier, N., Kolmerer, B., Ruegg, J.C., Labeit, S., 1996. Towards a molecular understanding of the elasticity of titin. J. Mol. Biol. 261, 62–71.

Linke, W.A., Rudy, D.E., Centner, T., Gautel, M., Witt, C., Labeit, S., Gregorio, C.C., 1999. I-band titin in cardiac muscle is a three-element molecular spring and is critical for maintaining thin filament structure. J. Cell Biol. 146, 631–644.

Lu, H., Isralewitz, B., Krammer, A., Vogel, V., Schulten, K., 1998. Unfolding of titin immunoglobulin domains by steered molecular dynamics simulation. Biophys. J. 75, 662–671.

Lu, H., Schulten, K., 1999. Steered molecular dynamics simulations of force-induced protein domain unfolding. Proteins. 35, 453–463.

Ludwig, M., Rief, M., Schmidt, L., Li, H., Oesterhelt, F., Gautel, M., Gaub, H.E., 1999. AFM, a tool for single-molecule experiments. Appl. Phys. A 68, 173–176.

Marko, J.F., Siggia, E.D., 1995. Stretching DNA. Macromolecules 28, 8759–8770.

Marszalek, P.E., Pang Y.P., Li, H., El Yazal, J., Oberhauser, A.F., Fernandez, J.M., 1999a. Atomic levers control pyranose ring conformations. Proc. Natl.Acad. Sci. USA 96, 7894–7898.

Marszalek, P.E., Lu, H., Li, H., Carrion-Vazquez, M., Oberhauser, A.F., Shulten, K., Fernandez, J.M., 1999b. Mechanical unfolding intermediates in titin modules. Nature 402, 100–103.

Marszalek, P.E., Oberhauser, A.F., Pang, Y.P., Fernandez, J.M., 1998. Polysaccharide elasticity governed by chair-boat transitions of the glucopyranose ring. Nature 396, 661–664.

Minajeva, A., Ivemeyer, M., Fernandez, J.M., Linke, W.A., Unfolding of titin domains explains the viscoelastic behavior of skeletal myofibrils. Submitted for publication.

Murzin, A.G., Brenner, S.E., Hubbard, T., Clothia, C., 1995. scop: a structural classification of proteins database for the investigation of sequences and structures. J. Mol. Biol. 247, 536–540.

Oberhauser, A.F., Marszalek, P.E., Erickson, H.P., Fernandez, J.M., 1998. The molecular elasticity of tenascin, an extracellular matrix protein. Nature 393, 181–185.

Oberhauser, A.F., Marszalek, P.E., Carrión-Vázquez, M., Fernández, J.M., 1999. Single protein misfolding events captured by atomic force microscopy. Nat. Struct. Biol. 6, 1025–1028.

Oesterhelt, F., Oesterhelt, D., Pfeifer, M., Engel, A., Gaub, H.E., Miller, D.J., 2000. Unfolding pathways of individual bacteriorhodopsius. Science 288, 143–146.

Ohashi, T., Kiehart, D.P., Erickson, H.P., 1999. Dynamics and elasticity of the fibronectin matrix in living cell culture visualized by fibronectin-green fluorescent protein. Proc. Natl. Acad. Sci. USA. 96, 2153–2158.

Perin, M.S., Fried, V.A., Mignery, G.A., Jahn, R., Sudhof, T.C., 1990. Phospholipid binding by a synaptic vesicle protein homologous to the regulatory region of protein kinase C. Nature 345, 260–263.

Pfanner, N., Meijer, M., 1995. Pulling in the proteins. Curr. Biol. 5, 132–135.

Plaxco, K.W., Simons, K.T., Baker, D., 1998. Contact order, transition state placement and the refolding rates of single domain proteins. J. Mol. Biol. 277, 985–994.

Politou, A.S., Gautel, M., Pfuhl, M., Labeit, S., Pastore, A., 1994a. Immunoglobulin-type domains of titin: same fold, different stability?. Biochemistry 33, 4730–4737.

Politou, A.S., Gautel, M., Joseph, C., Pastore, A., 1994b. Immunoglobulin-like domains of titin are stabilized by amino-terminal extension. FEBS Lett. 352, 27–31.

Politou, A.S., Thomas, D.J., Pastore, A., 1995. The folding and stability of titin immunoglobulin-like modules, with implications for the mechanism of elasticity. Biophys. J. 69, 2601–2610.

Politou, A.S., Gautel, M., Improta, S., Bangelista, L., Pastore, A., 1996. The elastic I-band region of titin is assembled by weakly interacting Ig-like domains. J. Mol. Biol. 255, 604–616.

Rief, M., Gautel, M., Oesterhelt, F., Fernandez, J.M., Gaub, H.E., 1997a. Reversible unfolding of individual titin immunoglobulin domains by AFM. Science 276, 1109–1112.

Rief, M., Gautel, M., Schemmel, A., Gaub, H.E., 1998. The mechanical stability of immunoglobulin and fibronectin III domains in the muscle protein titin measured by atomic force microscopy. Biophys. J. 75, 3008–3014.

Rief, M., Oesterhelt, F., Heymann, B., Gaub, H.E., 1997b. Single molecule force spectroscopy on polysaccharides by atomic force microscopy. Science 275, 1295–1297.

Rief, M., Pascual, J., Saraste, M., Gaub, H.E., 1999. Single molecule force spectroscopy of spectrin repeats: low unfolding forces in helix bundles. J. Mol. Biol. 286, 553–561.

Rothman, J.E., Kornberg, R.D., 1986. An unfolding story of protein translocation. Nature 322, 209–210.

Sambrook, J., Fritsch, E.F., Maniatis, T., 1989. Molecular Cloning: A Laboratory Manual, 2nd Edition. Cold Spring Harbor Laboratory Press, Cold Spring Harbor, NY.

Shao, X., Davletov, B.A., Sutton, R.B., Sudhof, T.C., Rizo, J., 1996. Bipartite Ca^{2+}-binding motif in C_2 domains of synaptotagmin and protein kinase C. Science 273, 248–251.

Shaub, A., 1999. Unraveling the extracellular matrix. Nature Cell Biol. 1, E173–E175.

Shtilerman, M., Lorimer, G.H., Englander, S.W., 1999. Chaperonin function: folding by forced unfolding. Science 284, 822–825.

Smith, B.L., Schaffer, T.E., Viani, M., Thompson, J.B., Frederick, N.A., Kindt, J., Belcher, A., Stucky, G.D., Morse, D.E., Hansma, P.K., 1999. Molecular mechanistic origin of the toughness of natural adhesives, fibres and composites. Nature 399, 761–763.

Strick, T.R., Allemand, J.-F., Bensimon, D., Croquette, V., 2000. Stress-induced structural transitions in DNA and proteins. Annu. Rev. Biophys. Biomed. Struct. 29, 523–543.

Studier, F.W., Rosemberg, A.H., Dunn, J.J., Dubendorff, J.W., 1990. Use of T7 RNA polymerase to direct expression of cloned genes. Meth. Enzymol. 185, 60–89.

Sudhof, T.C., Rizo, J., 1996. Synaptotagmins: C2-domain proteins that regulate membrane traffic. Neuron 17, 379–388.

Sutton, R.B., Davletov, B.A., Berghuis, A.M., Sudhof, T.C., Sprang, S.R., 1995. Structure of the first C2 domain of synaptotagmin I: a novel Ca^{2+}-phospholipid binding fold. Cell 80, 929–938.

Trinick, J., Tskhovrebova, L., 1999. Titin: a molecular control freak. Trends Cell Biol. 9, 377–380.

Wallace, R.W., Tallant, E.A., Cheung, W.Y., 1980. Assay, preparation, and properties of calmodulin. In: Calcium and Cell Function, Vol. 1. Academic Press, New York, pp. 13–40.

Witt, C.C., Olivieri, N., Centner, T., Kolmerer, B., Millevoi, S., Morell, J., Labeit, D., Labeit, S., Jockusch, Pastore, A., 1998. A survey of the primary structure and the interspecies conservation of I-band titin's elastic elements in vertebrates. J. Struct. Biol. 122, 206–215.

Yang, G., Cecconi, C., Baase, W.A., Vetter, I.R., Breyer, W.A., Haack, J.A., Matthews, B.W., Dahlquist, F.W., Bustamante, C., 2000. Solid-state synthesis and mechanical unfolding of polymers of T4 lysozyme. Proc. Natl. Acad. Sci. USA 97, 139–144.

Biology at the Single Molecule Level
S.H. Leuba an J. Zlatanova (Eds)

Review

The importance of molecular structure and conformation: learning with scanning probe microscopy

Bettye L. Smith*

Departments of Physics and Marine Science Institute, University of California, Santa Barbara, CA 93106, USA

Abstract

Molecular structure holds a key to understanding Nature's intricate design mechanisms and blueprints. If we can understand her blueprints and basic materials, perhaps we can begin to mimic her beautiful products more cost effectively and with less detrimental environmental consequences. Higher resolution instrumentation has allowed us to study single molecules. Indeed, many stellar contributions to the field have come forth in the last couple of years. We can measure the forces required to unravel individual domains of biological molecules such as titin or DNA to a few picoNewtons resolution. This review will attempt to provide a general overview of the field of single molecule analysis using scanning force microscopy. © 2000 Elsevier Science Ltd. All rights reserved.

Keywords: Atomic force microscope; Review; Scanning force microscopy; Biomineralization; DNA; Single molecule dynamics

Contents

*Corresponding author. Tel.: +1-805-893-4544; fax: +1-805-893-8315.
E-mail address: bettye@mac.com (B.L. Smith).

1. Introduction

As recently as 1987, it was possible to review all of scanning probe microscopy because there had only been about 60 papers written in the field. If coverage is restricted to only two scanning probe microscopes, the atomic force microscope (AFM), and the scanning ion conductance microscope, (SICM) and to only one sub-field, biological applications, there are still roughly 200 papers just within the last year. Most, if not all, of these elegant papers include single-molecule imaging or manipulation. Additionally, single-molecule analysis, from a biological perspective has even expanded into other disciplines, including engineering and materials research (e.g. bioengineering, biomaterials, etc). Reviewing a field of this size is indeed a daunting task.

This review, therefore, will include brief overview of single-molecule applications concerning biomaterials and organic polymers associated with these materials. It has been necessary to modify existing AFMs and build new microscopes to study the structure and activity of single molecules. Thus, a discussion of recent advances in AFM design and implementation along with recent progress in biophysics of single molecules and their activity is warranted. The abalone shell will be used as an example of an exquisite biomaterial that makes intriguing adhesives and fibers, because it is an example, which the author is most familiar. The structure of single biopolymers from the abalone, proteins and fibers at sub-molecular resolution has been examined and will be mentioned in relation to how Nature creates strong and tough fibers. Finally, other examples of Nature to be discussed are those studies that include monitoring the activity of single-molecule biological processes such as transcription and enzyme degradation. Because the field has grown exponentially, it is difficult to discuss every major contribution made with respect to AFM. Many laboratories produce phenomenal work. For the sake of simplicity, the author will focus mostly on contributions from a few selected groups.

2. Instrumentation

2.1. *Scanning probe microscopy*

Scanning probe microscopy (SPM) includes a family of microscopes spawned from the scanning tunneling microscope (Binnig et al., 1982). In general, SPMs analyze samples by probing the surface with a tip and measuring the interaction between tip and sample. Surface properties such as physical topography, charge, magnetic field and temperature can be discerned. The AFM is the most widely used SPM for Biophysics and Molecular Biological Research. For this reason, the AFM and related studies will be the focus of this general review.

The AFM was invented in 1986 (Binnig et al., 1986), and has dramatically evolved since its invention (for a review see Rugar and Hansma, 1990). The first AFMs operated in contact mode. The tip, which is mounted onto the end of a flexible cantilever, scans over the sample surface in a raster pattern. The cantilever deflects as the tip contacts the surface; the deflections are monitored and used to reveal the sample surface (Fig. 1a). Samples can be analyzed in liquids, air, or vacuum. Analyzing the samples in liquids or vacuum generally yields higher resolution, since strong capillary forces due to a thin liquid film present on all samples in air, are absent (Rugar and Hansma, 1990). Müller and Engel measured the contour of a film of native proteins at a lateral resolution of ~0.5 nm and a vertical resolution of ~0.1 nm using contact mode (Müller et al., 1999). Single biological molecules, however, are not easily imaged in contact mode, because they are weakly immobilized to the surface and thus easily scrapped away by contact mode AFM. Extreme care must be taken in sample preparation in order to get high-resolution images of soft biological materials (Czajkowsky and Shao, 1998; Mou et al., 1996; Möller et al., 1999). If the force is held constant, variations in sample compressibility yield inconsistent results and a true

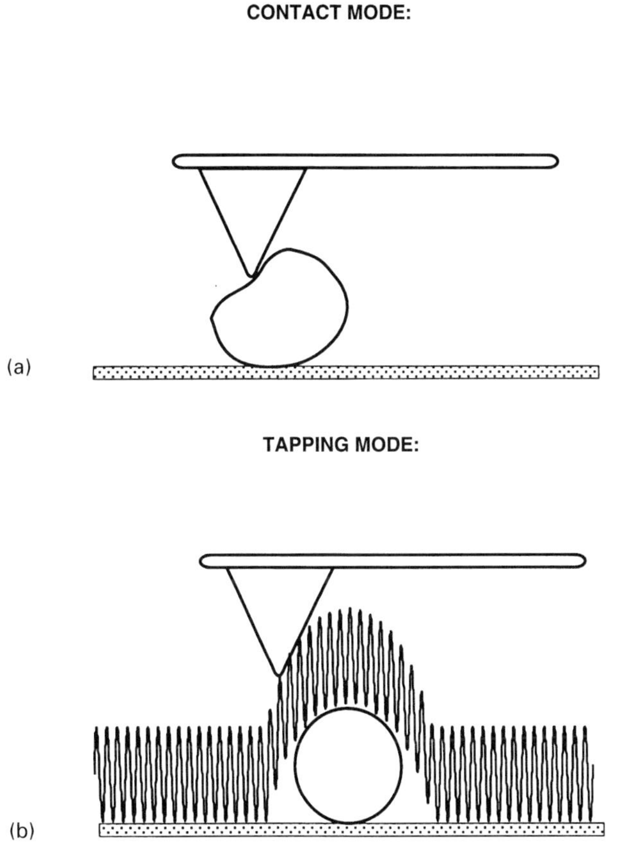

Fig. 1. Schematic of an atomic force microscope cantilever operating in (a) contact mode and (b) tapping mode in liquids. (a) In contact mode, soft samples are easily distorted by lateral forces. (b) In tapping mode in liquids the cantilever oscillates between 8 and 40 kHz at present, though this will increase as cantilever become smaller. Lateral forces are minimized and the loading force can be precisely controlled.

representation of the surface is not obtained. If the height of the sample is held constant, then the force applied to the sample increases with increasing cantilever deflection resulting in damage to the tip and/or the sample.

Non-contact mode was developed in an effort to more accurately image soft samples. In this case, the cantilever oscillates close to its resonant frequency a small distance (1–10 nm) above the surface (Martin et al., 1987). Long-range attractive forces induce changes in the amplitude, frequency and phase of the cantilever oscillations, and maintain a constant distance during scanning. Because the forces on the sample are much lower compared to contact mode, even the softest samples can be imaged without significant damage. One type of non-contact mode imaging is the dynamic force microscope (DFM) (Lüthi et al., 1994). In the DFM, interactions are sensed by the change in amplitude or phase of an oscillating tip. It has been demonstrated that low-amplitude DFM senses changes in the elastic and viscous properties of the surface at nanometer distances (O'Shea et al., 1999; Lantz et al., 1999; Lindsay, 1999).

The development of Tapping Mode® AFM (Zhong et al., 1993) has greatly enhanced imaging of soft biological samples. In this mode, the cantilever oscillates at its resonant frequency, but unlike non-contact mode, the cantilever gently taps the surface during scanning (Fig. 1b). Lateral forces, which move or damage the sample, are greatly reduced. Tapping mode in fluids was developed in the following year (Hansma et al., 1994a). In the first implementation of tapping mode in fluids, the sample, which sits on a piezoelectric scanner oscillates up and down and taps the tip at the apex of each oscillation cycle. The amplitude of the piezoelectric is set manually at the beginning of the run, and the tapping force is held constant by a feedback loop. Thus, the main difference between tapping in fluids and non-contact mode in air or vacuum is that in non-contact or tapping in air, a stiff cantilever is oscillated at its resonant frequency. The changes in cantilever amplitude, phase or resonant frequency is measured. Tapping in fluids modulates the tip–sample separation and measures the response of the cantilever. Additionally, for tapping in fluids, very soft cantilevers with nominal spring constants on the order of 0.1 N/m can be used, whereas the cantilevers for tapping in air have spring constants on the order of 1–100 N/m. This mode is especially suitable for imaging biological samples weakly adsorbed to the surface, and has recently been used to study active biological molecules at single-molecule resolution (Kasas et al., 1997a; Thomson et al., 1999). Near atomic resolution is still unavailable on single molecules. However, Möller et al. (1999) recently compared tapping-mode AFM with contact-mode AFM of the protein layers of the purple membrane (*Deinococcus raiodurans*) at high resolution. They demonstrated that in protein layers, submolecular resolution can be obtained with tapping mode under fluids. They report the ability to image individual peptide loops within the purple membrane. Thus, for "thin films" of biological materials, tapping mode has reached the sensitivity previously only seen in contact mode. These authors speculate that new instrumentation that would automatically adjust the setpoint would produce lower force/higher resolution images of static systems.

Dynamic biological systems would require AFMs capable of fast scanning in order to maintain the resolution mentioned above. Currently, commercial AFMs are often limited by scan rates slower than reaction rates of active molecules being studied. Increasing the scanning speed often damages delicate biological samples and/or increases signal noise from the AFM (Schäffer et al., 1996). Many reactions occur in times on the order of milliseconds, but scanning speeds must be much slower in order to maintain biological viability, usually about 3 lines per second (Butt et al., 1993; Kasas et al., 1997a). So, for an image resolution of 256 pixels, it would take 85 s to complete a single scan. Thus, currently, the only method available to image active complexes requires significantly slowing the reaction conditions. This greatly enhances the difficulty of the experiment and often leads to artifacts or inactive molecules.

One obvious way around this problem is to increase scanning speed of the microscope without damaging soft biological samples. However, increasing the scanning speed of commercial AFM increases the force applied to the sample, which inevitably results in damage and degradation. Rapid imaging using contact mode has been studied extensively (Manalis et al., 1996), but the problems associated with imaging biological molecules in contact mode arise again. Clearly, rapid imaging using tapping mode in fluids is an ideal method to image biological processes in action. One way to improve imaging speed is to use cantilevers with higher resonant frequencies (Walters et al., 1996). Advanced biological applications require low-noise AFMs for cantilevers with spring constants around 0.1 N/m. In order to keep the spring constant fixed, while raising the resonant frequency, the mass of the cantilever must be decreased. Cantilevers of the order of 9–40 μm in length which possess resonant frequencies an order of magnitude higher than commercially available cantilevers have been recently designed and fabricated (Fig. 2c) (Schäffer et al., 1997b). The resonant frequency of the smallest cantilever shown in Fig. 2c was 2.5 MHz in air and 0.94 MHz in water.

One study showed that rapid imaging of a growing calcite crystal in situ can be obtained using smaller cantilevers and a modified commercial AFM (Paloczi et al., 1998). The cantilevers were patterned using conventional photolithography. A 26 μm long, 13 μm wide and 0.3 μm thick cantilever with a spring constant of 0.66 N/m was used. The tapping drive frequency used for this experiment was ~100 kHz. In this method, scan speeds of 104 lines per second were reached. This corresponded to an image every 1.2 s, allowing high-resolution images of a growing screw dislocation on a calcite crystal. From this observation, the critical length a step must reach before another step can form was determined. This would not have been possible with a conventional AFM. In a related study using rapid imaging, Walters et al. (1997b), observed protein binding to growing calcite steps. Scanning at 79 lines per second, or 1.6 s per frame, they showed proteins binding to growing calcite and remaining bound for up to 20 frames.

As mentioned above, the development of smaller cantilevers allowed faster scanning which meets the demands of advanced biological applications. Decreasing the mass of the cantilever increased the resonant frequency while maintaining a constant spring constant. The spring constant must remain on the order of 0.1 N/m or less in order to image biological samples (Walters et al., 1997b). Smaller cantilevers also lower force on the samples and with that smaller cantilever deflection which leads to higher detection noise in the microscope. Therefore, it is important to have a good, low-noise detection system, such as one that utilizes the optical beam deflection method (Meyer and Amer, 1988; Alexander et al., 1989). This method uses angular changes of an optical beam, reflected off the cantilever, to measure the deflection. The optical beam deflection scheme is elegant and simple. Remote sensing with the optical beam that is focused to a spot on the cantilever and reflected off it physically separates the detector from the cantilever-sample environment (Hansma et al., 1994b). The cantilever can easily be submerged in transparent fluids, and cantilevers can be modified (size, composition, shape) to meet specific experimental needs (Drake et al., 1989).

In order to meet these demands, however, it was necessary to develop an AFM with optics able to use a wide range of cantilever sizes at optimum detection sensitivity. Hansma's group developed, characterized, and optimized a detection method for use with small cantilevers of various sizes, from ~5 to 40 μm (Schäffer and Hansma, 1998; Schäffer et al., 1997b). They used an

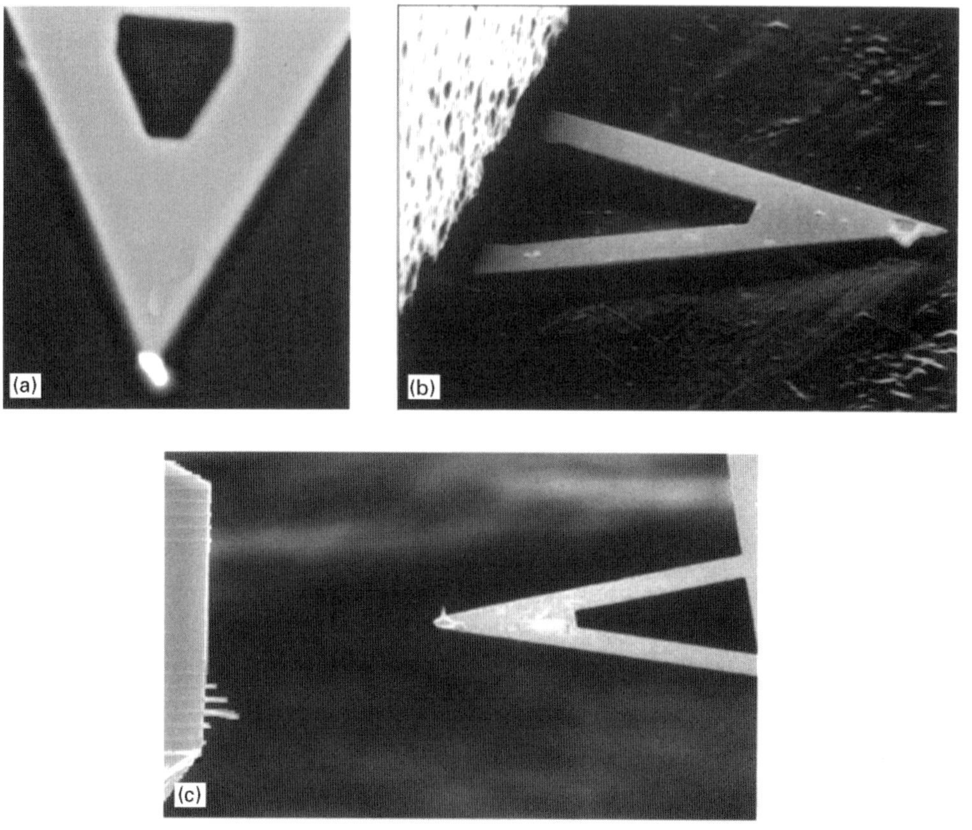

Fig. 2. Tips and cantilevers used for AFM. (a) One of the first tips used a piece of diamond glued to the end of the cantilever. (b) Today, the most common cantilevers are microfabricated out of silicon or silicon nitride with an integrated pyramidal tip. These cantilevers are generally of the order of 100–200 μm long and have nominal spring constants of ~0.1 N m^{-1} for biological applications. (c) A fabricated and characterized small metal cantilevers 4–12 μm long, 1–4 μm wide, and 60–300 nm thick. The small cantilevers shown above are made of nickel and have resonant frequencies >1 MHz and nominal spring constant of <0.1 N m^{-1} (Viani et al., 1999a).

adjustable aperture in the incident beam path of an AFM for small cantilevers. The focused spot size is 1.6 μm in diameter and is implemented with a variable focus adjustment that allows the user to re-focus on each cantilever. Different cantilever sizes require different aperture widths for maximum detection sensitivity (Schäffer and Hansma, 1998). The adjustable aperture optimizes the detection sensitivity when using the AFM with different size cantilevers. Single DNA molecules were recently imaged with at scanning speeds of less than a second per image. Structural continuity of the molecules was maintained (Viani et al., 1999b). Experiments to measure biological activity using rapid imaging are currently underway (Viani, unpublished results).

2.2. *Combined AFM/SICM*

The scanning ion conductance microscope (SICM) was one of the first offspring developed from the AFM (Hansma et al., 1989; Prater et al., 1990). This SICM was based on a pipette pulled down to a narrow aperture. An electrical system was used to monitor the ionic currents flowing through the aperture at the end of the pipette as it scanned over the surface. Proksch et al. (1996) recently improved upon the SICM by using bent probes that could also be used for tapping mode AFM. The position feedback is provided by the AFM. The SICM signal is simultaneously measured, but not used for feedback. The combined SICM/AFM thus measures surface topography and ion conductance simultaneously.

Schäffer et al. used this microscope to measure ion conductance through pores of organic interlamellar sheets isolated from abalone nacre (the "mother-of-pearl" region of the shell) (Schäffer et al., 1997a). The abalone shell is a complex composite of calcium carbonate and organic biopolymers secreted by the animal (Smith, 1998). The calcium carbonate appears in the shell in two distinct crystal forms, calcite and aragonite. The rough, outer portion of the abalone shell contains calcite plus organic biopolymers, while the lustrous nacreous part of the shell is composed of highly oriented aragonite arranged in a "stack of coins" plus organic polymers (Fig. 3). The interlocking tablets are approximately 400 nm thick and 5–10 μm wide surrounded

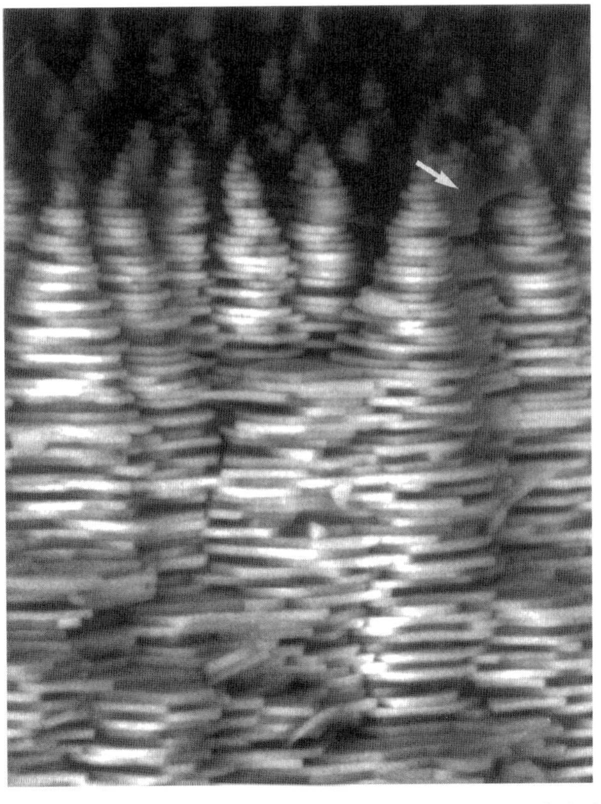

Fig. 3. Scanning electron micrograph of aragonite tablets found within the nacre of abalone shells. The interlocking tablets are approximately 400 nm thick and 5–10 μm wide surrounded on all sides by thin sheets of organic matrix approximately 30 nm thick. Interspersed between the aragonite "coins" or tablets are the interlamellar sheets (depicted by the arrow) (see also Smith, 1998). Published with kind permission of Deron Waters and Monika Fritz.

Fig. 4. Flat pearl. A glass coverslip was deposited between the shell and mantle of an adult red abalone. After 14 days, the coverslip was carefully removed and kept in sterilized seawater. This flat pearl is described by Fritz et al. (1994).

on all sides by thin sheets of organic matrix approximately 30 nm thick. Interspersed between the aragonite "coins" or tablets are the interlamellar sheets.

Because neither the SICM nor the AFM require rigorous sample preparation, the enzymatic, biochemical and structural composition of these organic sheets could be studied. The organic interlamellar sheets were isolated from "flat pearls" (Fig. 4), a biofabricated model system used to study abalone shell growth (Fritz et al., 1994). The existence of the pores in the interlamellar sheets was confirmed by this method. These pores corresponded stochastically to the lateral offset of the aragonite tablets from one layer to the next. The size and the spacing of the pores also correlates with protrusions found on aragonite tablets. As the aragonite grows through the pores, mineral bridges are formed between one layer and the next. Thus, it has been suggested that the aragonite grows through pores in the interlamellar sheets as a single crystal, and maintains its crystallographic alignment not only as it grows upward, but also as it grows outward. This notion has led to a new paradigm for shell growth (Fig. 5). Successive tablets grow from mineral bridges through the interlamellar sheets. The mineral bridges connect the tablets as a single crystal. As the tablets grow laterally, more mineral bridges might form between the tablets, allowing the tablets to grow together in more than one place.

Shortly after the invention of the SICM, microfabricated probes were developed (Prater et al., 1991). Though these early probes had their problems such as difficulties in aperture formation and placement and sensitivity to sample tilt, they give us a glimpse of the future of the field. The key will be wafer-scale probe production of probes suitable for simultaneous AFM and SICM as discussed above for bent pipettes.

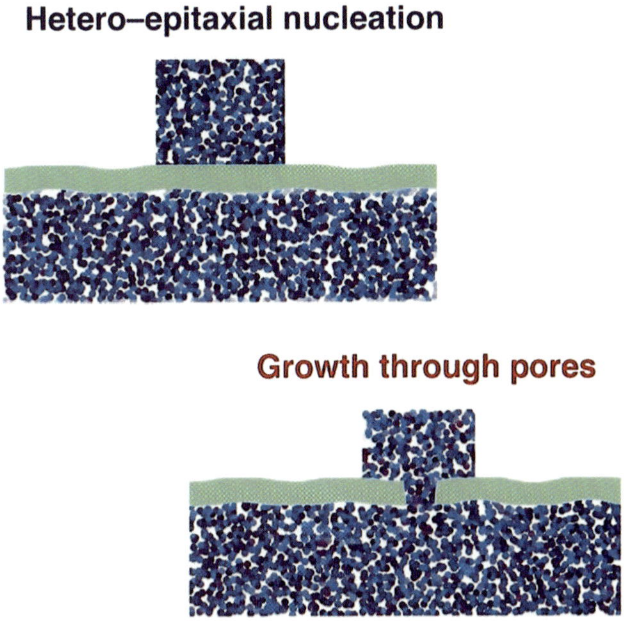

Fig. 5. A new paradigm for shell growth. Two competing paradigms for nacre formation are shown. The top figure represents the old paradigm in which nacre tablets are hetero-epitaxially nucleated on the surface of the interlamellar organic sheets (shown in green). In the new paradigm, successive tablets grow from mineral bridges *through* the interlamellar sheets as depicted by the bottom figure. The mineral bridges connect the tablets as a single crystal. As the tablets grow laterally, more mineral bridges might form between the tablets, allowing the tablets to grow together in more than one place. Thus, an inter-locking pattern of tablets emerges.

3. Examples from nature

3.1. Advice from the abalone shell

The discovery of nacre growth through mineral bridges represents a major advance for researchers attempting to understand the process of shell formation (Smith, 1998). The new model explains how the aragonite tablets are arranged as a single crystal and how they grow through successive interlamellar sheets. Intrinsic to the overall picture of shell formation, however are the organic biopolymers. The interlamellar sheets and the other families of soluble proteins are controlled by a genetic switch that dictates whether insoluble interlamellar sheets will form, or soluble proteins designed to control crystal growth and polymorph type will form. Thus, two general classes of organic biopolymers are found in abalone shells: (1) water-insoluble polymers found in the interlamellar sheets, and (2) water-soluble proteins that control crystal growth and polymorph selection. Single-molecule analysis of these biopolymers could greatly facilitate the elucidation of shell formation, which would ultimately lead to advancement of analogous synthetic materials.

The soluble protein families dictate which crystal structure is formed and when it is deposited. They are mainly polyanionic proteins, the bulk of which represents a mixture of glycoproteins.

The family of calcite directing proteins was isolated from the calcite portion of the abalone shell and studied by AFM to examine the nanometer-scale effects of shell proteins on a growing calcite surface in situ (Walters et al., 1997a). The interaction between abalone shell proteins and a growing calcite surface was directly visualized. The proteins induced rounding of the step edges on that surface, which is consistent with the macroscopic habit modification these proteins induce (Belcher et al., 1996). A roughening of the terraces upon addition of the family of calcite proteins was also observed. These phenomena were not observed with bovine serum albumin, suggesting that the soluble proteins from the outer region of the shell do indeed direct crystal growth.

When these experiments were repeated with the soluble proteins extracted from the nacreous region of the shell, a different morphological change was observed. The step edges were severely kinked and the entire surface was roughened (Walters et al., 1997a). High-resolution AFM analysis of these calcite surface changes has revealed that the atomic lattice of the original crystal is converted into aragonite (J. B. Thompson, unpublished results). Individual proteins from the aragonite family have been separated and purified (Michenfelder and Morse, unpublished). The effects of these individual proteins on a growing calcite surface in situ resulted in radical differences in the interactions of individual fractions upon the surface (Smith and Paloczi, unpublished results). Some of the proteins bind specifically to the step edges and irreversibly inhibit further growth, whereas others bind only to the terraces and do not effect the growth of a step. The most acidic protein seemed to exhibit the most drastic effect on alteration of the growing calcite surface.

The interlamellar sheets are usually made up of structural proteins and/or polysaccharides. Shen et al. (1997) resolved the complete amino acid sequence of one of the insoluble nacre matrix proteins, which was named Lustrin A. The complementary DNA and translated protein sequence reveal that the structure of this protein consists of about 10 alternating and highly conserved cysteine- and proline-rich domains, demonstrating that the structure is highly modular (Fig. 6).

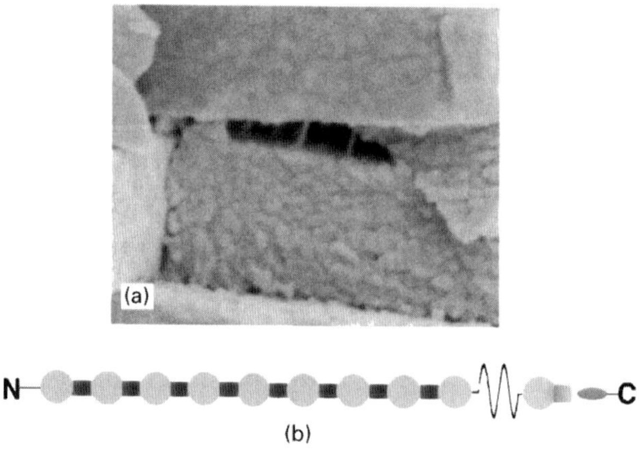

Fig. 6. Elastic fibers of abalone nacre. (a) Scanning electron micrograph of freshly cleaved abalone nacre under three-point mechanical stress. Fibers are easily revealed between consecutive mineral tablets, which are approximately 400 nm thick. Immunohistochemical analysis of these fibers revealed Lustrin A to be a component of the organic material between the mineral tablets (Belcher and Morse, unpublished). (b) Cartoon depicting the modular structure of Lustrin A based on its sequence. The balls represent cysteine-rich domains and the rectangular region represents the proline-rich regions.

The main function of this protein is thought to hold the aragonite tablets together. Lustrin A is considered an elastic protein, due in part to glycine loops within the sequence. Indeed, adhesive ligaments found between successive tablets have been observed by SEM and TEM (Smith et al., 1999). Immunohistochemical analysis of these fibers revealed Lustrin A to be a component of the adhesive between the nacre mineral tablets (A. Belcher and D. E. Morse, unpublished results).

3.2. Adhesives and fibers

Aside from Lustrin, Nature makes a plethora of materials that are renowned for their strength, toughness and elasticity. Examples are nearly endless. Spider dragline silk has a breakage energy per unit weight two orders of magnitude greater than high tensile steel (Heslot, 1998). The tough tendons of mussel byssus tightly glues mussels to rocks in strong currents (Qin et al., 1997). The abalone shell is roughly 3000 times more fracture-resistant than geologically formed calcium carbonate (Jackson et al., 1988). The organic material locked within the abalone shell is thought to be the key to the overall toughness of the shell, despite its relative paucity as a filler (just a few percent by weight). Yet, to date no one has been able to adequately reproduce some of Nature's composites. Almqvist et al. (1999) showed that synthetic materials made of interlocking ceramic tablets together with ordinary adhesives do not have a toughness comparable to nacre. The list of tough composites and fibers goes on and on. What is the common thread tying all of these strong materials together?

A mechanism behind the toughness of organic adhesives and fibers found in Nature was recently reported (Smith et al., 1999). An elegant technique first described by Rief et al. was deployed to examine adhesive properties of single biopolymers (Rief et al., 1997a, b). A molecule is attached between a flat surface and a cantilever of an AFM and extended. By plotting the force versus extension curves, Rief et al. measured the force required to unravel a single subunit in its modular structure. As the molecule stretched, the force–extension curve revealed a sawtooth pattern (Fig. 7a). They demonstrated that each tooth corresponded to a single-domain unfolding. This technique affords a unique opportunity to examine some of Nature's fibers and hopefully unravel some of its mysteries in building modular elastic fibers.

The abalone shell represents a strong composite model. Recently, this pulling technique was used to study the molecules holding the shell together (Smith et al., 1999). A cantilever of an AFM was touched to the surface of freshly cleaved nacre until a molecule was grabbed. A sawtooth-like pattern in the force versus extension curves was observed that was repeatable through many pulls without touching the surface again with the cantilever, suggesting that some kind of bonds are breaking and reforming. Breaking forces of the order of 100–400 pN were observed in these curves (Fig. 7b). Since Lustrin is present on this surface, it is possible that the authors observed unfolding of this protein, though, it should be noted that the mechanism behind the strength and toughness of the adhesive is revealed by the force–extension curves. It *does not* depend on the identification of the specific molecules involved. This behavior may reflect the successive opening of intra-chain loops or folded domains within a single molecule (Fig. 8), or the successive release of sacrificial inter-chain bonds holding a crosslinked multi-chain matrix together (Smith et al., 1999).

A modular adhesive or fiber that would contain intermediate-strength sacrificial bonds would display a force versus extension curve that rises to a large force quickly, but then maintains that force over large extensions. It would consist of long molecules with many domains that are either

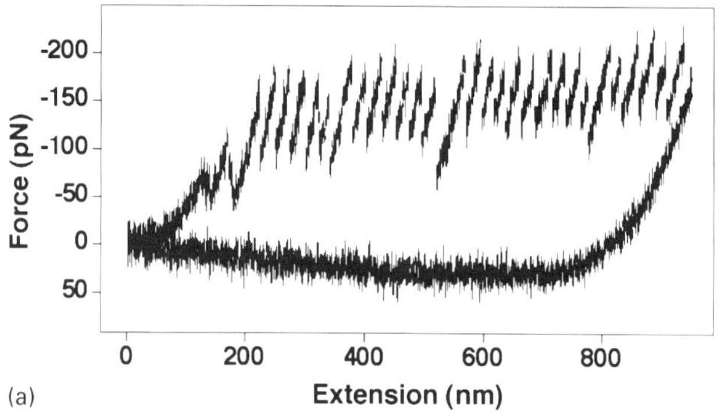

(a)

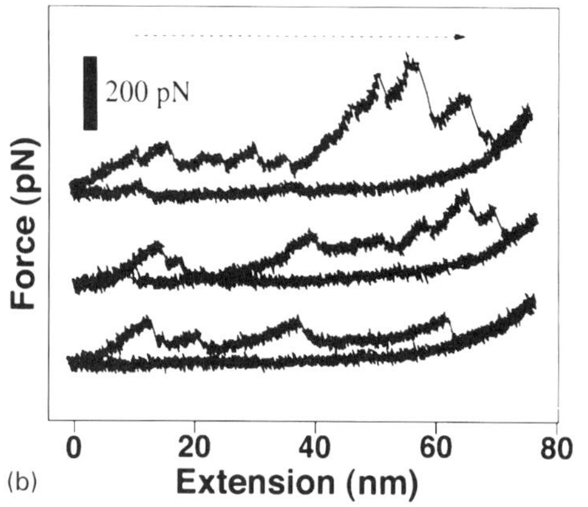

(b)

Fig. 7. Force versus extension curves. (a) One end of a single molecule of titin is attached to a mica surface and the other to a cantilever. Each peak in the pattern represents one subunit unfolding (Rief et al., 1997b). A single titin molecule was pulled on up to 50 times without touching the surface with the cantilever confirming earlier studies that single domains are indeed unfolding and refolding. (b) Three consecutive force versus extension curves from pulling on freshly cleaved nacre surface (Smith et al., 1999). A similar sawtooth appearance is visible in each curve. The cantilever did not touch the surface between pulls suggesting that some refolding took place, possibly of Lustrin A. The hysteresis observed shows that energy is dissipated upon unfolding.

folded or looped, with intermediate-strength bonds to compact the molecule. These intermediate-strength bonds are stronger than hydrogen bonds or van der Waals forces, but weaker than the covalent bonds of the backbone of the molecule. Alternatively, the cumulative effect of multiple weak bonds may act in concert to produce the same effect. Thus, for such a molecule, when it is initially stretched, the force rises quickly to a significant fraction of the force required to break the backbone (e.g. a strong bond). However, before the backbone breaks, a domain unfolds or a loop is opened, thereby avoiding breaking the strong bonds of the backbone. If the molecule is pulled even more, the force rises again to a significant fraction of a strong bond, but another domain or loop unfolds before the backbone can break. If the tension is released, the domains refold adding strength to the molecule. But, if the molecule is finally stretched until all domains unfold, then the molecule finally breaks. The net result is to sustain a large force over the pulling extension making

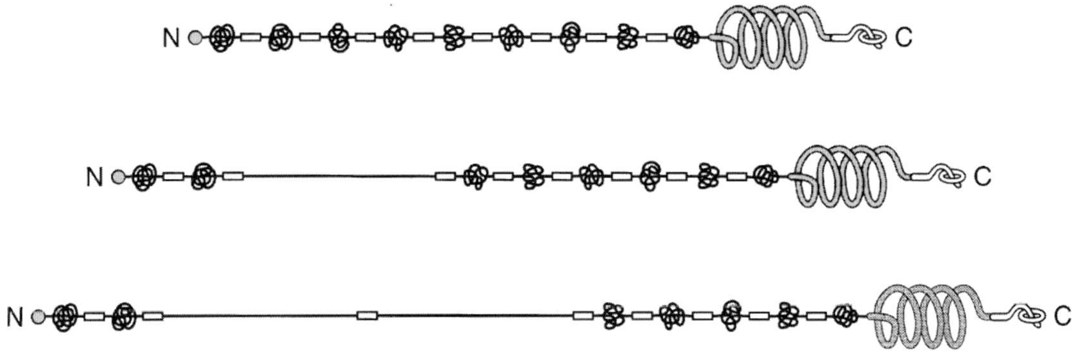

Fig. 8. Long polymers behaving as entropic springs. The first molecule is compacted with many domains that are held together by intermediate-strength sacrificial bonds. As the molecule is stretched, some of the intermediate-strength sacrificial bonds break as depicted in the second and third molecule. Before the backbone of the molecule can break (of the order of a nN) more domains unfold. Repeated unfolding allows stretching up to large extensions and requires significant energy. Therefore, a long molecule with domains containing intermediate-strength sacrificial bonds combines both high tensile strength and great toughness.

the molecule "strong", while producing a relatively large area under the force versus extension curve, making it "tough" as well (Smith et al., 1999).

3.3. *Single DNA molecules*

DNA has become one of the most studied molecules via AFM and has been imaged on several surfaces in air and in a variety of solutions (Hansma et al., 1992a,b; 1993a,b; Bezanilla et al., 1993; Lyubchenko et al., 1992). When DNA was immersed in alcohol, adhesion forces were dramatically reduced, enhancing the specimen's stability and improving the reproducibility of the technique (Hansma et al., 1993a; 1992b; Lyubchenko et al., 1993).

Imaging in water and aqueous buffer has been much more challenging because of the weakened DNA adsorption to the surface. The problem has been circumvented by the discovery that divalent cations (Hansma and Laney, 1996) play a critical role in DNA adhesion to mica surfaces and the development of tapping mode in liquids (Hansma et al., 1994a). Chemical modification of surfaces also aides in adsorbing DNA to the surface (Bezanilla et al., 1995; Thundat et al., 1992). High-resolution images of DNA have been taken by Shao et al. (1996), in which they deposited DNA on a cationic lipid bilayer and imaged in contact mode under liquid. They report observing the right-handed double helix of a double-stranded piece of DNA, which requires 2–3 nm resolution. The motion and enzymatic degradation of DNA by restriction endonuclease has also been observed using AFM (Bezanilla et al., 1994).

Aside from qualitative observations of DNA in various buffers and on specific substrates, quantitative observations of DNA using AFM have been made (Lee et al., 1994). The unbinding force between complementary strands of DNA have been measured by dynamic force spectroscopy (Strunz et al., 1999). These studies measured the force required to mechanically separate a single strand of DNA as a function of loading force. Under these conditions, the unbinding forces varied between 20 and 50 pN, and scaled logarithmically with the loading rate.

Persistence length measurements of moving and stationary DNA molecules, along with contour lengths and changes in end-to-end distance have been measured (Hansma et al., 1997; Guthold et al., 1999). Intercalators and carcinogen binding to DNA has been mapped quantitatively by directly measuring the contour length and end-to-end distances via AFM (Coury et al., 1996; Smith and MacLeod, 1998). Recent results that analyze a potent carcinogen, benzo(a)pyrene diol epoxide by AFM has shown that this molecule causes a strand kinking in DNA, consistent with indirect measurements (Pietrasanta et al., 2000). This and prior studies suggest that AFM is a novel technique that will eventually make its way into the clinical research field (Bottomley et al., 1996; Coury et al., 1996). Indeed, AFM has even used to measure the extent of DNA condensation for gene therapy (Hansma et al., 1998b).

3.4. Dynamic biological systems

Proteins are dynamic entities changing shapes and moving around on a millisecond basis (Kasas et al., 1997b). Their motion often involves changes in length of several angstroms for a time scale that can be detected by AFM. The study of these motions is an integral key to understanding the processes involved in enzymatic reactions.

For example, Radmacher et al. (1994b) measured height fluctuations of lysozyme upon addition of substrate. In the presence of inhibitor, the height fluctuations decreased to the level without substrate. The height fluctuations corresponded to conformational changes upon activation by substrate. Thomson et al. (1996) also measured the height fluctuations over three different proteins, immunoglobulin G, urease, and microtubules using AFM under liquids using tapping mode. In this case, a protein tracking system was used, where the AFM tip was periodically repositioned above a single protein molecule as thermal drift occurred. Data were taken in 1–2 s time series with the tip positioned over the molecule, and compared to data taken on just the surface. The measured fluctuations were consistently higher when the tip was positioned over the protein, compared to the surface on which the protein was adsorbed. The increase in noise was shown to result from motion of the proteins.

Cooperative protein binding has also been observed by AFM. Poly(A) binding protein (PABP) was shown to bind cooperatively to poly(A) RNA (Smith et al., 1997). It has been previously shown that PABP binds cooperatively with a packing density of one PABP molecule per 25 adenosine molecules. When directly observed by AFM, following binding to poly(A) RNA, PABP appeared to be present in variable size complexes that bound length wise along the RNA. This complex size appeared to be maximal, suggesting that PABP binding to poly(A) RNA is self-limiting. Thus, AFM provided potential new insights into the complexes formed by this RNA binding protein.

Recently, a single RNA polymerase transcribing a strand of DNA was observed (Kasas et al., 1997a). Stalled ternary complexes of the polymerase, double stranded DNA, and nascent RNA transcript were adsorbed on mica and imaged under continuously flowing buffer. On introduction of the necessary ribonucleoside triphosphates (rNTP), DNA was pulled through the polymerase (Fig. 9). In general, some of the complexes were inactive, some released the DNA and some transcribed it efficiently. The rate of transcription was measured to be 0.5–2 bases/s at an rNTP concentration of 5 μm. It was necessary to use such a low concentration of rNTPs in order to observe the polymerase transcribing DNA. The nascent RNA transcripts did not bind to the

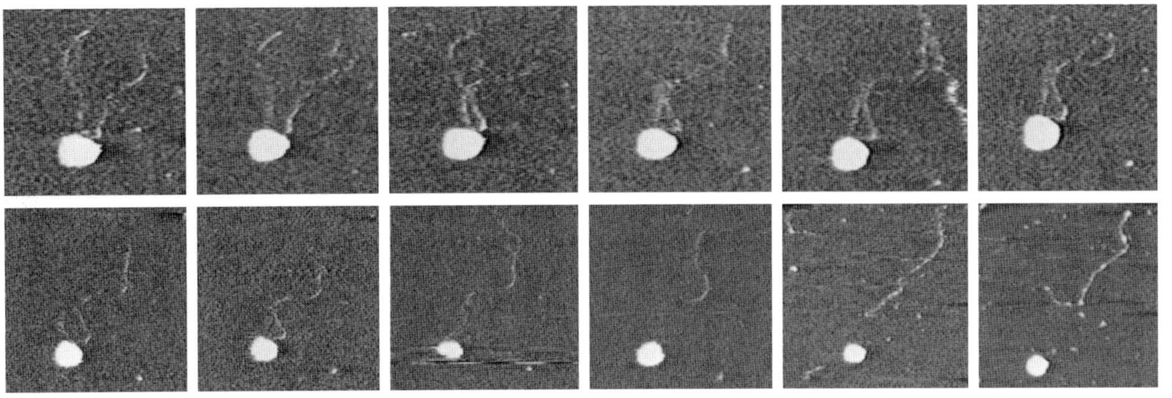

Fig. 9. RNA polymerase transcribes a strand of DNA. This time-lapse series shows transcription of a double-stranded 1-kilobase template of DNA by RNA polymerase. The first four images (left to right, top row) show that DNA moves around on the surface before addition of nucleotide triphosphates (rNTPs). rNTPs arrive in the image by the fifth panel of the top row. The remaining sequential images show that one arm of the DNA template becomes progressively shorter until it is finally released. Reprinted with permission from Kasas et al. (1997a), Copyright (1997) American Chemical Society.

surface well enough in fluids to be seen. However in experiments using a small single-stranded circular DNA template (i.e. a rolling circle), transcripts up to 1–2 μm could be observed with AFM once the samples were dried and imaged in air. Thus, ternary complexes adsorbed to mice were still active, in agreement with RNA polymerase immobilized to beads (Kashlev et al., 1993). Furthermore, Hansma et al. (1998a) observed that RNA polymerase adsorbs to mica in a left-handed orientation. The DNA binds to the side rather than through the middle of the polymerase, lending credence to the notion that the complexes are active when adsorbed to substrate.

Nevertheless, knowing the orientation of active molecules on the surface would be very useful. Making the active site more accessible would prevent topological constraints from impeding data acquisition and interpretation, and would greatly facilitate measuring the activity and motion of enzymes. Properly orienting a molecule on a surface should increase the activity of that molecule, and therefore increase the probability of observing events related to activity as well as conformational or structural changes within the molecule. Recombinant *E. coli* RNA polymerase containing histidine tags was oriented onto ultra-flat gold surface functionalized with alkanethiols (Thomson et al., 1999) (Fig. 10). One alkanethiol was terminated in an ethylene-glycol (EG) group, which rejects protein adsorption, and the other was terminated in an *N*-nitrilotriacetic acid (NTA) group, which binds the histidine tag of recombinant proteins through two coordination sites with nickel or cobalt (Wagner et al., 1996; Whitesides and Gorman, 1995). The NTA-terminated alkanethiol phase segregated, and the recombinant proteins bound specifically to the phase-segregated moieties. Unfortunately, the DNA was refractory to binding the EG-terminated alkanethiol as well, so in this method the DNA was not observed as it threaded through the polymerase. However, an active complex exhibited a large conformational shift upon addition of rNTPs (Fig. 11). Thus, orienting molecules on a surface opens doors for measuring their activity and conformational changes. Rapid imaging should facilitate these observations on time-scales upon which these occurrences transpire naturally.

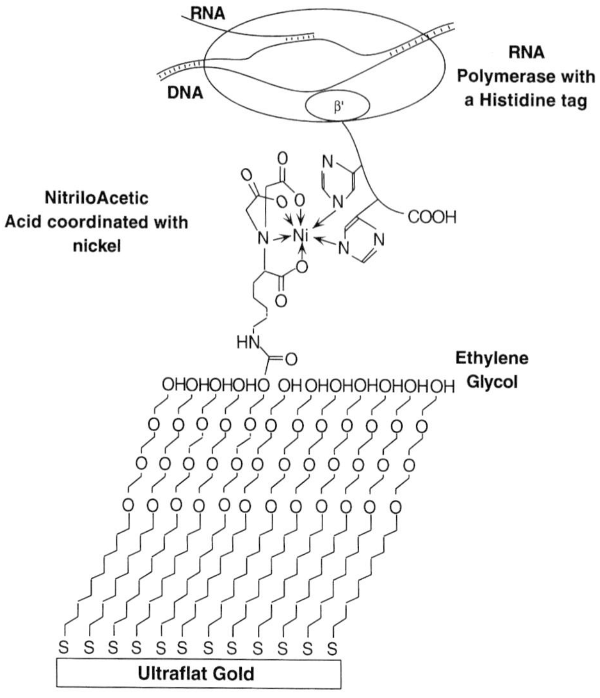

Fig. 10. Schematic of an RNA polymerase molecule oriented on a surface. Gold was sputtered or evaporated onto mica and glued onto steel disks gold-side down. The mica was peeled off and self-assembled monolayers were formed by incubating the surface into a mixture of the two alkane-thiols. Prior to use the surface was treated with nickel to form coordination site for histidine tag binding (Thomson et al., 1999).

4. Sample–tip interactions

Adhesion can be a menace or a blessing depending upon the experimental questions and conditions. In order to understand tip–sample interactions, many studies have explored adhesion forces on substrates and surfaces. Mapping forces between tip and sample is relatively straightforward. The deflection of the cantilever is recorded while the sample is approached, brought into contact with the tip, and then withdrawn. Force curves contain information about surface properties such as topography, adhesion, elasticity, and interactions between single molecules.

Hoh et al. (1992b) observed that a silicon nitride cantilever tip strongly adheres to glass in pure water. Interestingly, the adhesive forces were greatly dependent on pH. At high pH the force curve was repulsive, suggesting that both surfaces became strongly charged and presented a substantial double-layer force. The forces observed were larger than would have been expected for van der Waals interactions; thus, it seemed that the Nature of the interaction was due to hydrogen bonding (Hoh et al., 1992a). Since the AFM can be used to manipulate structures such as Langmuir–Blodgett films (Hansma et al., 1991), this study shows that AFM

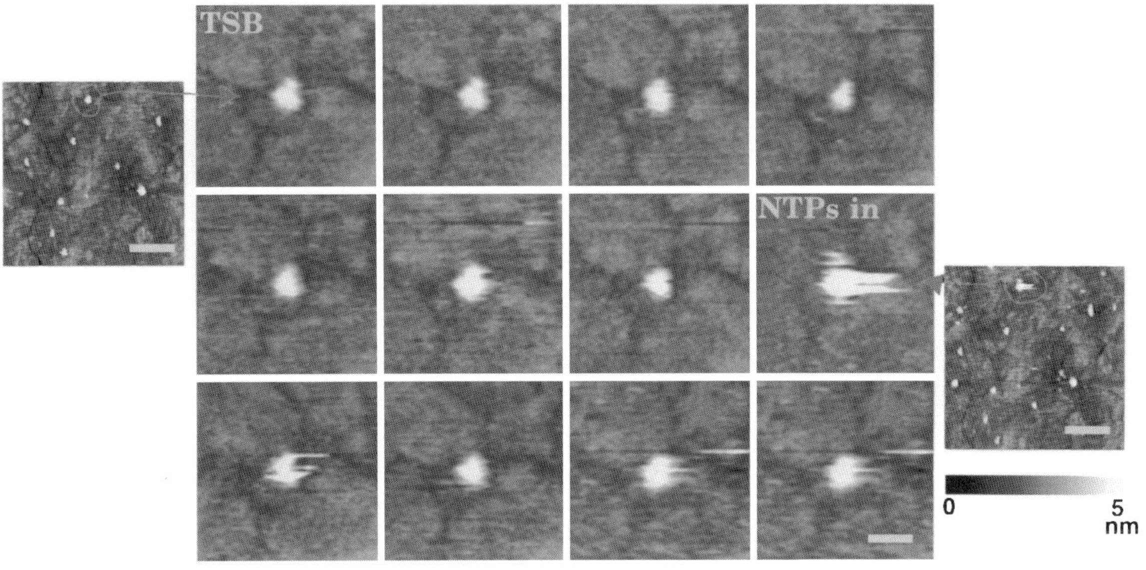

Fig. 11. RNA polymerase oriented on a functionalized gold surface. A specifically oriented RNA polymerase is adsorbed to a functionalized gold surface via a histidine tag of the recombinant protein. Smith, Almqvist and Thomson scanned a 130 nm area (outset) that displayed approximately 10 molecules. We zoomed in on one molecule (circled by the green arrow) that showed a dramatic shape change upon addition of rNTPs. The top panel shows the molecule as we imaged in ordinary transcription buffer (TSB). The fourth image of the middle panel underwent an extraordinary shape change upon addition of rNTP, suggesting that this protein was transcriptionally active and underwent a conformational shift. Scale bar of the zoomed molecule is 30 nm.

could be a useful application for dissecting biological molecules, or perhaps analyzing electrical impulses.

Because the AFM can measure very small forces, it can be used to delineate properties of structured surfaces, such as microtubules or tubulin (Radmacher et al., 1995), or lysozyme (Radmacher et al., 1994a). In the case of lysozyme, it was found that pH greatly affected lysozyme not only binding to mica, but also with itself (Radmacher et al., 1994a). At pH 6.4 adsorption was weak and only aggregates of lysozyme could be imaged with AFM. But, at pH 4.0, the protein becomes highly charged and binds to mica as monomers. In the latter case, the protein could be imaged and force maps taken. From these studies, the viscoelastic properties were deduced. In the case of tubulin (Radmacher et al., 1995), it was found that the observed adhesion is larger than expected by only van der Waals forces, suggesting that additional, complex interactions determine the adhesion between the tip and the protein.

Using tip adhesion can be an advantage. Many studies have functionalized tips to specially interact with molecules on the surface. The most common functionalization involves streptavidin/avidin binding (Florin et al., 1994). In this way, the AFM can be used as a "force sensor" to measure the bond strength between two molecules such as ligand and receptors. Such experiments have even made their way into cardiovascular research (Arnsdorf and Xu, 1996). Indeed, the area has become so popular that the term "chemical force microscopy" has been coined, in which

AFM is used to measure the mechanics of molecular recognition between the surface and a functionalized cantilever tip (Mazzola et al., 1999).

5. Conclusions

AFM continues to grow and expand into a plethora of fields and subfields since its invention over a decade ago. With respect to imaging single biological molecules, the field has progressed from simply observing the shape of single molecules to measuring biological activity to quantifying conformational changes and unfolding on a submolecular level.

The invention of faster AFMs with smaller cantilevers allows the measurement of forces necessary to unravel individual protein domains. With the help of the abalone shell and these faster microscopes, one of Nature's mechanisms in making strong and tough fibers and adhesives has already been elucidated (Smith et al., 1999). Thus, we can use the knowledge that we have gained from these studies and attempt to mimic Nature's design of strong fibers, biocomposites, and even nano-motors (protein machines). As AFMs continue to improve, the ability to measure single-molecule conformational dynamics on time-scales equivalent to actual changes that occur within the molecule becomes a real possibility.

Acknowledgements

I apologize for omitting many stellar contributions to the field of AFM, due to the specific focus of this very limited review. I chose to concentrate on the work of a few selected groups because of the broad scope of their research (instrumentation and biophysics), and because I am most familiar with their contributions. I thank Mario Viani for use of the small cantilevers figure. Very special thanks go to Timothy Maddux and Faye Smith for infinite support and unique discussions. I gratefully acknowledge the support of the Army Research Office (Grant No. DAAH04-96-1-0443), the Office of Naval Research (Grant No. N00014-93-10584), and the Materials Research Laboratory Program of the National Science Foundation (Grant No. DMR96-32716). The U.S. Government is authorized to reproduce and distribute copies for governmental purposes.

References

Alexander, S., Hellemans, L., Marti, O., Schneir, J., Elings, V., Hansma, P.K., Longmire, M., Gurley, J., 1989. J. Appl. Phys. 65, 164–167.

Almqvist, N., Thomson, N.H., Smith, B.L., Stucky, G.D., Morse, D.E., Hansma, P.K., 1999. Mater. Sci. Eng. 7, 37–43.

Arnsdorf, M.F., Xu, S., 1996. J. Cardiovasc. Electrophysiol. 7, 639–652.

Belcher, A.M., Wu, X.H., Christensen, R.J., Hansma, P.K., Stucky, G.D., Morse, D.E., 1996. Nature 381, 56–58.

Bezanilla, M., Brake, B., Nudler, E., Kashlev, M., Hansma, P.K., Hansma, H.G., 1994. Biophys. J. 67, 2454–2459.

Bezanilla, M., Bustamante, C., Hansma, H.G., 1993. Scanning 7, 1145–1148.

Bezanilla, M., Manne, S., Laney, D.E., Lyubchenko, Y.L., Hansma, H.G., 1995. Langmuir 11, 655–659.
Binnig, G., Quate, C.F., Gerber, C., 1986. Phys. Rev. Lett. 56, 930–933.
Binnig, G., Rohrer, H., Gerber, C., 1982. Phys. Rev. Lett. 49, 57–61.
Bottomley, L.A., Coury, J.E., First, P.N., 1996. Anal. Chem. 68, R185–R230.
Butt, H.-J., Siedle, P., Seifert, K., Fendler, K., Seeger, T., Bamberg, E., Weisenhorn, A.L., Goldie, K., Engel, A., 1993. J. Micro. 169, 75–84.
Coury, J.E., McFail-Isom, L., Williams, L.D., Bottomley, L.A., 1996. Proc. Natl. Acad. Sci. 93, 12283–12286.
Czajkowsky, D.M., Shao, Z., 1998. FEBS Lett. 430, 51–54.
Drake, B., Prater, C.B., Weisenhorn, A.L., Gould, S.A., Albrecht, T.R., Quate, C.F., Cannell, D.S., Hansma, H.G., Hansma, P.K., 1989. Science 243, 1586–1589.
Florin, E.-L., Moy, V.T., Gaub, H.E., 1994. Science 264, 415–417.
Fritz, M., Belcher, A.M., Radmacher, M., Walters, D.A., Hansma, P.K., Stucky, G.D., Morse, D.E., Mann, S., 1994. Nature 371, 49–51.
Guthold, M., Zhu, X.S., Rivetti, C., Yang, G.L., Thomson, N.H., Kasas, S., Hansma, H.G., Smith, B., Hansma, P.K., Bustamante, C., 1999. Biophys. J. 77, 2284–2294.
Hansma, H.G., Benzanilla, M., Nudler, E., Hansma, P.K., Hoh, J., Kashlev, M., Firouz, N., Smith, B.L., 1998a. Probe Micro 1, 127–134.
Hansma, H.G., Bezanilla, M., Zenhausern, F., Adrian, M., Sinsheimer, R.L., 1993a. Nucl. Acids Res. 21, 505–512.
Hansma, H.G., Golan, R., Hsieh, W., Lollo, C.P., Mullen-Ley, P., Kwoh, D., 1998b. Nucl. Acids Res. 26, 2481–2487.
Hansma, H.G., Gould, S.A.C., Hansma, P.K., Gaub, H.E., Longo, M.L., Zasadzinski, J.A.N., 1991. Langmuir 7, 1051–1054.
Hansma, H.G., Kim, K.J., Laney, D.E., Garcia, R.A., Argaman, M., Allen, M.J., Parsons, S.M., 1997. J. Struct. Biol. 119, 99–108.
Hansma, H.G., Laney, D.E., 1996. Biophys. J. 70, 1933–1939.
Hansma, H.G., Sinsheimer, R.L., Groppe, J., Bruice, T.C., Elings, V., Gurley, G., Bezanilla, M., Mastrangelo, I.A., Hough, P.V.C., Hansma, P.K., 1993b. Scanning 15, 296–299.
Hansma, H.G., Sinsheimer, R.L., Li, M.Q., Hansma, P.K., 1992a. Nucl. Acids Res. 20, 3585–3590.
Hansma, H.G., Vesenka, J., Siegerist, C., Kelderman, G., Morrett, H., Sinsheimer, R.L., Elings, V., Bustamante, C., Hansma, P.K., 1992b. Science 256, 1180–1184.
Hansma, P.K., Cleveland, J.P., Radmacher, M., Walters, D.A., Hillner, P.E., Bezanilla, M., Fritz, M., Vie, D., Hansma, H.G., Prater, C.B., Massie, J., Fukunaga, L., Gurley, J., Elings, V., 1994a. Appl. Phys. Lett. 64, 1738–1740.
Hansma, P.K., Drake, B., Grigg, D., Prater, C.B., Yashar, F., Gurley, G., Elings, V., Feinstein, S., Lal, R., 1994b. J. Appl. Phys. 76, 796–799.
Hansma, P.K., Drake, B., Marti, O., Gould, S.A.C., Prater, C.B., 1989. Science 243, 641–643.
Heslot, H., 1998. Biochimie (Paris) 80, 19–31.
Hoh, J.H., Cleveland, J.P., Prater, C.B., Revel, J.P., Hansma, P.K., 1992a. J. Am. Chem. Soc. 114, 4917–4918.
Hoh, J.H., Revel, J.P., Hansma, P.K., 1992b. Nanotechnology 3, 119–122.
Jackson, A.P., Vincent, J.F., Turner, R.M., 1988. Proc. Roy. Soc. London B 234, 415–440.
Kasas, S., Thomson, N.H., Smith, B.L., Hansma, H.G., Zhu, X., Guthold, M., Bustamante, C., Kool, E.T., Kashlev, M., Hansma, P.K., 1997a. Biochemistry 36, 461–468.
Kasas, S., Thomson, N.H., Smith, B.L., Hansma, P.K., Miklossy, J., Hansma, H.G., 1997b. Int. J. Imaging Systems Technol. 8, 151–161.
Kashlev, M., Martin, E., Polyakov, A., Severinov, K., Nikiforov, V., Goldfarb, A., 1993. Gene 130, 9–14.
Lantz, M.D., Liu, Y.Z., Cui, X.D., Tokumoto, H., Lindsay, S.M., 1999. Sur Interfac. Anal. 27, 354–360.
Lee, G.U., Chrisey, L.A., Colton, R.J., 1994. Science 266, 771–773.
Lindsay, S.M., 1999. The scanning probe microscope in biology. In: Bonnell, D. (Ed.), Scanning Tunneling Microscopy and Related Techniques. Wiley, New York.
Lüthi, R., Meyer, E., Howald, L., Haefke, H., Anselmetti, D., Dreier, M., Rüetschi, M., Bonner, T., Overney, R.M., Frommer, J., Güntherodt, H.-J., 1994. J. Vac. Sci. Technol. B 12, 1673–1676.

Lyubchenko, Y.L., Gall, A.A., Shlyakhtenko, L.S., Harrington, R.E., Jacobs, B.L., Oden, P.I., Lindsay, S.M., 1992. J. Biomol. Struct. Dyn. 10, 589–606.
Lyubchenko, Y.L., Oden, P.I., Lampner, D., Lindsay, S.M., Dunker, K.A., 1993. Nucl. Acids Res. 21, 1117–1123.
Manalis, S.R., Minne, S.C., Atalar, A., Quate, C.F., 1996. Rev. Sci. Instr. 67, 3294–3297.
Martin, Y., Williams, C.C., Wickramasinghe, H.K., 1987. J. Appl. Phys. 61, 4723–4729.
Mazzola, L.T., Frank, C.W., Fodor, S.P.A., Mosher, C., Lartius, R., Henderson, E., 1999. Biophys. J. 76, 2922–2933.
Meyer, G., Amer, N.M., 1988. Appl. Phys. Lett. 53, 1045–1047.
Möller, C., Allen, M., Elings, V., Engel, A., Müller, D.J., 1999. Biophys. J. 77, 1150–1158.
Mou, J., Sheng, S., Ho, R., Shao, Z., 1996. Biophys. J. 71, 2213–2221.
Müller, D.J., Fotiadis, D., Scheuring, S., Müller, S.A., Engel, A., 1999. Biophys. J. 76, 1101–1111.
O'Shea, S.J., Lantz, M.A., Tokumoto, H., 1999. Langmuir 15, 922–925.
Paloczi, G.T., Smith, B.L., Hansma, P.K., Walters, D.A., Wendman, M.A., 1998. Appl. Phys. Lett. 73, 1658–1660.
Pietrasanta, L. I., Smith, B. L., MacLeod, M. C., 2000. Chem. Res. Tox. 13, 351–355.
Prater, C.B., Drake, B., Gould, S.A.C., Hansma, H.G., Hansma, P.K., 1990. Scanning 12, 50–52.
Prater, C.B., Hansma, P.K., Tortonese, M., Quate, C.F., 1991. Rev. Sci. Instr. 62, 2634–2638.
Proksch, R., Lal, R., Hansma, P.K., Morse, D., Stucky, G., 1996. Biophys. J. 71, 2155–2157.
Qin, X.X., Coyne, K.J., Waite, J.H., 1997. J. Biol. Chem. 272, 32623–32627.
Radmacher, M., Fritz, M., Allersma, M.W., Schmidt, C., Hansma, P.K., 1995. SPIE 2384, 136–143.
Radmacher, M., Fritz, M., Cleveland, J.P., Walters, D.R., Hansma, P.K., 1994a. Langmuir 10, 3809–3814.
Radmacher, M., Fritz, M., Hansma, H.G., Hansma, P.K., 1994b. Science 265, 1577–1579.
Rief, M., Gautel, M., Oesterhelt, F., Fernandez, J.M., Gaub, H.E., 1997b. Science 276, 1109–1112.
Rief, M., Oesterhelt, F., Heymann B and Gaub, H., 1997a. Science 275, 1295–1297.
Rugar, D., Hansma, P.K., 1990. Phys. Today 43, 23–30.
Schäffer, T.E., Cleveland, J.P., Ohnesorge, F., Walters, D.A., Hansma, P.K., 1996. J. Appl. Phys. 80, 3227–3622.
Schäffer, T.E., Hansma, P.K., 1998. J. Appl. Phys. 84, 4661–4666.
Schäffer, T.E., IonescuZanetti, C., Proksch, R., Fritz, M., Walters, D.A., Almqvist, N., Zaremba, C.M., Belcher, A.M., Smith, B.L., Stucky, G.D., Morse, D.E., Hansma, P.K., 1997a. Chem. Mater. 9, 1731–1740.
Schäffer, T.E., Viani, M., Walters, D.A., Drake, B., Runge, E.K., Cleveland, J.P., Wendman, M.A., Hansma, P.K., 1997b. SPIE 3009, 48–52.
Shao, Z., Mou, J., Czajkowsky, D.M., Yang, J., Yuan, J.Y., 1996. Adv. Phys. 45, 1–86.
Shen, X., Belcher, A.M., Hansma, P.K., Stucky, G.D., Morse, D.E., 1997. J. Biol. Chem. 272, 32472–32481.
Smith, B.L., 1998. Chem. Ind. Aug 17(16), 649–653.
Smith, B.L., Gallie, D.R., Le, H., Hansma, P.K., 1997. J. Struct. Biol. 119, 109–117.
Smith, B.L., MacLeod, M.C., 1998. Proc. Am. Assoc. Can. Res. 39, 286.
Smith, B.L., Schäffer, T.E., Viani, M., Thompson, J.B., Frederick, N.A., Kindt, J., Belcher, A., Stucky, G.D., Morse, D.E., Hansma, P.K., 1999. Nature 399, 761–763.
Strunz, T., Oroszlan, K., Schafer, R., Guntherodt, H.-J., 1999. Proc. Natl. Acad. Sci. 96, 11277–11282.
Thomson, N.H., Fritz, M., Radmacher, M., Cleveland, J.P., Schmidt, C.F., Hansma, P.K., 1996. Biophys. J. 70, 2421–2431.
Thomson, N.H., Smith, B.L., Almqvist, N., Schmitt, L., Kashlev, M., Kool, E.T., Hansma, P.K., 1999. Biophys. J. 76, 1024–1033.
Thundat, T., Allison, D.P., Warmack, R.J., Brown, G.M., Jacobson, K.B., Schrick, J.J., Ferrell, T.L., 1992. Scanning Micro 6, 911–918.
Viani, M.B., Schäffer, T.E., Chand, A., Rief, M., Gaub, H.E., Hansma, P.K., 1999a. J. Appl. Phys. 86, 2258–2262.
Viani, M.B., Schäffer, T.E., Paloczi, G.T., Pietrasanta, L.I., Smith, B.L., Thompson, J.B., Richter, M., Rief, M., Gaub, H.E., Plaxco, K.W., Clelena, A.N., Hansma, H.G., Hansma, P.K., 1999b. Rev. Sci. Instr. 70, 4300–4303.
Wagner, P., Hegner, M., Kernen, P., Zaugg, F., Semenza, G., 1996. Biophys. J. 70, 2052–2066.
Walters, D.A., Cleveland, J.P., Thomson, N.H., Hansma, P.K., Wendman, M.A., Gurley, G., Elings, V., 1996. Rev. Sci. Instr. 67, 3583–3590.

Walters, D.A., Smith, B.L., Belcher, A.M., Paloczi, G.T., Stucky, G.D., Morse, D.E., Hansma, P.K., 1997a. Biophys. J. 72, 1425–1433.
Walters, D. A., Viani, M., Paloczi, G. T., Schäffer, T. E., Cleveland, J. P., Wendman, M. A., Gurley, G., Elings, V., Hansma, P. K., 1997b. SPIE 3009, pp. 43–47.
Whitesides, G.M., Gorman, C.G., 1995. In: Hubbard, A.T. (Ed.), Handbook of Surface Imaging and Visualisation. CRC Press, Boca Raton, FL, pp. 713–733.
Zhong, Q., Inniss, D., Kjoller, K., Elings, V.B., 1993. Surf. Sci. Lett. 290, L888–L692.

Biology at the Single Molecule Level
S.H. Leuba an J. Zlatanova (Eds)

Review

Twisting and stretching single DNA molecules

Terence Strick[a,*], Jean-Francois Allemand[a,b], Vincent Croquette[a], David Bensimon[a]

[a] *LPS, ENS, UMR 8550 CNRS, 24 rue Lhomond, 75231 Paris Cedex 05, France*
[b] *Département de Chimie, ENS, UMR 8640 CNRS, 24 rue Lhomond, 75231 Paris, Cedex 05, France*

Abstract

The elastic properties of DNA are essential for its biological function. They control its bending and twisting as well as the induction of structural modifications in the molecule. These can affect its interaction with the cell machinery. The response of a single DNA molecule to a mechanical stress can be precisely determined in single-molecule experiments which give access to an accurate measurement of the elastic parameters of DNA.

Contents

*Corresponding author.

1. Introduction

The past seven years have witnessed the emergence of a wealth of new techniques and tools for the study of single-molecule biophysics. Methods as diverse as optical and magnetic tweezers, microfibers and atomic force microscopy are now used in many labs to manipulate (displace, stretch or twist) single biomolecules (DNA, proteins, carbohydrates, etc.). In parallel optical methods based on fluorescence (by energy transfer or directly with evanescent wave, two-photons or confocal configurations) have also been developed to study biochemical processes at the single-molecule level. Currently, many groups are actively trying to combine both aspects, i.e. to visualize the displacement and activity of a single-molecule (myosin, RNA-polymerase, etc.) under stress.

In this paper we shall focus on the manipulation of single DNA molecules and on the measurement of their elastic properties in particular. After a brief review of the relevance of these properties to the biology of DNA, we shall describe the techniques involved in the manipulation of DNA. First, the anchoring of the molecule's ends to appropriate surfaces and the use of a meniscus to stretch and align a DNA molecule on a surface. This technique known as molecular combing has found many applications in the field of genomics. We shall then review various techniques for the manipulation of DNA and in particular the magnetic trap which allows one to both pull and twist a single-molecule. We will then present the various models used to describe DNA under tension and discuss their adequacy with the experimental results. The stretching of twisted DNA will be reviewed with an emphasis on the measurement of the torsional constant of the molecule and a description of the torque-induced structural phase transitions. Finally we shall review some of the experiments studying DNA/protein interactions at the single-molecule level.

DNA is one of the longest molecule in nature. A human chromosome for example is a few centimeters long. To squeeze such a lengthy molecule in a micron-size nucleus DNA is strongly bent and wrapped around histones, forming the bead on a string structure of chromatin, itself further compactified by extensive coiling. The bending and torsional properties of DNA (and chromatin) are therefore essential to an understanding of its compactification in the nucleus.

DNA is a polymer, i.e. a linear chain made of repeating structural units. These consist of a ribose-phosphate to which four different groups can be linked: adenine (A), guanine (G), cytosine (C) or thymine (T). DNA differs from most polymers in that it is formed by the winding around each other of two ribose-phosphate polymer chains (a DNA strand) locked by hydrogen bonding between their complementary bases: adenine (guanine) on one strand with thymine (cytosine) on the other. This double helical structure prevents the relaxation of torsional stress by rotation about a single covalent bond as common with man-made polymers. Moreover, the stacking of the bases on top of each other confers unto DNA an unusually large flexional rigidity.

This structure also poses some formidable mechanical problems to the cellular machinery which has to read, transcribe and replicate the instructions of the genetic code buried inside the double helix. To make the code accessible to the DNA or RNA polymerase enzymes, the molecule has to be unwound and the two strands separated. Thus, as an RNA polymerase proceeds along the molecule the DNA upstream of the transcription complex is overwound, whereas downstream it is underwound (Liu and Wang, 1987). The regulation of the winding and torsional stresses involved in those processes is performed by a battery of enzymes known as topoisomerases (Wang et al., 1998). To study the function of these DNA-associated molecular motors one has to first understand the mechanical response of DNA under stress.

2. Single molecule micromanipulation

There are by now many new techniques to manipulate single molecules: optical (Simmons et al., 1996) or magnetic tweezers (Amblard et al., 1996; Gosse and Croquette, 1999) and traps (Smith et al., 1996; Strick et al., 1996), atomic force cantilevers (Florin et al., 1994), microfibers (Ishijima et al., 1991; Cluzel et al., 1996) and hydrodynamic drag (Smith et al. 1992). In all these techniques, a DNA molecule (but also a protein or some other polymer) is first anchored to a surface at one end and to a force sensor at the other. The force sensor is usually a trapped micron-sized bead or a cantilever whose displacements are used to measure the force, see Fig. 1. Different force range and measurement time scales are afforded by these techniques. Magnetic traps and microfibers further allow twisting of the molecule by rotating the magnets or the fibers where the bead is attached.

The upper bound for force measurements in micromanipulation experiments is the tensile strength of a covalent bound, on the order of eV/Å or about 1000 pN (1 pN = 10^{-12} N). The smallest measurable force is set by the Langevin force which is responsible for the Brownian motion of the sensor. Because of its random nature, the Langevin force is a noise density in force which is simply written as $f_n = \sqrt{4k_BT6\pi\eta r}$ (η is the viscosity of the medium, r is the radius of the particle). For a 1−μm diameter bead in water, $f_n \sim 0.017\ \text{pN}/\sqrt{\text{Hz}}$. In between those two extremes lies the forces typical of the molecular scale, which are of order $k_BT/\text{nm} \sim 4$ pN. This is roughly the stall force of a single-molecular motor such as myosin (4 pN; Finer et al., 1994) or RNA-polymerase (15–30 pN; Yin et al., 1995; Wang et al., 1998). It is also the typical force needed to unpair the DNA bases (about 15 pN; Essevaz-Roulet et al., 1997).

2.1. *End-specific anchoring of DNA and molecular combing*

The first step in any DNA manipulation experiment is to anchor the DNA (preferentially via its extremities) to appropriately treated surfaces. Many different methods have been developed to achieve specific DNA binding to surfaces. They have found useful applications, from gene mapping, sequencing and analysis (Chee et al., 1996) to the development of very sensitive

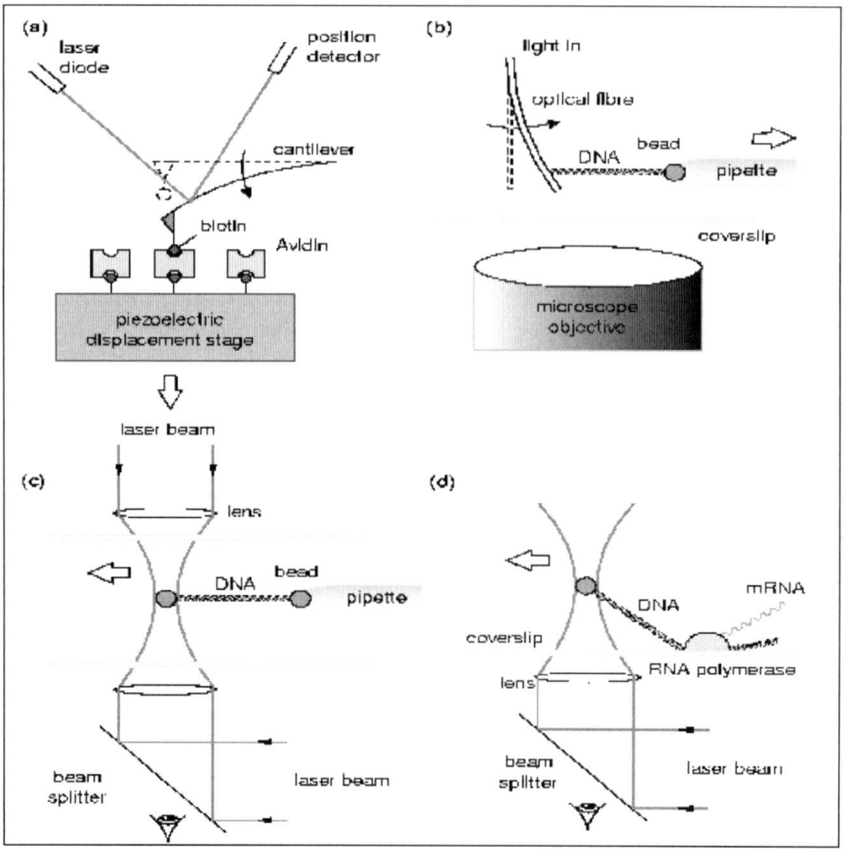

Fig. 1. Example of forces transducers used in various experiments. (a) An AFM cantilever is often used as a force transducer during intermolecular force measurements (Florin et al., 1994) and protein unfolding experiments (Rief et al., 1997). Its deflection upon pulling is detected by the displacement of a laser beam reflected from the cantilever. (b) In some of the experiments involving DNA pulling (Cluzel et al., 1996; Léger et al., 1998) and unzipping (Essevaz-Roulet et al., 1997), the force transducer is an optical fiber, which deflection is detected optically (by measuring the displacement either of the fiber directly on a microscope stage or of a light beam emitted from its pulled end). (c) In the experiment of Bustamante and coworkers (Smith et al., 1996; Kellermayer et al., 1997), the force transducer is an optical trap consisting of two co-axial counter-propagating and focused laser beams. The displacement of the bead in the trap is observed with a microscope and together with the trap stiffness is used to assess the trapping force. (d) A force transducer often used to characterize molecular motors (Simmons et al., 1996; Finer et al., 1994; Yin et al., 1995) is the optical tweezers which consists of a single strongly focused laser beam holding a bead in its focal point. Force measurement is carried out as in (c). Figure, courtesy of *Structure* (Bensimon, 1996).

immunological assays (immuno-PCR) (Sano et al., 1992). Most of these applications achieve the required binding specificity via biochemical reactions between a (possibly modified) DNA molecule and an appropriately treated surface. For example, the extremity of the molecule can be functionalized with biotine (or digoxigenine) which can interact specifically with streptavidine (or an antibody to digoxigenine) bound to a surface (Smith et al., 1992). Similarly, surfaces coated with oligonucleotides can be used to recognize the complementary extremity of DNA molecules.

Finally, there exist a large range of chemical methods to anchor (with various degrees of specificity) the extremities of DNA to surfaces bearing reactive groups (e.g. primary or secondary amine, carboxyl or thiol moieties).

An interesting alternative which does not require any modification of the molecule relies on the specific adsorption of DNA by its ends on hydrophobic surfaces at a pH $\sim$5.5 (Allemand et al., 1997). On many different hydrophobic materials (teflon, polystyrene, graphite, silanised glass, etc.) DNA was observed to adhere strongly and non-specifically at low pH and weakly or not at all at high pH. In between there exists a narrow pH range (pH $= 5.5 \pm 0.2$) where DNA binds to the surface by its extremities only.

Once DNA is anchored to a surface by its end(s), a very easy way to stretch it is to drain the solution (e.g. by pulling the surface out of it or by letting it evaporate), see Fig. 2. In that process, known as molecular combing, the anchored DNA molecules are aligned on the surface by the receding meniscus as are algae on the shore by the receding tide. It has been shown that the force applied on the molecule by the receding air/water interface (estimated to be $\sim$160 pN) is large enough to stretch it but not to break its bond(s) with the surface (Bensimon et al., 1994, 1995). Because this force is acting locally (at the interface), DNA is stretched uniformly (Allemand et al., 1997): $\sim$1 μm for every 2 kbytes. Fluorescent in situ hybridization (FISH) on combed molecules (Bacs, Yacs or even whole genomes) allows one to obtain very accurate ($\sim$1 kbps) genomic maps (Weier et al., 1995; Michalet et al., 1997). The ordering, orientation and distance between genes, the existence of genomic rearrangements (e.g. deletions) can thus be determined.

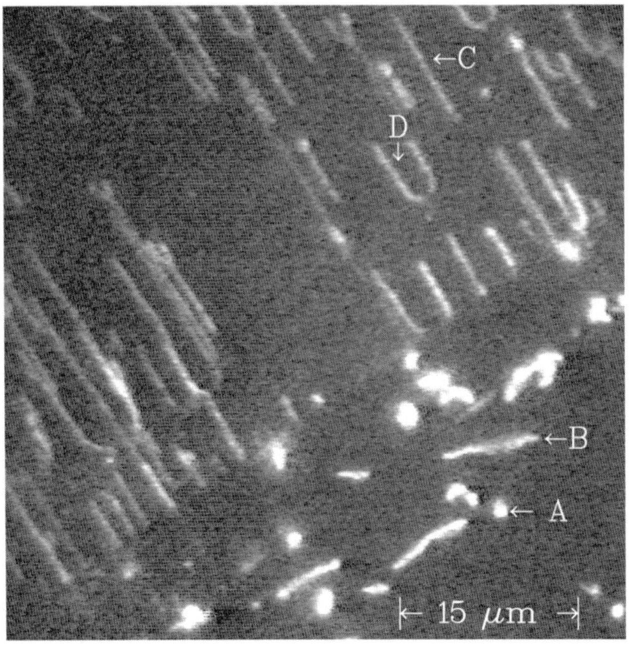

Fig. 2. Molecular combing of λ–DNA stained with YOYO1. The molecules in solution (bottom) are bound at one (A) or both (B) ends. The meniscus extends across the image from the lower left to upper right. The extended molecules left behind the meniscus are visible as straight segments (C) if bound at one end, or loops (D) if bound at both ends.

Observation of the combing of DNA molecules grafted at both ends to a surface is instructive, see Fig. 3. First as the meniscus moves past the anchoring points of the molecule it stretches its two anchored segments (legs) *perpendicular* to the interface. The portion of the molecule in solution decreases until it spans the distance between the two legs. It is then stretched *parallel* to the contact line, its length diminishing as the meniscus recedes thus forming a loop. Then as the tension in the molecule increases the loop breaks. The length of the DNA molecule at the breaking point l_b can be measured and compared with the unstretched length $l_{b,0}$, see Fig. 3. It turns out that $l_b/l_{b,0} = 2.14 \pm 0.2$ (Bensimon et al., 1995): DNA can be extended to more than twice its length before breaking! As we shall see below, this huge deformation of DNA (it is similar to the ratio of the interphosphate distance to the distance between base-pairs) has to be accommodated by a profound restructuring of the Watson–Crick double helix.

2.2. DNA manipulation and force measurements

Once DNA has been bound to a surface, it can be manipulated by displacing the anchoring object. The first manipulations of DNA molecules involved translating with the help of optical tweezers a small streptavidine-coated bead to which a single biotynilated DNA had been anchored (Chu, 1991; Perkins et al., 1994a,b). The molecule stained with a fluorescent dye was stretched by the hydrodynamic drag and its relaxation in various conditions was studied. In particular when stretched in a dilute solution of other DNA molecules the relaxation of the molecule provided the first direct observation of the reptation dynamics of an entangled polymer.

By anchoring a DNA at both ends to different surfaces (typically a coverslip, a microbead or a microneedle), a force can be exerted on the molecule by displacing the surfaces relative to each other. Thus, DNA can be anchored to an AFM cantilever (or microfiber) at one end and to a

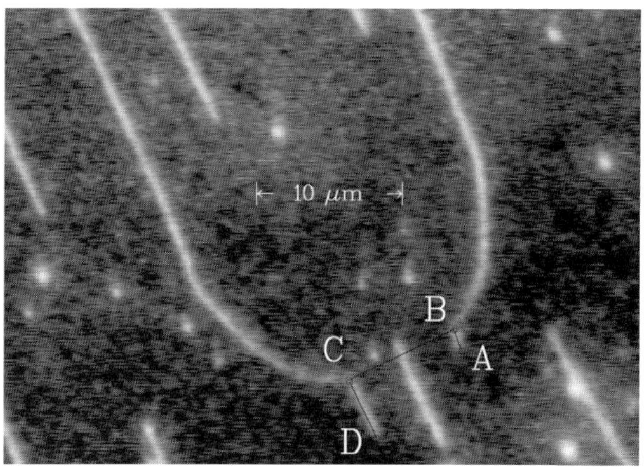

Fig. 3. Fragment of the genomic DNA, of one of the authors combed on a hydrophobic surface. The DNA bound at both ends form a typical broken loop. From measurements on such figures we deduce the extension of DNA at the breaking point: the stretched molecule length is $l_b = \overline{BC}$, its unstretched length $l_{b,0}$ is deduced by dividing the length $\overline{AB} + \overline{DC}$ by the extension factor observed for a straight molecule combed on an hydrophobic surface (1 μm for 2 kbytes).

treated surface at the other (Engel et al., 1999; Rief et al., 1999, 1997; Florin et al., 1994). By displacing the surface, see Fig. 1, the molecule is stretched and pulls on the cantilever. Knowing its spring constant, the force on the molecule can be measured. Very high forces (many thousands of pN) can be achieved and measured in these experiments with a precision of $\sim$10 pN and a spatial resolution of $O(0.1\ \text{nm})$.

By using two co-axial counter-propagating laser beams a small transparent bead can be trapped, see Fig. 1 with a force of $O(100\ \text{pN})$ and a spatial resolution $\sim$10 nm. The force exerted on the bead can be deduced from the displacement of the trapping beam due to its refraction in the bead, i.e. by directly measuring the momentum transfer (Smith, 1998). This absolute measure bypasses the need for a calibration of this optical trap.

Another popular technique involves the stretching by an intensely focused laser beam (optical tweezers (Simmons et al., 1996)) of a transparent bead of radius r of $O(1\ \mu\text{m})$ anchored by a DNA molecule to a surface, see Fig. 1. Forces $F < 50$ pN can be achieved and measured with a precision of $\sim$0.1 pN by following the displacement δx of the trapped bead from its equilibrium (zero force) position: $F = k_{\text{trap}}\delta x$, where k_{trap} is the elastic stiffness of the optical trap. δx can be measured with a precision of $O(1\ \text{nm})$ (Simmons et al., 1996), enough to resolve a single step of myosin (Finer et al., 1994; Saito et al., 1994) or kinesin (Schnitzer and Block, 1997; Hua et al., 1997). k_{trap} has to be determined prior to any force measurement, for example by pulling on the bead with a known force such as the hydrodynamic drag F_{s} of a fluid (of viscosity η) flowing with velocity v around the bead: $F_{\text{s}} = 6\pi\eta r v$. Alternatively, one may determine k_{trap} by measuring the intensity of the Brownian fluctuations $\langle\delta x^2\rangle$ of the trapped bead. By the equipartition theorem they have to satisfy (Simmons et al., 1996; Einstein, 1956; Reif, 1965):

$$\frac{k_{\text{trap}}\langle\delta x^2\rangle}{2} = \frac{k_{\text{B}}T}{2}. \tag{1}$$

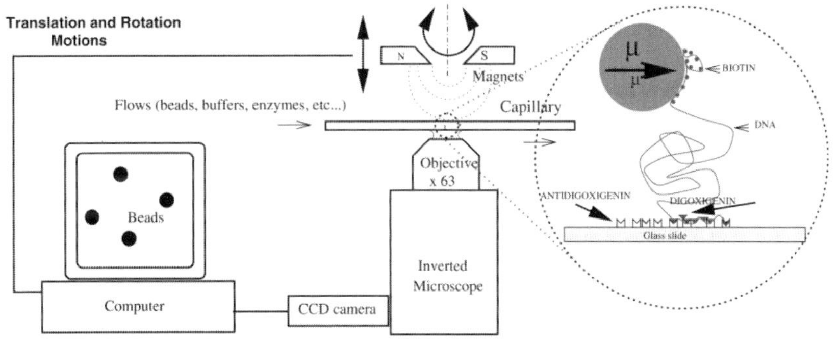

Fig. 4. Schematic view of the apparatus used to twist and stretch single DNA molecules. DNA molecules were first prepared with *biotin* attached to one end and *digoxigenin* (*dig*) bound to the other. These end-labeled DNA molecules are incubated with *streptavidin*-coated magnetic beads and then flowed into a square glass capillary coated with an antibody to *dig*, *antidig*. The DNA molecules bind specifically to the bead via *biotin/streptavidin* coupling and to the glass via *dig/antidig* coupling. The capillary is placed above an inverted microscope. Magnets are placed above the capillary. By approaching the magnets we increase the stretching force on the bead and thus on the molecule. By rotating the magnets the molecule is twisted at constant force. A frame grabber installed in a PC allows for tracking of the Brownian fluctuations $\langle\delta x^2\rangle$ of the bead. The determination of $\langle\delta x^2\rangle$ and of the molecule's extension l leads to a measure of the stretching force $F = k_{\text{B}}Tl/\langle\delta x^2\rangle$.

To twist and stretch a DNA molecule and study its interactions with proteins, a magnetic trapping technique (Strick et al., 1996) has proved particularly convenient. Briefly, it consists in stretching a single DNA molecule bound at one end to a surface and at the other to a magnetic micro-bead (1–4.5 μm in diameter), see Fig. 4. Small magnets, whose position and rotation can be controlled, are used to pull on and rotate the micro-bead and thus stretch and twist the molecule. This system allows one to apply and measure forces ranging from a few fN (10^{-3}pN) to nearly 100 pN (see Strick et al., 1998b) with a relative accuracy of $\sim 10\%$.

In contrast with other techniques, this force measurement is absolute and does not require a calibration of the sensor. It is based on the analysis of the Brownian fluctuations of the tethered bead, which is completely equivalent to a damped pendulum of length $l = \langle z \rangle$ pulled by a magnetic force F (along the z-axis). Its longitudinal ($\delta z^2 = \langle z^2 \rangle - \langle z \rangle^2$) and transverse δx^2 fluctuations are characterized by effective rigidities $k_{||} = \partial_z F$ and $k_\perp = F/l$. By the equipartition theorem they satisfy (Einstein, 1956; Reif, 1965)

$$\delta z^2 = \frac{k_B T}{k_{||}} = \frac{k_B T}{\partial_z F}, \tag{2}$$

$$\delta x^2 = \frac{k_B T}{k_\perp} = \frac{k_B T l}{F}. \tag{3}$$

Thus from the bead's Brownian fluctuations ($\delta x^2, \delta y^2$) one can extract the force pulling on the molecule (the smaller the fluctuations the greater F) and from δz^2 one obtains its first derivative, $\partial_z F$. This measurement method can be used with magnetic (but not optical) traps because the variation of the trapping gradients occurs on a scale ($O(1\,\text{mm})$) much larger than the scale on which the elasticity of the molecule changes ($O(0.1\,\mu\text{m})$). In other words, the stiffness of the optical trap is very large compared to F/l. A further bonus of the magnetic trap technique is that measurements on DNA at constant force are trivial (just keep the position of the magnets fixed). With cantilevers or optical tweezers to work at constant force requires an appropriate feedback to ensure that the displacement of the sensor is kept constant. However, because its stiffness depends on the force, the magnetic trap technique has at weak forces (<1 pN) a lower spatial resolution of $O(10\,\text{nm})$, than the other manipulation methods. Finally notice that by using electro-magnets, a faster and more versatile magnetic tweezers system has recently been developed (Gosse and Croquette, 1999).

3. Models of polymer elasticity

Just like any polymer in solution, free DNA adopts a random coil conformation which maximizes its entropy (de Gennes, 1979). Pulling on the molecule reduces this entropy and costs energy. The associated entropic forces result from a reduction of the number of possible configurations of the system consisting of the molecule (be it a polymer, DNA or a protein) and its solvent (water, ions), so that at full extension there is but one configuration left: a straight polymer linking both ends. To reach that configuration work has to be done against entropy, a force has to be applied. The entropic forces are rather weak, typically <10 pN. Beyond this regime and up to about 70 pN DNA stretches like any spring: it is in an enthalpy dominated regime.

3.1. The Kratky–Porod model

For simplicity, let us first consider a polymer chain with no torsional stress. Such a chain is often described by the Kratky–Porod model (Cantor and Schimmel, 1980): a succession of N segments of length b and orientation vector $\mathbf{t}_i$, see Fig. 5. The energy $\mathscr{E}_{\mathrm{KP}}$ of a given chain configuration (the ensemble of segment orientations $\{\mathbf{t}_i\}$) is the sum of the bending energies of successive segments:

$$\mathscr{E}_{\mathrm{KP}} = -\frac{B}{b}\sum_{i=2}^{N}\mathbf{t}_i \cdot \mathbf{t}_{i-1} = -\frac{B}{b}\sum_{i=2}^{N}\cos\theta_i, \tag{4}$$

where θ_i is the angle between successive orientation vectors and B is the bending modulus. (Notice the analogy between the statistical mechanics of a Kratky–Porod chain and that of a classical one-dimensional magnetic (spin) system (Fisher, 1964).) This model has been solved exactly (Fisher, 1964). The angular correlation decays exponentially with distance along the chain:

$$\langle \mathbf{t}_i \cdot \mathbf{t}_j \rangle = \mathrm{e}^{-b|i-j|/\xi_T}, \tag{5}$$

where $\xi_T = B/k_{\mathrm{B}}T$ is the decay length of the angular correlation. It reflects the stiffness of the chain and is known as the persistence length. The chain end-to-end mean square distance R_g satisfies

$$R_g^2 \equiv \langle \mathbf{R}^2 \rangle = \left(b\sum_{i=1}^{N}\mathbf{t}_i \right)^2 \simeq 2Nb\xi_T = 2l_0\xi_T, \tag{6}$$

where $l_0 = Nb$ is the chain length. A DNA molecule in solution thus adopts a fluctuating random coil configuration of typical size R_g, known as the gyration radius. For many years, the measurement of R_g by various means (sedimentation, light scattering, etc. (Cantor and Schimmel, 1980; Hagerman, 1988) was the only way to estimate the persistence length of DNA (or any polymer).

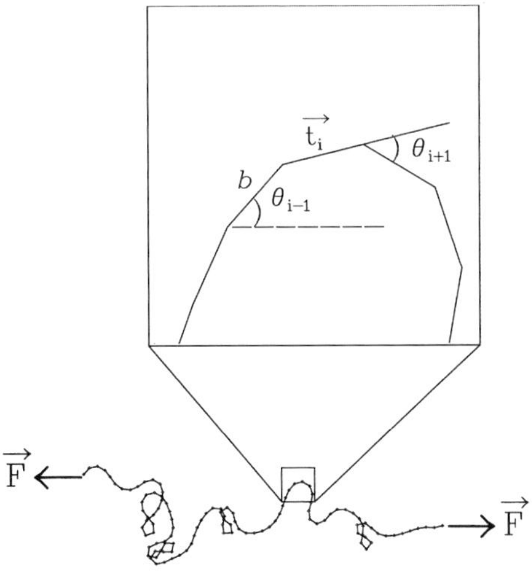

Fig. 5. A continuous polymer chain can be simulated by a chain of freely rotating segments of size b and orientation vector $\mathbf{t}_i$. The direction of the stretching force F defines the z-axis.

The stretching of a single DNA molecule now provides a much more precise way of measuring ξ_T. To model the behavior of a polymer chain under tension, it suffices to add to Eq. (4) a term representing the work $W = -\mathbf{F} \cdot \mathbf{R} = -Fb \sum t_{i,z} = -Fb \sum \cos \Theta_i$ done by a force F acting on the chain along the z−axis (Θ_i is the angle between $\mathbf{t}_i$ and the z−axis):

$$\mathscr{E}_{\mathrm{KP}} = -\frac{B}{b}\sum_{i=2}^{N} \mathbf{t}_i \cdot \mathbf{t}_{i-1} - Fb \sum_{i=1}^{N} t_{i,z} = -\frac{B}{b}\sum_{i=2}^{N} \cos \theta_i - Fb \sum_{i=1}^{N} \cos \Theta_i. \tag{7}$$

Unfortunately, this model can be solved only for small forces, where the mean extension of the chain $l \ll l_0$ (Fisher, 1964) is

$$l = \frac{2F\xi_T}{3k_{\mathrm{B}}T}\, l_0 = \frac{F}{3k_{\mathrm{B}}T}\, R_g^2. \tag{8}$$

To compute the elastic response of a chain at higher forces one has to resort to numerical calculations (e.g. transfer matrix methods) or to various approximations of the Kratky–Porod model.

3.2. *The freely jointed chain model*

An interesting limit is the freely jointed chain (FJC) model, which consists in setting $B = 0$ in Eq. (7). It models a chain whose segments are unrestricted in their respective orientation and corresponds to a discretization of a polymer with segments of length $b = 2\xi_T$ (the so-called Kuhn length). In the FJC model the energy of a given chain configuration $\{\mathbf{t}_i\}$ is thus $\mathscr{E}_{\mathrm{FJC}} = W = -Fb \sum \cos \Theta_i$. The partition function $\mathscr{Z}$ is

$$\mathscr{Z} = \sum_{\mathbf{t}_i} \mathrm{e}^{-\mathscr{E}_{\mathrm{FJC}}/k_{\mathrm{B}}T} = \sum_{\mathbf{t}_i} \prod_{i=1}^{N} \mathrm{e}^{Fb \cos \Theta_i / k_{\mathrm{B}}T} \tag{9}$$

$$= \left[\int \mathrm{d}\Omega \, \mathrm{e}^{Fb \cos \Theta / k_{\mathrm{B}}T}\right]^N = \left[\frac{2\pi k_{\mathrm{B}}T}{Fb} \sinh \frac{Fb}{k_{\mathrm{B}}T}\right]^N. \tag{10}$$

From the free energy $\mathscr{F} = -k_{\mathrm{B}}T \log \mathscr{Z}$, one can compute the mean extension of the chain l

$$l = -\frac{\partial \mathscr{F}}{\partial F} = l_0 \left(\coth \frac{Fb}{k_{\mathrm{B}}T} - \frac{k_{\mathrm{B}}T}{Fb}\right). \tag{11}$$

Notice that at small forces one recovers our previous result, Eq. (8). However, as shown in Fig. 6, the FJC model is too crude and is *not* a good approximation of the elastic behavior of a DNA molecule at large extensions ($l > R_g$).

3.3. *The worm like chain model*

A much more precise description is afforded by the worm like chain (WLC) model, the continuous ($b \to 0$) limit of Eq. (7):

$$\frac{\mathscr{E}_{\mathrm{WLC}}}{k_{\mathrm{B}}T} = \frac{\xi_T}{2} \int_0^{l_0} \left(\frac{\mathrm{d}\mathbf{t}}{\mathrm{d}s}\right)^2 \mathrm{d}s - \frac{F}{k_{\mathrm{B}}T} \int_0^{l_0} \cos \Theta(s)\, \mathrm{d}s, \tag{12}$$

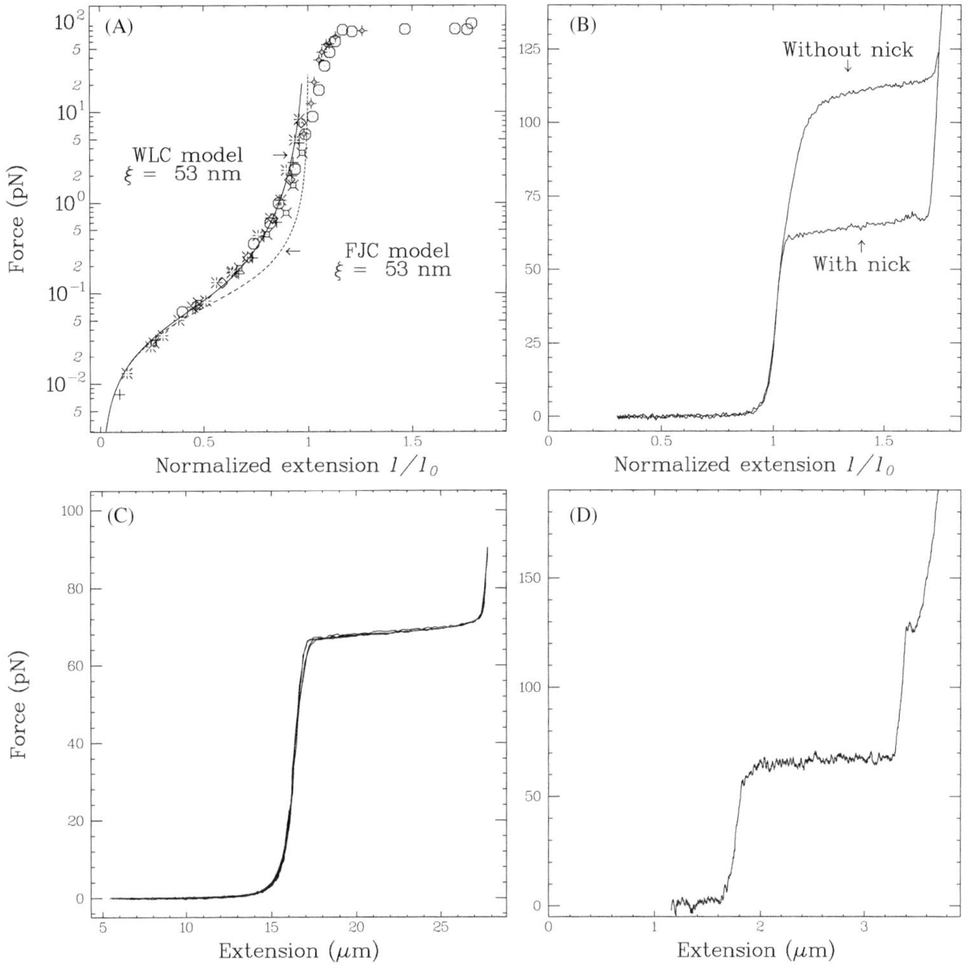

Fig. 6. Force versus extension curves of single DNA molecules obtained by different groups. (A) The dots correspond to several experiments performed over a wide range of forces. The force was measured using the Brownian fluctuation technique (Strick et al., 1996). The full line curve is a best fit to the WLC model for forces smaller than 5 pN. The dashed curve is the result of the FJC model with the same persistence length (it is clearly a worse description of the behavior of DNA under stress than the WLC model). At high forces, the molecule first elongates slightly, as would any material in its elastic regime. Above 70 pN, the length abruptly increases, corresponding to the appearance of a new structure called S-DNA. (B) The same transition observed by J.-F. Léger and D. Chatenay using a glass needle deflection on a nicked molecule and an unnicked molecule (the transition occurs for a higher force). (C) The transition is also observed by S. Smith and C. Bustamante using optical tweezers. (D) Finally, H. Clausen-Schaumann and H. Gaub observe also the transition using an AFM. (We thank C. Bustamante, D. Chatenay, H. Clausen-Schaumann, H. Gaub, J.-F. Léger and S. Smith for sharing their data).

where s is the curvilinear coordinate along the chain. The calculation of the partition function $\mathscr{Z}$ and the free energy $\mathscr{F}$ of that model calls upon an analogy with the quantum mechanical problem of a dipole in an electric field, which is beyond the scope of this paper. It has however been solved by Marko and Siggia (Fixman and Kovac, 1973; Bustamante et al., 1994; Vologodskii, 1994; Marko and Siggia, 1995a,b). Though there is no analytic formula equivalent to Eq. (11) for the force vs. extension behavior of a WLC, a simple and efficient numerical solution was recently provided by Bouchiat et al. (1999), who gave an approximation better than 0.1%:

$$F = (k_B T/\xi_T) g(x), \tag{13}$$

where $x = l/l_0$ and

$$g(x) = x - \frac{1}{4} + \frac{1}{4(1-x)^2} + \sum_{i=2}^{7} a_i x^i \tag{14}$$

with $a_2 = -0.5164228$, $a_3 = -2.737418$, $a_4 = 16.07497$, $a_5 = -38.87607$, $a_6 = 39.49944$, $a_7 = -14.17718$ (Bouchiat et al., 1999). Notice that at small relative extensions $x \ll 1$, as for the FJC model we recover Eq. (8). However, when compared over the whole extension range, the WLC model is a much better description of the behavior of DNA than the FJC model. As shown in Fig. 6 the WLC model fits extremely well the measured data and allows a very precise estimation of the DNA's persistence length: $\xi_T = 52 \pm 2$ nm in physiological conditions (10 mM phosphate buffer (pH = 7.5), 10 mM NaCl).

3.4. Self-avoidance effects

In the theoretical models described previously, the fact that a real polymer cannot intersect itself was not taken into consideration. For example in the computation of the partition function $\mathscr{Z}$, Eq. (10), we included these unrealistic configurations. In that section we shall try to justify this approach. A treatment of the self-avoidance effects of a stretched polymer exists only at low extensions ($l \ll l_0$), where a heuristic argument due to Flory (1975) works remarkably well. Due to self-avoidance one expects a real chain to occupy a larger volume than the so-called Gaussian (intersecting or phantom) chains considered previously, which occupied a volume R_g^3. That of course cost some entropic energy which is, however, compensated by a reduction in the probability of interaction. In this approach the free energy of a FJC polymer is (recall that at low extensions all models are equivalent):

$$\frac{\mathscr{F}}{k_B T} = \frac{3l^2}{2R_g^2} + \frac{vN^2}{l^3}. \tag{15}$$

The first term on the right describes the energy cost associated with a swelling of the polymer ($\sim \int F \, dl$, with F given by Eq. (8)). The second term accounts for the self-avoidance. It is proportional to the probability that two segments of the FJC of volume $v = \pi r^2 b$ will share the same location in the space l^3 occupied by the chain (the segments are assumed cylindrical with a radius r and length b). That repulsive term decreases as the chain swells. By minimizing $\mathscr{F}$ with respect to l, one obtains the equilibrium (Flory's) radius of a self-avoiding polymer:

$$R_F = (vb^2)^{1/5} N^{3/5}. \tag{16}$$

Excluded volume interactions will become important when in the volume explored by one monomer, typically b^3 there is another monomer, i.e. when the monomer concentration $c = N/R_F^3 = 1/b^3$. Self-avoidance thus becomes non-negligible when $N > (b/r)^{3/2}/\pi^{3/4}$. For a DNA molecule with $r = 1$ nm and $b = 100$ nm, this corresponds to $N \sim 400$, i.e. a molecular length: $l_{0,F} = Nb \sim 40\,\mu$m. Almost all manipulations of DNA molecules so far have been in a regime $l_0 < l_{0,\mathrm{F}}$, where self-avoidance is totally negligible. That is except in torsionally constrained experiments where self-avoidance is crucial as it stabilizes plectonemic loops (see Section 4 below).

3.5. *Beyond the entropic regime*

Beyond the entropic regime, i.e. from ~6 pN to about 70 pN, DNA behaves like an elastic rod with stretch modulus $EA \sim 1000$ pN (Smith et al., 1996; Wang et al., 1997) (where E is the Young modulus of DNA and A its effective cross-sectional area (Hogan and Austin, 1987)). Neglecting entropic contributions, the force vs. extension curve follows a simple Hookean law (as any elastic material): $F = EA(x - 1)$ (with $x = l/l_0 > 1$). Some ad hoc formulas exist, interpolating between the entropic and Hookean regimes, e.g. replacing the term $(1 - x)^2$ in Eq. (14) by $(1 - x + F/EA)^2$ (Wang et al., 1997).

Finally at about 70 pN a surprising transition has recently been discovered where DNA stretches to about 1.7 times its crystallographic length (Smith et al., 1996; Cluzel et al., 1996). As characteristic of the first-order transitions in nature (e.g. boiling) that transition is highly cooperative: a small change in force results in a large change in extension (see Fig. 6). A phenomenological description of that transition has been proposed (Cluzel et al., 1996; Cizeau and Viovy, 1997; Ahsan et al., 1998; Marko, 1998), where the force plays the same role as the magnetic field in a ferromagnetic context. In this model, the observed sharpness of the transition (its high cooperativity) is associated with a large interfacial energy between the phases, suggesting that the typical domain size is about 100 bases long (Cizeau and Viovy, 1997).

To address the possible structural modification in the molecule resulting from pulling on it, a numerical energy minimization of DNA under stress was performed by R. Lavery and collaborators (Cluzel et al., 1996; Lebrun and Lavery, 1996). Its results reveal the existence of a new conformation called S-DNA indeed 70% longer than B-DNA, whose exact structure depends on which extremities of the DNA are being pulled (3′–3′ or 5′–5′). If both 3′ extremities are being pulled the double helix unwinds upon stretching. The final structure resembles a ladder. If both 5′ ends are pulled a helical structure is preserved. It is characterised by a strong base pair inclination, a narrow minor groove and a diameter roughly 30% less than that of B-DNA. In both cases the rupture of the molecule (by unpairing of the bases) occurs as observed during molecular combing (Bensimon et al., 1994, 1995) when its extension is more than twice that of B-DNA (Wilkins et al., 1951).

These numerical results are supported by experiments done almost 50 years ago by Wilkins et al. (1951) before the double helix structure of DNA was even proposed. Those suggested that stretched DNA fibers indeed undergo a transition to a structure with tilted bases about twice longer than the relaxed molecule. However, recent experiments by Léger et al. (1999) stretching torsionally constrained single molecules have shown that the S-DNA phase has a helical pitch of 22 nm with 38 bases per right-handed turn, which would make it look more like a slightly twisted ladder.

The existence of a new stable form of DNA at high extension might have considerable interest for the study of DNA/protein interactions. Thus RecA is known to induce a 60% extension of B-DNA (Stasiak and Di Capua, 1982; Stasiak et al., 1983) and facilitate the formation of a triple helix, a putative intermediate during recombination. Smith et al. (1996) calculated that the existence of an extended S-form of DNA reduced the energetics of RecA binding to DNA by as much as $15k_BT$ (9 kcal/mol) per complex. Recent experiments (Léger et al., 1998; Shivashankar et al., 1999; Hegner et al., 1999) have indeed shown that the polymerisation of RecA on a dsDNA was facilitated by stretching the molecule. This implies that the barriers to nucleation and accretion of a RecA fiber on DNA is lowered by the presence of a S-DNA sub-phase. Numerical modeling further suggest that its structure is closer to the RecA/DNA complex than regular B-DNA (Lebrun and Lavery, 1997).

3.6. Elasticity of heterogeneous polymers

In the preceding discussion, the polymer properties were assumed homogenous, i.e. we did not consider possible sequence specific effects on the elastic behavior of DNA. The Kratky–Porod model allows one to treat those to some extent. It is thus easy to introduce some local preferred bending (i.e. orientational angle ψ_i between successive segments) by replacing the term $\cos\theta_i$ in Eq. (7) by $\cos(\theta_i - \psi_i)$. The case where the ψ_i's are fixed randomly along the chain, thus defining a static (disorder induced) angular correlation ξ_d, has been analysed by two groups using slightly different models (Bensimon et al., 1998; Nelson, 1998). Surprisingly, it turns out that the elastic behavior of such a heterogeneous chain is to all practical purposes identical to that of an homogenous WLC with an effective persistence length ξ_{eff} which is a model-dependent function of ξ_d and ξ_T. Therefore, one does not expect sequence specific effects to alter drastically the results of the WLC model described previously. It is worth emphasizing the theoretical point that sequence inhomogeneities must be treated as 'quenched', i.e. the free energy of the system should be averaged over the disorder (Binder and Young, 1986). Some results claiming to describe sequence disorder effects (Nelson, 1998; Trifonov et al., 1987; Schellman and Harvey, 1995) actually employ an 'annealed' averaging of the partition function over disorder which is mathematically simpler but is irrelevant to experiment where the sequence is fixed.

4. DNA under torsion

4.1. Topological properties of coiled DNA

To describe DNA under torsional stress it is first necessary to introduce some topological concepts. The first is the twist (Tw), the number of helical turns along the molecule. For a torsionally unconstrained B-DNA, $\mathrm{Tw} = \mathrm{Tw}_0 = N/h$ where N is the number of base-pairs and $h = 10.4$ is the number of base-pairs per turn of the helix. The second topological quantity of interest is the writhe (Wr) of the molecule. Wr is a measure of the coiling of the DNA axis about itself, as a twisted phone cord which forms interwound structures in order to relieve its torque. If the DNA molecule is torsionally constrained, then the total number of times the two strands of the helix cross each other (either by twist or writhe) is a topological invariant of the system called the linking number $\mathrm{Lk} = \mathrm{Tw} + \mathrm{Wr}$ (White, 1969). For relaxed linear DNA molecules, assuming the absence of any spontaneous local curvature, $\mathrm{Lk} = \mathrm{Lk}_0 = \mathrm{Tw}_0$. The relative difference in linking number between the supercoiled and relaxed forms of DNA is called the degree of supercoiling, σ:

$$\sigma = (\mathrm{Lk} - \mathrm{Lk}_0)/\mathrm{Lk}_0 = \Delta\mathrm{Lk}/\mathrm{Lk}_0. \tag{17}$$

The value of σ for most circular molecules isolated from cells or virions is roughly -0.06. A notable exception occurs in hyperthermophilic archeabacteria (Woese and Fox, 1977; Woese et al., 1990) who have a positively coiled DNA (Kikuchi and Asai, 1984; Forterre, 1996). In the experiments described here, provided that the anchoring of the DNA molecule is achieved at multiple points at both ends, a torsional constraint can be applied on the molecule by simply rotating the magnets. As one turn of the magnets implies a change of one turn of the molecule, we have simply $\Delta\mathrm{Lk} = \pm n$, where $\pm n$ is the number of turns by which the magnets are made to rotate. Note that at fixed Lk the ratio Tw/Wr will depend on the force stretching the molecule, the writhe being suppressed by high forces. As a consequence, pulling on a coiled molecule increases the torque twisting it, until its writhe reaches zero.

4.2. The mechanical buckling instability

Twisting DNA leads to a torsional buckling instability analogous to that observed on telephone cords or rubber tubes. This instability leads to the formation of interwound structures known as supercoils or plectonemes. Of course, a DNA molecule is also animated by very strong thermal fluctuations which play an important role. However, it is instructive to first consider the purely mechanical (zero-temperature) instability of a rubber tube of length l and torsional modulus C. If we firmly hold one end of the tube while simultaneously rotating and pulling on the second end with a force F, we observe the following phenomenon (see Fig. 7): when the twist constraint is small, the associated torque Γ increases linearly with the twist angle θ, $\Gamma = C\theta/l$ and the tube remains straight. As the tube is further twisted, a critical twist angle $\theta_{\mathrm{c,b}}$ and torque $\Gamma_{\mathrm{c,b}}$ are reached where the tube ceases to be straight: it locally buckles and forms a small loop of radius $R_{\mathrm{c,b}}$. The torsional energy thus gained is $2\pi\Gamma_{\mathrm{c,b}}$, whereas the energy cost (due to bending and work against F) is (see Eq. (12)): $\mathscr{E} = \pi B/R + 2\pi RF$. The cost is minimized for a loop of radius $R_{\mathrm{c,b}} = \sqrt{B/2F}$. The critical torque for the formation of plectonemes is controlled by the balance between energy gain and cost, i.e. by the stretching force:

$$\Gamma_{\mathrm{c,b}} = \mathscr{E}_{\min}/2\pi = \sqrt{2BF}. \tag{18}$$

As we twist the tube further, we increase the length of the plectonemes but the torque in the tube remains basically fixed at its critical value $\Gamma_{\mathrm{c,b}}$.

4.3. DNA under torsion: the rod like chain model

For DNA, the picture is pretty much the same (Bouchiat and Mézard, 1998; Moroz and Nelson, 1998; Vologodskii and Marko, 1997). The thermal fluctuations which will be most important near the mechanical instability at $\theta_{\mathrm{c,b}}$ will tend to round it off. Hence as one is coiling a DNA molecule under fixed force F, one observes the following behavior (see Fig. 7): at low degrees of supercoiling $|\sigma|$ the molecule's extension varies little. Beyond a critical value σ_{c} (which depends on the force), the molecule shortens continuously as it is twisted further. A theoretical treatment of this behavior is given by the RLC model developed by Bouchiat and Mézard (1998,

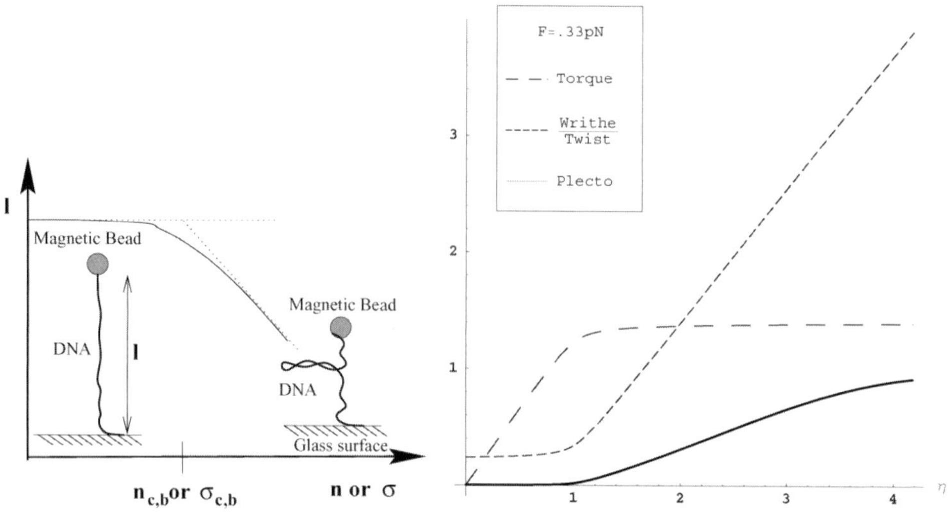

Fig. 7. LEFT: schematic view of the buckling transition for a twisted rubber tube (dotted line) or a DNA molecule (solid line). Below a critical number of turns $n_{c,b}$ the rubber tube's torque increases linearly as it stores twisting energy. When $n_{c,b}$ turns have been added the system abruptly exchanges twisting energy for bending energy and plectonemes begin to form. The plectonemes grow linearly with subsequent twisting and the torque remains constant thereafter. In the case of DNA the same picture holds, except for the fact that thermal fluctuation round off the transition which takes place at $n_{c,b}$. RIGHT: Results from the RLC model corresponding to a stretching force of $F = 0.33$ pN. The x-axis represents the supercoiling variable $\eta = 2\pi n\xi_T/l_0 \simeq 95\sigma$ (Bouchiat and Mézard, 1998), and the y-axis is in arbitrary units. The long-dash curve represents the torque acting on the DNA: as decribed above, it increases linearly until $\eta_{c,b} \sim 1$ ($\sigma \sim 0.01$) and remains essentially constant thereafter. The short-dash curve represents the ratio of writhe to twist: note that the writhe is never zero and increases rapidly as $\eta > 1$. Finally, the full line measures the fraction of plectonemes in DNA: stable supercoiled structures only appear after the torsional buckling transition has been passed.

1999) which consists in describing a DNA molecule not as a chain free to rotate but as a rod with a finite torsional modulus C. The energy of this rod like chain (RLC) model $\mathscr{E}_{\mathrm{RLC}}$ is obtained by adding to the energy of the WLC model, Eq. (12), an energy of twist $\mathscr{E}_{\mathrm{T}}$:

$$\mathscr{E}_{\mathrm{RLC}} = \mathscr{E}_{\mathrm{WLC}} + \mathscr{E}_{\mathrm{T}} = \mathscr{E}_{\mathrm{WLC}} + \frac{C}{2}\int_0^{l_0} \mathrm{d}s\, \Omega^2(s), \tag{19}$$

where $\Omega(s)$ is the local twist of the chain. A detailed solution of this model which is beyond the scope of the present paper can be found in Bouchiat and Mézard (1998, 1999) and Moroz and Nelson (1998a,b). A particular mathematical subtely of this model is that the twist integrant in Eq. (19) is singular. That singularity has to be regularized either by a truncation of its series expansion (Moroz and Nelson, 1998a,b) or by the reintroduction of the short length cut-off b of the Kratky–Porod model (Bouchiat and Mézard, 1998; Bouchiat et al., 1999). Although the RLC model neglects the chain's self-avoidance, which stabilizes the plectonemes, its predictions fit remarkably well with our observations at low forces (<0.4 pN) where as the molecule is coiled its end to end distance decreases, see Fig. 8. From a fit of the RLC predictions to our data, one can extract a value for the torsional modulus C of DNA: $C/k_{\mathrm{B}}T = 86 \pm 5$ nm (Bouchiat and Mézard,

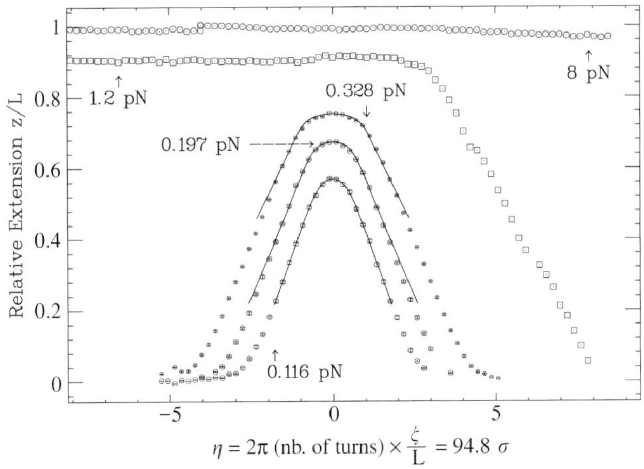

Fig. 8. Relative extension of a DNA molecule vs. the degree of supercoiling $\eta = 2\pi n\xi_T/l_0 \simeq 95\sigma$ for various stretching forces. For the three curves obtained at low force, the behavior is symmetrical under $\sigma \to -\sigma$. The shortening corresponds to the formation of plectonemes upon writhing. For these low forces the comparison between the experimental data (points) and the rod like chain model with $C/k_BT = 86$ nm (full-line) is very good. When the force is increased above 0.5 pN, the curve becomes asymmetric: supercoils still form for positive coiling while local denaturation adsorbs the torsional stress for negative σ. At forces larger than 3 pN no plectonemes are observed: the torsional stress is adsorbed not by writhe but in local structural changes of the molecule.

1998). This result depends weakly on the magnitude of b which best-fitted value of ~6 nm is roughly equal to twice the DNA pitch.

A number of groups have attempted to go beyond the model defined in Eq. (19) by introducing a coupling between the stretch on the double helix and its twist (Moroz and Nelson, 1998b; Marko, 1997; Nelson et al., 1997). The approach is sensible, but the comparison with the experimental data is very problematic due to the existence of structural transitions in DNA (see below) and the smallness of the predicted effect. Much theoretical work has also been done on the coupling between the intrinsic curvature and the twisting of the molecule (Schlick and Olson, 1992; Olson, 1996; Chirico and Langowski, 1996; Garrivier and Fourcade, 2000). As many regulating factors are supposed to bend the molecule, a coupling to its twist could modulate the interaction of distant sites along the DNA. This very interesting and important problem, deserves to be studied with the new tools now at our disposal. Finally, the dynamical aspects related to relaxation or transport of torsional stress along DNA (Nelson, 1999) need also to be addressed experimentally.

4.4. *Torque induced transitions in DNA*

The torsional buckling instability just described treats the DNA molecule as a continuous elastic tube. It ignores the underlying double-helical structure of the molecule, and its relevance is therefore limited to very low forces ($F < 0.4$ pN) or low degrees of supercoiling ($-0.015 < \sigma < 0.037$). For higher forces and degrees of supercoiling, the buildup of torque in the molecule can be large enough to actually modify its internal structure (Strick et al., 1999). This is evidenced by breaking of the $\sigma \to -\sigma$ symmetry in the extension vs. supercoiling curves, see Fig. 8. As a critical force is reached (ipso facto a critical torque), the molecule undergoes a transition from a contracted state (plectonemic B-DNA) to a highly extended one (Strick et al., 1998). This state is characterized by the coexistence of B-DNA with denatured DNA for $\sigma < 0$ and with a new phase called P-DNA for $\sigma > 0$ (Allemand et al., 1998). This new P-DNA structure has an intrinsic degree of coiling $\sigma_p = 3$, i.e. it has 2.6 bps/turn (Allemand et al., 1998; Léger et al., 1999). Numerical simulations, supported by chemical reactivity studies, suggest that in P-DNA the phosphate backbone is winding inside the structure and the bases are exposed in solution.

These results show that at very low forces and low degrees of supercoiling DNA can locally undergo major structural transitions. These transitions might be relevant to the activity of RNA-polymerase, which is known to exert forces as high as 35 pN and to under(over)-wind the molecule up(down)-stream. It is furthermore worthwhile to notice that a structure very similar to P-DNA has apparently been observed in the packaging of DNA in the Pf1 virus. There this unusual structure is stabilised by the proteins of the virus' coat (Liu and Day, 1994).

4.5. Twisting rigidity measured through the critical torque of denaturation

The denaturation transition offers a second way to evaluate the elastic torsional persistence length C using a very simple model (Strick et al., 1999). It consists in measuring the difference in work done while stretching a single DNA molecule wound either positively or negatively by the same number of turns.

Fig. 8 shows the molecular extension as a function of supercoiling for various forces. At a low force ($F \sim 0.2$ pN) the elastic behavior of DNA is symmetric under $\sigma \to -\sigma$. Pulling on the molecule removes the writhe and thus increases the twist and the torque on the molecule. For underwound molecules (in 10 mM phosphate buffer) above the critical force ($F_c \sim 0.5$ pN) and its associated critical torque ($\Gamma_c \sim 9$ pN nm) writhing becomes energetically unfavorable. The molecule elongates, see Fig. 8 as plectonemes (which are used to absorb twist) are converted locally into melted (denatured) regions of DNA. For positive supercoilings the critical force and torque are significantly higher. Thus, we can easily find values of $|\sigma|$ such that denaturation occurs at $-|\sigma|$ whereas plectonemes remain at $|\sigma|$.

In the following, we shall use the force vs. extension measurements on DNA supercoiled by $\pm n$ turns, i.e. with the same $|\sigma|$, to estimate the torsional constant, C, the critical torque at denaturation Γ_c and the energy of denaturation per base pair (bp), ε_d.

Consider the case of a DNA of contour length l_0 at an initial extension $l_{A^+}(< l_0)$. Let us coil the molecule by $n > 0$ turns to state A^+, (Fig. 9), requiring a torsional energy T_{A^+} ($F < 0.2$ pN) and then extend it to state B^+ ($F = 4$ pN), so as to pull out its plectonemes and eliminate its writhe. Alternatively, state B^+ could be reached by first stretching the torsionally relaxed DNA and then twisting it. In that case its torsional energy T_{B^+} is purely twist. The free energy of a stretched coiled DNA being a state variable, the mechanical works performed on the molecule by stretching it from thermodynamic state A^+ to B^+ along these two different paths should be equal:

$$T_{A^+} + \Delta W_{AB^+} = T_{B^+} = \frac{C}{2l_0}(2\pi n)^2. \tag{20}$$

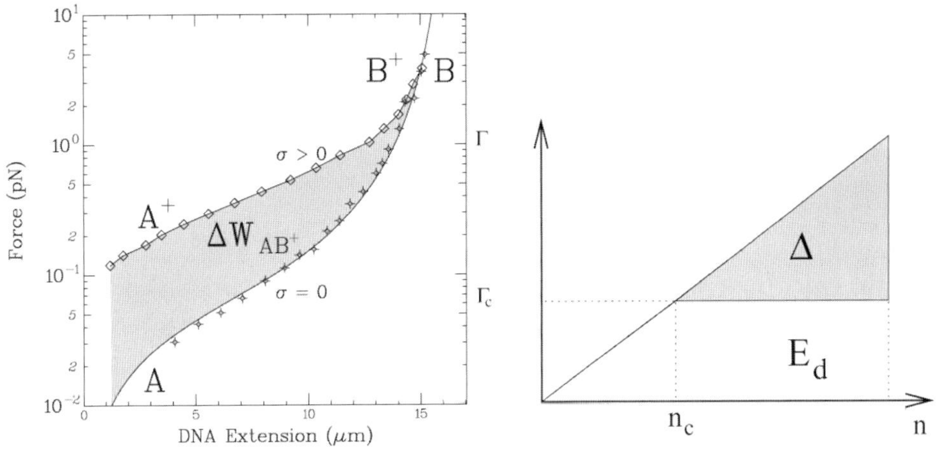

Fig. 9. LEFT: The extra work performed while stretching an overwound DNA. The molecule is overwound from points A to A^+ and then stretched along the $\sigma > 0$ curve to point B^+. The extra work performed while stretching is the shaded area between the $\sigma > 0$ and the $\sigma = 0$ curves. RIGHT dependence of the torque on the twist (number of turns). In a DNA molecule as in a twisted rod the torque increases linearly with the twist angle (number of turns). If the molecule melts because of torsional stress as expected when underwound, the torque stabilizes at a value Γ_c as it does in a rod which undergoes a torsional buckling instability. The difference Δ in the work of over and under-twisting is the shaded area shown here and in Fig. 10 left.

Here ΔW_{AB^+} is the extra work performed in stretching a coiled molecule from A^+ to B^+, the shaded area in Fig. 9. For the sake of simplicity we neglect the correction to the bare torsional constant C_0 due to the thermal fluctuations (Bouchiat and Mézard, 1998; Moroz and Nelson, 1998a). We shall see later that this approximation ($C \approx C_0$) is justified. Consider now the case in which DNA is underwound by $-n$ turns to state A^- and then stretched to state B^-. By the same reasoning as above we may write

$$T_{A^-} + \Delta W_{AB^-} = T_{B^-}. \tag{21}$$

Since when underwound the molecule partially denatures as it is pulled from A^- to B^-, the torsional energy T_{B^-} will consist of twist energy and energy of denaturation. We can nevertheless estimate T_{B^-} by considering the alternative pathway for reaching B^- by first stretching the molecule and then twisting it. In this case as the molecule is underwound, the torque Γ initially rises as in a twisted rod:

$$\Gamma = \frac{C}{l_0} 2\pi n. \tag{22}$$

When Γ reaches a critical value Γ_c after $-n_c$ turns, the molecule starts to denature. Any further increase in n enlarges the denaturation region, without affecting the torque in the molecule which stabilizes at $\Gamma = \Gamma_c$. Since mixing entropy can be neglected (Strick et al., 1998a,c; Allemand et al., 1998) the energy of denaturation is simply, see Fig. 9:

$$E_d = 2\pi(n - n_c)\Gamma_c \quad \text{while} \quad T_{B^-} = \frac{C}{2l_0}(2\pi n_c)^2 + E_d. \tag{23}$$

Note that at $F \sim 4\,\text{pN}$ the extension of dsDNA and partially denatured DNA is the same (Allemand et al., 1998). Thus, no extra work is performed against the force when the molecule partially melts. At low force the elastic behavior of a DNA molecule is symmetric under $n \to -n$: $T_{A^+} = T_{A^-}$. Thus subtracting Eq. (21) from Eq. (20) yields

$$\Delta \equiv \Delta W_{AB^+} - \Delta W_{AB^-} = T_{B^+} - T_{B^-} = \frac{2\pi^2 C}{l_0}(n - n_c)^2, \tag{24}$$

where Δ is the measured difference between the work performed while stretching an overwound molecule and the work done while pulling on an underwound one, see shaded area in Fig. 10 (left). Plotting the value of $\sqrt{\Delta}$ vs. n, one obtains a straight line (see Fig. 10 (right)), the slope of which allows one to determine the value of the torsional constant: $C/k_B T = 86 \pm 10\,\text{nm}$. The intercept of that line with the n-axis yields $n_c = 66$ turns, from which one can estimate the critical torque $\Gamma_c = 9\,\text{pN nm}$ and denaturation energy per bp $\varepsilon_d = E_d/10.5(n - n_c) = 1.35 k_B T$. Although the error bar on the measurement of C is still rather large, this method can be improved to yield a more precise value of C. It is nevertheless at least consistent with the current very imprecise estimate of $C/k_B T = 75 \pm 25\,\text{nm}$.

We may now estimate the validity of our approximation neglecting the correction to C_0 due to the thermal fluctuations (Moroz and Nelson, 1998a). At high force these renormalize C as

$$\frac{1}{C} = \frac{1}{C_0} + \frac{k_B T}{4B\sqrt{BF}}. \tag{25}$$

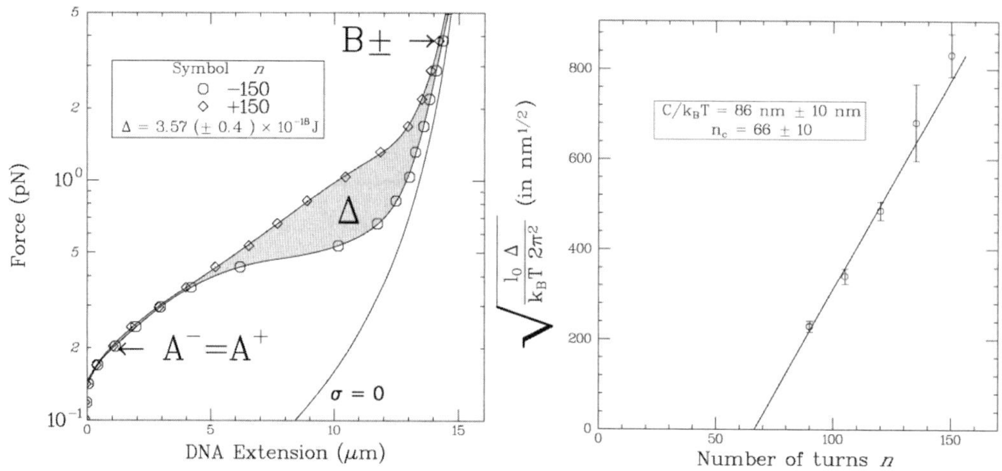

Fig. 10. LEFT: Difference in the work of stretching over and underwound DNA. (∘): DNA unwound by $n = -150$ turns. (◇) DNA overwound by $n = 150$ turns. The solid curves are polynomial fits to the force-extension data. The bottom curve is the theoretical (worm-like chain) fit to the data obtained for this molecule at $\sigma = 0$: $l_0 \sim 15.7\,\mu\text{m}$ and $\xi_p = B/k_B T \sim 48\,\text{nm}$ (Strick et al., 1999). The shaded surface between the $\sigma > 0$ and $\sigma < 0$ curves represents the work difference Δ. In both cases, point A^+ (respectively, A^-) is reached by overwinding (underwinding) the DNA which is initially at low extension (point A, not shown). Point B^+ (B^-) is reached by stretching the molecule along the appropriate $\sigma > 0$ ($\sigma < 0$) curve. RIGHT: Plot of the square root of the work difference Δ vs. the number of turns n the molecule is over or underwound. The straight line is a best fit through the experimental points. From its slope we extract the value $C/k_B T = 86 \pm 10\,\text{nm}$ and from its intercept with the n-axis, $n_c = 66$ turns, we infer $\Gamma_c = 2\pi n_c C/l_0 = 9\,\text{pN nm}$.

The last term on the right implies a correction of only 5% to the value of C at $F = 4\,\text{pN}$, smaller than our experimental uncertainty. It is interesting to note that the value of C determined here is in good agreement with the one obtained from the measurement of the molecular extension versus σ at constant force (Bouchiat and Mézard, 1998), a totally independent measurement based on the model of a rod like chain polymer, see above.

5. DNA–protein interactions

The understanding gained on the manipulation and mechanical properties of a stretched and coiled DNA, allows one to use these measurements as a tool to probe DNA/protein interactions at the level of a single molecule, e.g. chromatin (Chatenay et al., 1997; Cui and Bustamante, 2000).

For example, the lactose repressor-mediated loop formation in a single DNA molecule was followed by studying the decrease in the Brownian motion of a DNA tethered bead (Finzi and Gelles, 1995), as its molecular leash gets shorter. Subjecting that bead to a force F should yield the free energy δG for the formation of the complex repressor/DNA. Indeed when the formation and breakdown of the loop are equiprobable: $\delta G = F\delta l$ (Marko and Siggia, 1997), where δl is the loop length.

By using the stretching force as a control parameter the polymerization of RecA on a double-stranded DNA has been induced (Léger et al., 1998; Shivashankar et al., 1999; Hegner et al., 1999). As we have seen previously, stretching a molecule with a force of $\sim 70\,\text{pN}$ induces a transition to S-DNA, a structure of DNA presumably similar to the one adopted by the double-stranded molecule when interacting with RecA (Cluzel et al., 1996) or the TATA box-binding protein (Lebrun et al., 1997). By applying a force (even if less than 70 pN) on the molecule, one increases the probability of nucleation of an S-DNA region in the regular B-DNA helix. This S-DNA bubble has a higher affinity for RecA than B-DNA and serves as a nucleation center for the growth of a RecA fiber. As the complex RecA/DNA is about 60% longer than B-DNA, the kinetics of RecA polymerization can be followed as a function of time (at various forces) by monitoring the change in the molecular extension. Whether this stretched induced polymerization of RecA on dsDNA actually occurs in vivo remains an interesting suggestion.

The progression of an *E. coli* RNA-polymerase on a dsDNA can be followed by tethering the molecule to a small bead (Schafer et al., 1991; Yin et al., 1994) and anchoring the molecule at its other end to a RNA-pol bound to a surface. As the DNA is transcribed, the "leash" binding the bead to the surface gets shorter. The extent of its fluctuations is thus reduced and can be used to monitor the progression of the enzyme (Schafer et al., 1991). By trapping the bead with an optical tweezers, a force can be applied on a single polymerase. Its speed, pauses and the force to stall it (about 35 pN) can thus be determined (Wang et al., 1998; Yin et al., 1995).

Similarly, the replication of a single-strand DNA by a single DNA-polymerase could be observed by following the change in extension of a stretched template as a new strand is synthesized (Wuite et al., 2000; Maier et al., 2000). From the variation of the replication rate with the applied force, one could argue that a few bases (2 for T7-DNA polymerase, 4 for the Klenow fragment of polI) had to be fit between the two strands in the replication site for the enzyme to proceed. These results are in agreement with structural data and support the "induced-fit kinetic mechanism" for replication (Wong et al., 1991).

Finally, the relaxation of DNA supercoiling by a single topoisomerase (Wang, 1998) could be monitored (Strick et al., 2000) and individual enzymatic cycles observed at low ATP concentrations. From the distribution of the time intervals between successive cycles we could deduce that a single ATP was apparently burned per turnover. By averaging over many single enzymatic reactions, we could regain the kinetic behavior of topoII known from bulk measurements. Surprisingly, stretching the supercoiled DNA resulted in a lower enzymatic activity, indicating that the DNA gate religation step might be rate limiting. Finally in absence of ATP, the enzyme was observed to stabilise DNA crossovers.

These experiments open the way to a study of single mechanical enzymes, with a possibility of a detailed analysis of their cycle. Future experiments will certainly combine these manipulation techniques with single molecule observations (Kitamura et al., 1999; Harada et al., 1999; Funatsu et al., 1995; Yasuda et al., 1998) using fluorescence polarisation (Sase et al., 1997) or energy transfer (Weiss, 1999) to monitor the enzyme's structural deformation and energy consumption during its cycle.

Acknowledgements

We thank C. Bustamante, D. Chatenay, H. Clausen-Schaumann, H. Gaub, J.-F. Léger and S. Smith for sharing their data. We also thank C. Bouchiat, B. Maier, M. Mézard, for stimulating discussions.

References

Ahsan, A., Rudnick, J., Bruinsma, R., 1998. Elasticity theory of the B-DNA to S-DNA transition. Biophys. J. 74, 132–137.

Allemand, J.-F., Bensimon, D., Jullien, L., Bensimon, A., Croquette, V., 1997. pH-dependent specific binding and combing of DNA. Biophys. J. 73, 2064–2070.

Allemand, J.-F., Bensimon, D., Lavery, R., Croquette, V., 1998. Stretched and overwound DNA form a Pauling-like structure with exposed bases. Proc. Natl. Acad. Sci. USA 95, 14,152–14,157.

Amblard, F., Yurke, B., Pargellis, A., Leibler, S., 1996. A magnetic manipulator for studying local rheology and micromechanical properties of biological system. Rev. Sci. Instrum. 67 (2), 1–10.

Bensimon, A., 1996. Force: a new structural control parameter? Structure 4, 885–889.

Bensimon, D., Dohmi, D., Mézard, M., 1998. Stretching a heteropolymer. Europhys. Lett. 42, 97–102.

Bensimon, A., Simon, A.J., Chiffaudel, A., Croquette, V., Heslot, F., Bensimon, D., 1994. Alignment and sensitive detection of DNA by a moving interface. Science 265, 2096–2098.

Bensimon, D., Simon, A.J., Croquette, V., Bensimon, A., 1995. Stretching DNA with a receding meniscus: experiments and models. Phys. Rev. Lett. 74, 4754–4757.

Binder, K., Young, A.P., 1986. Spin glasses: experimental facts, theoretical concepts and open questions. Rev. Mod. Phys. 58 (4), 801–976.

Bouchiat, C., Mézard, M., 1998. Elasticity theory of a supercoiled DNA molecules. Phys. Rev. Lett. 80, 1556–1559.

Bouchiat, C., Mézard, M., 1999. Elasticity rod model of supercoiled DNA molecules. condmat/9904018.

Bouchiat, C., Wang, M.D., Block, S.M., Allemand, J.-F., Strick, T.R., Croquette, V., 1999. Estimating the persistence length of a worm-like chain molecule from force-extension measurements. Biophys. J. 76, 409–413.

Bustamante, C., Marko, J.F., Siggia, E.D., Smith, S., 1994. Entropic elasticity of λ-phage DNA. Science 265, 1599–1600.

Cantor, C.R., Schimmel, P.R., 1980. Biophysical Chemistry, Part III: The Behaviour of Biological Macromolecules. W.H. Freemann, San Francisco.

Chatenay, D., Houchmanzadeh, B., Marko, J.F., Libchaber, A., 1997. Elasticity and structure of eukaryote chromosomes studied by micromanipulation and micropipette aspiration. J. Cell. Biol. 139, 1–12.

Chee, M., Yang, R., Hubbell, E., Berno, A., Huang, X.C., Stern, D., Winkler, J., Lockhart, D.J., Morris, M.S., Fodor, S.P.A., 1996. Accessing genetic information with high-density DNA arrays. Science 274, 610–614.

Chirico, G., Langowski, J., 1996. Brownian dynamics simulations of supercoiled DNA with bent sequences. Biophys. J. 71, 955–971.

Chu, S., 1991. Laser manipulation of atoms and particles. Science 253, 861–866.

Cizeau, P., Viovy, J.L., 1997. Modeling extreme extension of DNA. Biopolymers 42, 383–385.

Cluzel, P., Lebrun, A., Heller, C., Lavery, R., Viovy, J.-L., Chatenay, D., Caron, F., 1996. DNA: an extensible molecule. Science 271, 792–794.

Cui, Y., Bustamante, C., 2000. Pulling a single chromatin fiber reveals the forces that maintain its higher order structure. Proc. Natl. Acad. Sci. (USA) 97, 127–132.

de Gennes, P.G., 1979. Scaling concepts in Polymer Physics. Cornell University Press, Ithaca.

Einstein, A., 1956. Investigation of the Brownian Theory of Movement. Dover Publication, New York.

Engel, A., Gaub, H.E., Müller, D.J., 1999. Atomic force microscopy: a forceful way with single molecules. Curr. Biol. 9, R133–R136.

Essevaz-Roulet, B., Bockelmann, U., Heslot, F., 1997. Mechanical separation of the complementary strands of DNA. Proc. Natl. Acad. Sci. (USA) 94, 11,935–11,940.

Finer, J.T., Simmons, R.M., Spudich, J.A., 1994. Single myosin molecule mechanics: piconewton forces and nanometre steps. Nature 368, 113–119.

Finzi, L., Gelles, J., 1995. Measurement of lactose repressor-mediated loop formation and breakdown in single DNA molecules. Science 267, 378–380.

Fisher, M.E., 1964. Magnetism in one-dimensional systems — the Heisenberg model for infinite spin. Am. J. Phys. 32, 343–346.

Fixman, M., Kovac, J., 1973. Polymer conformational statistics III: modified Gaussian models of the stiff chains. J. Chem. Phys. 58, 1564–1568.

Florin, E.L., Moy, V.T., Gaub, H.E., 1994. Adhesion force between individual ligand–receptor pairs. Science 264, 415–417.

Flory, P., 1975. Principles of Polymer Chemistry. Cornell University Press, Ithaca.

Forterre, P., 1996. A hot topic: the origin of hyperthermophiles (minireview). Cell 85, 789–792.

Funatsu, T., Harada, Y., Tokunaga, M., Saito, K., Yanagida, T., 1995. Imaging of single fluorescent molecules and individual ATP turnovers by single myosin molecules in aqueous solution. Nature 374, 555–559.

Garrivier, D., Fourcade, B., 2000. Twisting DNA with variable intrinsic curvature. Europhys. Lett. 49, 390–395.

Gosse, C., Croquette, V., 1999. Magnetic tweezers. Rev. Sci. Instrum., in preparation.

Hagerman, P.J., 1988. Flexibility of DNA. Ann. Rev. Biophys. Chem. 17, 265–268.

Harada, Y., Funatsu, T., Murakami, K., Nonoyama, Y., Ishihama, A., Yanagida, T., 1999. Single-molecule imaging of RNA polymerase-DNA interactions in real time. Biophys. J. 76, 709–715.

Hegner, M., Smith, S.B., Bustamante, C., 1999. Polymerization and mechanical properties of single RecA-DNA filaments. Proc. Natl. Acad. Sci. USA 96, 10,109–10,114.

Hogan, M.E., Austin, R.H., 1987. Importance of DNA stiffness in protein-DNA binding specificity. Nature 329, 263–266.

Hua, W., Young, E.C., Fleming, M.L., Gelles, J., 1997. Coupling of kinesin steps to ATP hydrolysis. Nature 388, 390–393.

Ishijima, A., Doi, T., Sakurada, K., Yanagida, T., 1991. Sub-piconewtown force fluctuations of actomyosin in vitro. Nature 352, 301–306.

Kellermayer, M.S.Z., Smith, S.B., Granzier, H.L., Bustamante, C., 1997. Folding–unfolding transition in single titin molecules characterized with laser tweezers. Science 276, 1112–1116.

Kikuchi, A., Asai, K., 1984. Reverse-gyrase — a topoisomerase which introduces positive superhelical turns into DNA. Nature 309, 677–681.

Kitamura, K., Tokunaga, M., Iwane, A.H., Yanagida, T., 1999. A single myosin head moves along an actin filament with regular steps of 5.3 nanometers. Nature 397, 129–134.

Lebrun, A., Lavery, R., 1996. Modelling extreme deformations of DNA. Nucl. Acids Res. 24, 2260.

Lebrun, A., Lavery, R., 1997. Unusual DNA conformations. Curr. Op. Struct. Biol. 7, 348–354.

Lebrun, A., Shakked, Z., Lavery, R., 1997. Local DNA stretching mimics the distortion caused by the TATA box-binding protein. Proc. Natl. Acad. Sci. USA 94, 2993–2998.

Léger, J.F., Robert, J., Bourdieu, L., Chatenay, D., Marko, J.F., 1998. RecA binding to a single double-stranded DNA molecule: a possible role of DNA conformational fluctuations. Proc. Natl. Acad. Sci. USA 95, 12,295–12,296.

Léger, J.F., Romano, G., Sarkar, A., Robert, J., Bourdieu, L., Chatenay, D., Marko, J.F., 1999. Structural transitions of a twisted and stretched DNA molecule. Phys. Rev. Lett. 83, 1066–1069.

Liu, D.J., Day, L.A., 1994. Pf1 virus structure: helical coat protein and DNA with paraxial phosphates. Science 265, 671–674.

Liu, L.F., Wang, J.C., 1987. Supercoiling of the DNA template during transcription. Proc. Natl. Acad. Sci. USA 84, 7024–7027.

Maier, B., Bensimon, D., Croquette, V., 2000. Replication by a single DNA-polymerase of a stretched single strand DNA. Proc. Natl. Acad. Sci. USA, accepted for publication.

Marko, J.F., 1997. Stretching must twist DNA. Europhys. Lett. 38, 183–188.

Marko, J.F., 1998. DNA under high tension: overstreching undertwisting and relaxation dynamics. Phys. Rev. E 57 (2), 2134–2149.

Marko, J.F., Siggia, E., 1995a. Statistical mechanics of supercoiled DNA. Phys. Rev. E 52 (3), 2912–2938.

Marko, J.F., Siggia, E., 1995b. Stretching DNA. Macromolecules 28 (26), 8759–8770.

Marko, J.F., Siggia, E.D., 1997. Driving proteins off DNA using applied tension. Biophys. J. 73, 2173–2178.

Moroz, J.D., Nelson, P., 1998a. Torsional directed walks, entropic elasticity and DNA twist stiffness. Proc. Natl. Acad. Sci (USA) 94, 14,418–14,422.

Moroz, J.D., Nelson, P., 1998b. Entropic elasticity of twist-storing polymers. Macromolecules 31, 6333–6347.

Michalet, X., Ekong, R., Fougerousse, F., Rousseaux, S., Schurra, C., Povey, S., Beckmann, J.S., Bensimon, A, 1997. Dynamic molecular combing: stretching the whole human genome for high resolution studies. Science 277, 1518–1523.

Nelson, P., 1998. Sequence-disorder effects on DNA entropic elasticity. Phys. Rev. Lett. 80, 5810–5812.

Nelson, P., 1999. Transport of torsional stress in DNA. Proc. Natl. Acad. Sci. 96, 14,342–14,347.

Nelson, P., Kamien, R.D., Lubensky, T.C., O'Hern, C.S., 1997. Direct determination of DNA twist-stretch coupling. Europhys. Lett. 38, 237–242.

Olson, W.K., 1996. Simulating DNA at low resolution. Curr. Op. Struct. Biol. 6, 242–256.

Perkins, T.T., Smith, D.E., Chu, S., 1994a. Direct observation of tube-like motion of a single polymer chain. Science 264, 819–822.

Perkins, T.T., Quake, S.R., Smith, D.E., Chu, S., 1994b. Relaxation of a single DNA molecule observed by optical microscopy. Science 264, 822–826.

Reif, F., 1965. Fundamentals of Statistical and Thermal Physics. McGraw-Hill, New York.

Rief, M., Clausen-Schaumann, H., Gaub, H.E., 1999. Sequence-dependent mechanics of single DNA molecules. Nature Struct. Bio. 6, 346–349.

Rief, M., Gautel, M., Oesterhelt, F., Fernandez, J.M., Gaub, H.E., 1997. Reversible unfolding of individual titin immunoglobulin domains by AFM. Science 276, 1109–1112.

Saito, K., Takaaki Aoki, Toshiaki Aoki, Yanagida, T., 1994. Movement of single myosin filaments and myosin step size on an actin filament suspended in solution by a laser trap. Biophys. J. 66, 769–777.

Sano, T., Smith, C.L., Cantor, C.R., 1992. Immuno-PCR: very sensitive antigen detection by means of specific antibody-DNA conjugates. Science 258, 120–122.

Sase, I., Miyata, H., Ishiwata, S., Kinosita, K., 1997. Axial rotation of sliding actin filaments revealed by single-fluorophore imaging. Proc. Natl. Acad. Sci. USA 94, 5646–5650.

Schafer, D.A., Gelles, J., Sheetz, M.P., Landick, R., 1991. Transcription by single molecules of RNA polymerase observed by light microscopy. Nature 352, 444–448.

Schellman, J.A., Harvey, S.C., 1995. Static contributions to the persistence length of DNA and dynamic contributions to DNA curvature. Biophys. Chem. 55, 95–114.
Schlick, T., Olson, W.K., 1992. Supercoiled DNA energetics and dynamics by computer simulation. J. Mol. Biol. 223, 1089–1119.
Schnitzer, M.J., Block, S.M., 1997. Kinesin hydrolyses one ATP per 8-nm step. Nature 388, 386–390.
Shivashankar, G.V., Feingold, M., Kritchevsky, O., Libchaber, A., 1999. RecA polymerization on double-stranded DNA by using single-molecule manipulation: the role of ATP hydrolysis. Proc. Natl. Acad. Sci. USA 96, 7916–7921.
Simmons, R.M., Finer, J.T., Chu, S., Spudich, J.A., 1996. Quantitative measurements of force and displacement using an optical trap. Biophys. J. 70, 1813–1822.
Smith, S.B., 1998. Stretch transitions observed in single biopolymer molecules (DNA or protein) using laser tweezers. Thesis, U. Twente, Netherlands.
Smith, S.B., Cui, Y., Bustamante, C., 1996. Overstretching B-DNA: the elastic response of individual double-stranded and single-stranded DNA molecules. Science 271, 795–799.
Smith, S.B., Finzi, L., Bustamante, C., 1992. Direct mechanical measurements of the elasticity of single DNA molecules by using magnetic beads. Science 258, 1122–1126.
Stasiak, A., Di Capua, E., 1982. The helicity of DNA in complexes with Rec-A protein. Nature 299, 185–187.
Stasiak, A., Di Capua, E., Koller, T., 1983. Unwinding of duplex DNA in complexes with Rec-A protein. Cold Spring Harbor Symposium on Quantitative Biology, Vol. 47, pp. 811–820.
Strick, T., Allemand, J.F., Bensimon, D., Bensimon, A., Croquette, V., 1996. The elasticity of a single supercoiled DNA molecule. Science 271, 1835–1837.
Strick, T., Allemand, J.-F., Bensimon, D., Croquette, V., 1998a. The behavior of supercoiled DNA. Biophys. J. 74, 2016–2028.
Strick, T., Allemand, J.-F., Croquette, V., Bensimon, D., 1998b. Physical approaches to the study of DNA. J. Stat. Phys. 93, 647–672.
Strick, T., Croquette, V., Bensimon, D., 1998c. Homologous pairing in streched supercoiled DNA. Proc. Natl. Acad. Sci. (USA) 95, 10,579–10,583.
Strick, T., Allemand, J.-F., Bensimon, D., Lavery, R., Croquette, V., 1999. Phase coexistence in a single DNA molecule. Physica A 263, 392–404.
Strick, T., Bensimon, D., Croquette, V., 1999. Micro-mechanical measurement of the torsional modulus of DNA. Genetica, Proceedings of NATO ARW on Structural Biology and Functional Genomics, 1998, Vol. 106, pp. 57–62.
Strick, T.R., Croquette, V., Bensimon, D., 2000. Single-molecule analysis of DNA uncoiling by a type II topoisomerase. Nature 404, 901–904.
Trifonov, E.N., Tan, R.K.-Z., Harvey, S.C., 1987. Static Persistence Length of DNA. Adenine Press.
Vologodskii, A.V., 1994. DNA extension under the action of an external force. Macromolecules 27, 5623–5625.
Vologodskii, A., Marko, J.F., 1997. Extension of torsionally stressed DNA by external force. Biophys. J. 73, 123–132.
Wang, J.C., 1998. Moving one DNA double helix through another by a type II DNA topoisomerase: the story of a simple molecular machine. Quart. Rev. Biophys. 31, 107–144.
Wang, M.D., Schnitzer, M.J., Yin, H., Landick, R., Gelles, J., Block, S., 1998. Force and velocity measured for single molecules of RNA polymerase. Science 282, 902–907.
Wang, M.D., Yin, H., Landick, R., Gelles, J., Block, S., 1997. Stretching DNA with optical tweezers. Biophys. J. 72, 1335–1346.
Weier, H.-U.G., Wang, M., Mullikin, J.C., Zhu, Y., Cheng, J.-F., Greulich, K.M., Bensimon, A., Gray, J.M., 1995. Quantitative DNA fiber mapping. Human Mol. Gen. 4, 1903–1910.
Weiss, S., 1999. Fluorescence spectroscopy of single biomolecules. Science 283, 1676–1683.
White, J.H., 1969. Self linking and the gauss integral in higher dimensions. Am. J. Math. 91, 693–728.
Wilkins, M.H.F., Gosling, R.G., Seeds, W.E., 1951. Nucleic acid: an extensible molecule? Nature 167, 759–760.
Woese, C.R., Fox, G.E., 1977. Phylogenetic structure of the prokaryotic domain: the primary kingdoms. Proc. Natl. Acad. Sci. USA 74, 5088–5090.
Woese, C.R., Kandler, O., Wheelis, M.L., 1990. Towards a natural system of organisms: proposal for the domain: archaea, bacteria and eucarya. Proc. Natl. Acad. Sci. USA 87, 4576–4579.

Wong, I., Patel, S.S., Johnson, K.A., 1991. An induced-fit kinetic mechanism for DNA replication fidelity: direct measurement by single-turnover kinetics. Biochemistry 30, 526–537.

Wuite, G.J.L., Smith, S.B., Young, M., Keller, D., Bustamante, C., 2000. Single-molecule studies of the effect of template tension on T7 DNA polymerase activity. Nature 404, 103–106.

Yasuda, R., Noji, H., Kinosita, K., Yoshida, M., 1998. F_1-ATPase is a highly efficient molecular motor that rotates with discrete 120° steps. Cell 93, 1117–1124.

Yin, H., Landick, R., Gelles, J., 1994. Tethered particle motion method for studying transcript elongation by a single RNA polymerase molecule. Biophys. J. 67, 2468–2478.

Yin, H., Wang, M.D., Svoboda, K., Landick, R., Block, S., Gelles, J., 1995. Transcription against a applied force. Science 270, 1653–1657.

Part II

Biology at the Single Molecule Level
S.H. Leuba an J. Zlatanova (Eds)
Published by Elsevier Science Ltd.

Review

Single molecule measurements of titin elasticity

Kuan Wang[a,*], Jeffrey G. Forbes[a,b], Albert J. Jin[a,c]

[a] *Laboratory of Physical Biology, NIAMS, NIH, Room 408, Building 6, Bethesda, MD 20892, USA*
[b] *Department of Chemical Engineering, University of Maryland, College Park, MD 20742-2111, USA*
[c] *Division of Bioengineering and Physical Science, ORS/OD, NIH, Room 3N18, Building 13, Bethesda, MD 20892, USA*

Abstract

Titin, with a massive single chain of 3–4 MDa and multiple modular motifs, spans the half-sarcomere of skeletal and cardiac muscles and serves important, multifaceted functions. In recent years, titin has become a favored subject of single molecule observations by atomic force microscopy (AFM) and laser optical trap (LOT). Here we review these single titin molecule extension studies with an emphasis on understanding their relevance to titin elasticity in muscle function. Some fundamental aspects of the methods for single titin molecule investigations, including the application of dynamic force, the elasticity models for filamentous titin motifs, the technical foundations and calibrations of AFM and LOT, and titin sample preparations are provided. A chronological review of major publications on recent single titin extension observations is presented. This is followed by summary evaluations of titin domain folding/unfolding results and of elastic properties of filamentous titin motifs. Implications of these single titin measurements for muscle physiology/pathology are discussed and forthcoming advances in single titin studies are anticipated. Published by Elsevier Science Ltd.

Keywords: Atomic force microscopy; AFM; Laser optical trap; LOT; Polymer elasticity; Biological forces; Muscle protein

Contents

*Corresponding author. Tel.: +1-301-496-4097; fax: +1-301-402-8566; http://www.nih.gov/niams/about/irp/lbpintro.htm.

E-mail address: wangk@exchange.nih.gov (K. Wang).

1. Introduction

Although striated muscle research has had hundreds of years of history, the classic two-filament sarcomere consisting of myosin thick filaments and actin thin filaments as the force-generating unit has only been known for less than 50 years (as reviewed e.g. in Huxley, 1990). The discovery of the giant protein titin (also known as connectin) over 20 years ago added a third filament system that is crucial to the sarcomeric elasticity, structure and function (Fig. 1) (Maruyama et al., 1981; Wang et al., 1979). Proteins of the titin family are the longest single polypeptide chains currently known and have a molecular mass in the range of 3–4 MDa. The complete cDNA sequence of human cardiac titin was first reported by Labeit and coworkers (1995), and intriguing tissue- and species-specific alternative splicing pathways of titin have now emerged (Freiburg et al., 2000; Labeit and Kolmerer, 1995) (Fig. 2A). Human titin gene has 234 exons, the largest number in a single human gene (Venter et al., 2001). Isolated titin is a monomer that is long, slender and flexible. Its conformational malleability can be seen in its EM images in Fig. 3. When the molecules are relaxed they take on a highly convoluted configuration (Fig. 3A). However, titin can be easily entangled and readily stretched by meniscus forces while preparing the EM samples (Fig. 3B–E). The knowledge of modular motifs in titin has accelerated more targeted studies via both the traditional myofibril mechanics/immunolocalization methods and various single molecule techniques, yielding an ever-increasing body of insights. These extensive studies in recent years have solidified our knowledge of the multifaceted roles that titin plays in the sarcomere [see e.g. recent reviews (Horowits, 1999; Maruyama, 1997; Trinick and Tskhovrebova, 1999; Wang, 1996, 1999)].

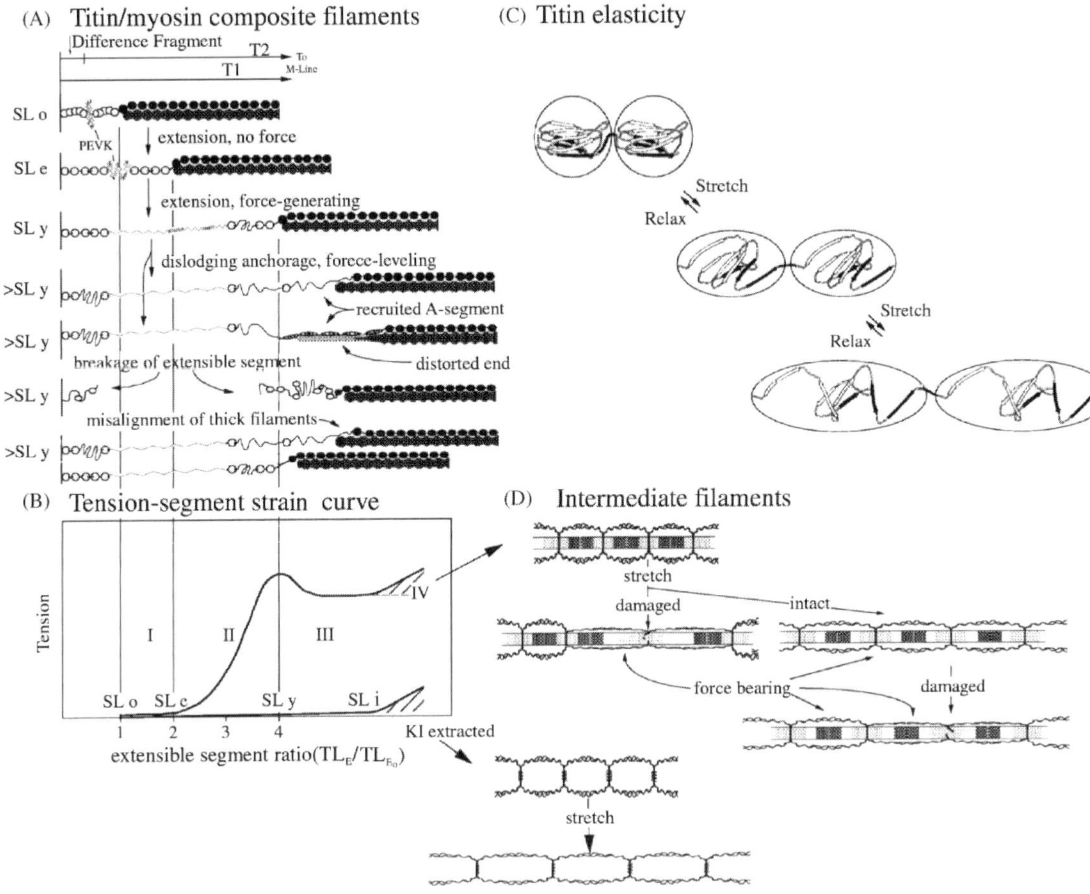

Fig. 1. Elasticity and segmental extension of sarcomere associated cytoskeletons. (A) Proposed structural changes in titin/myosin composite filaments upon stretching. (B) Resting tension–sarcomere length relationship of myofibrillar bundles containing both sarcomere matrix and intermediate filaments. Straightening of slack titin filaments at low tension (Zone I) is followed by the extension of the I band segment of titin/myosin composite filaments that gives rise to the exponential rise in tension (ERT) (zone II, between SL_e and SL_y). The detachment of part of the anchored titin segment at and beyond the elastic limit (yield point) of the sarcomere (Sl_y) contributes to the leveling of tension at higher degrees of stretch (zone III). Stretching of intermediate filament lattice contributes to the further increase in tension beyond 5 μm (zone IV). The resting tension curves of several types of muscles with different titin isoforms and elastic limits can be normalized to the same master curve by plotting resting tension as a function of the degree of stretch of the extensible segment of titin (extensible segmental ratio in B). (C) The mechanical unfolding or "yielding" of Ig and/or Fn3 domains that adjusts length and spring constant of the titin molecule. (D) The extension of the intermediate filaments that surround the sarcomeres and insert at the force-bearing structures at the Z line and M lines. The intermediate filaments provide mechanical continuity of the myofibril even when individual sarcomere is damaged. The segmental elasticity of titin and sarcomere is correlated closely to the extension of its rich molecular motifs such as the PEVK segment (marked in A as unfolded strings) and the Ig/Fn3 domains (marked in A as folded beads), see Fig. 2 and the text. (See Wang et al. (1993) for more details and the figure was adopted with permission from Biophysical Journal.)

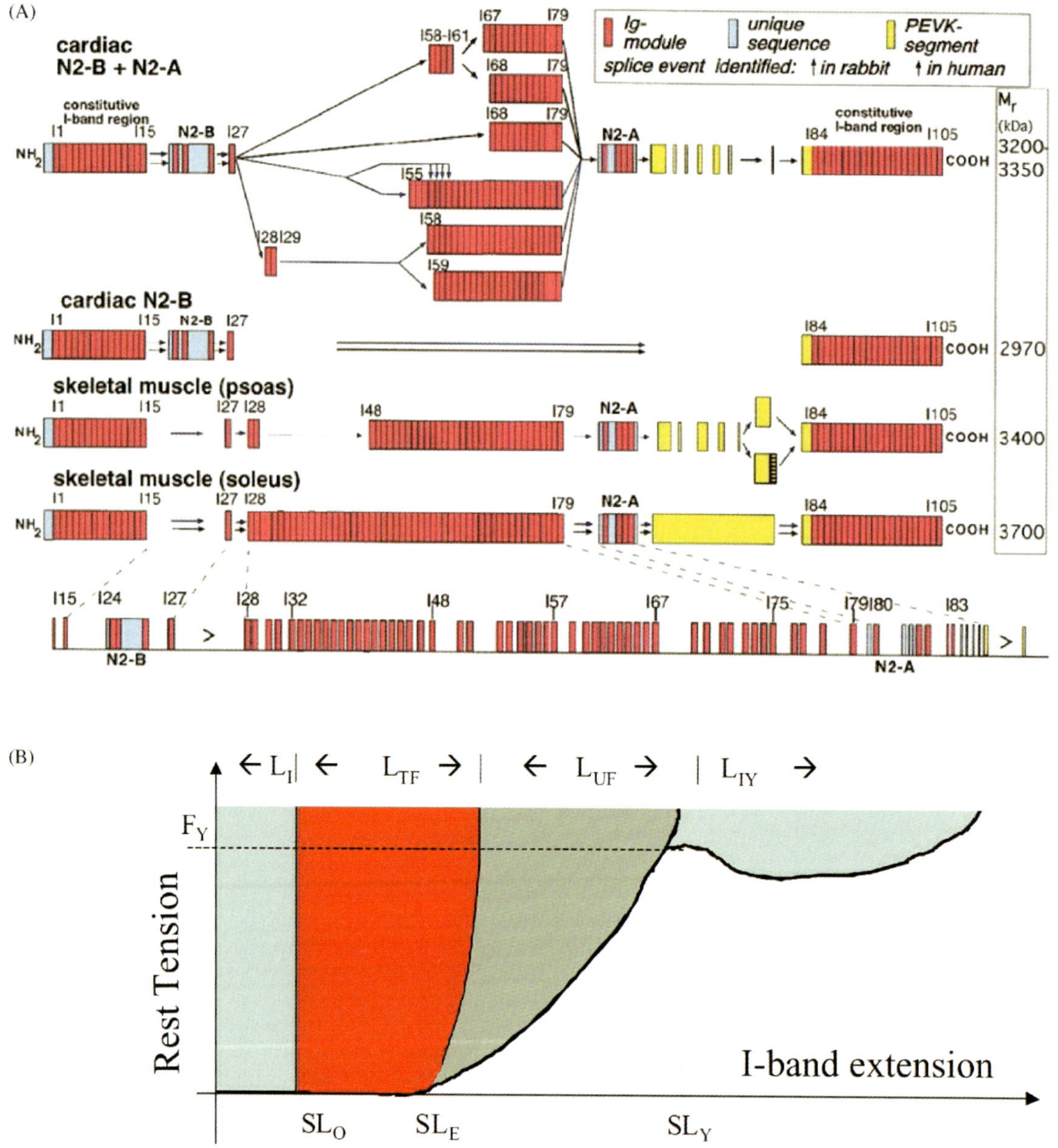

Fig. 2. Schematic diagrams of titin motifs and myofibril elasticity. (A) Differential motifs and splicing pathways of titin in the extensible I-band. Adopted from Freiburg et al. (2000) with permission from American Heart Association and Lippincott Williams & Wilkins. Identified splice routes are indicated by arrows, black for human and blue for rabbit. Predicted molecular weights of respective isoforms are given (right). All isoforms share the I27 exon and I84 exon (see bottom row). These alternative splicings for skeletal titins and cardiac N2-A titin are in combinations of additional 53 Ig domains (red), N2-B to N2-A (red-blue) substitution, and variable PEVK sequence length (yellow). All skeletal titins have the 53 Ig domain additions and variable PEVK length (with, e.g. 2174 residues in human soleus). Cardiac N2-A titin has a shorter extendable N2 insertion sequence, but a larger PEVK length. The details of Z-line, A-band and M-line motifs are not shown (see, e.g., Freiburg et al. (2000) and Labeit and Kolmerer (1995)). (B) Segmental extension of titin motifs (after Wang et al., 1991). Passive-force and sarcomere extension curves from most muscle types all scale to a multi-stage master relationship. The defining characteristics of the force curves are mainly the total lengths of the extensible Ig-motifs, which set the length SL_E–SL_O, and the unique sequences (PEVK plus N2-B and/or N2-A), which set the length SL_Y–SL_E and the strength of titin/myosin interaction which determines the yield point (SL_Y). The multistage passive tension vs. sarcomere I-band extension are attributed to the straightening of Ig-motifs (low force), the unique insertions such as PEVK and N2-A/B (exponential rising force), and the participation of newly recruited Fn3 motifs from the thick filament segment (beyond the yield point force, F_Y). See text for further details.

This review article focuses on titin as the source of passive tension (elasticity) in the sarcomere via its elastic properties as provided by single molecule investigations. Passive tension refers to the mechanical force exerted by the sarcomere against stretching, in contrast to the active force produced by actin/myosin interactions that is driven by ATP hydrolysis. Titin was quickly established to be the major filament that supports passive tension in the sarcomere by linking the tips of myosin filaments to the Z-line via several serially connected elastic segments with distinct extensibility (Fig. 1). The elasticity of titin filaments in situ was predicted at the time of its discovery as a major myofibrillar component (Maruyama et al., 1981; Wang et al., 1979). A strong correlation between the passive tension and the extension of titin was established by a number of studies using immunoelectron microscopy and the translocation of unique titin epitopes in sarcomeres (Horowits et al., 1989; Maruyama, 1994; Trombitas and Pollack, 1993; Wang et al., 1991). The pivotal role of titin in muscle elasticity was also supported by studies using protease digestion and low doses of ionizing radiation. Both of these techniques readily destroy the passive tension of muscle fibers in proportion to the degradation of the high-molecular-weight titin (Funatsu et al., 1990; Horowits et al., 1986). Furthermore, titin was found to contribute the majority of the passive tension in muscle compared with other force-bearing filaments (e.g. collagen and intermediate filaments) in the normal working range of muscle extension (Granzier and Irving, 1995; Granzier and Wang, 1993b). These results prove conclusively that, by acting as an elastic connector between the tip of thick filaments and the Z-line (i.e., the I band of the sarcomere), titin imparts elasticity and structural integrity to the sarcomere.

Quantitatively, sarcomere passive tension and in situ titin extensibility can be scaled across multiple muscle tissue types to a single master relationship by the length of such elastic connectors in the sarcomere, which in turn is determined by the length of the titin isoform(s) selectively expressed in each muscle (Fig. 2A). The segmental extension model of titin elasticity thus emerged from such intriguing correlation of titin size isoforms and the elastic limits (yield points) of each muscle tissue (Wang et al., 1991) (Fig. 2B). In the I-band region, where titin extends during passive force development, titin is comprised of serially-linked motifs of three types: (1) the folded 100-residue modular repeats of immunoglobin (Ig) or fibronectin III (Fn3) domains, (2) a unique PEVK motif consists of 70% of proline, glutamate, valine, and lysine residues, and (3) the N2-A or N2-B insert (Labeit and Kolmerer, 1995). Tissue- and species-specific alternative splicing extensively alters the expression of these motifs (Freiburg et al., 2000; Labeit and Kolmerer, 1995). The cardiac I-band titin is comprised of two sub-types, termed N2-B and N2-BA, both having the N2-B titin insert that is a few hundred amino acid residues longer than the N2-A titin insert and a PEVK motif that is as short as 163 amino acids. The skeletal titins, on the other hand, all have much longer poly-Ig motifs with up to 53 additional Ig modules, N2-A insertion, and much longer PEVK motifs with up to 2174 residues in human soleus muscle (Fig. 2A). During the stretching of sarcomeres, the poly-Ig motifs of titin first straighten and produce a very low restoring force. The unique segments, PEVK and N2-B or N2-A, then extend greatly, producing an essentially exponentially rising force. Upon reaching the yield point force, the inextensible titin segment in the A band dislodges from the thick filaments, becomes extensible, and then the passive force levels off (Figs. 1A and 2B). The major defining characteristics of each tissue-specific force–extension curve are the total length of the extensible Ig-motifs which sets the length SL_E–SL_O for initial low force–low extension region; the total length of the unique sequences (PEVK plus N2-B and/or N2-A), which sets the length SL_Y-SL_E; and the intrinsic extensibility of each sequence motifs of PEVK/N2 which sets the force scale for the exponential rising force region.

The strength of titin/thick filament interactions (F_Y) determines the elastic limit (SL_Y) and further stretching beyond the yield point only detaches an increasing number of the Ig and Fn3 domains that were originally adhered to the myosin filaments, without further increase of force. Thus the recruitment of titin domains from the thick filament region acts as a mechanical delimiter to prevent the massive unfolding of its Ig and Fn3 domains (Figs. 1 and 2B) while maintaining the structural continuity and preventing the detrimental fracturing of the sarcomere. The segmental extension and sequential recruitment of segments/motifs of titin in the sarcomere form a scaling base for unifying passive tension characteristics of skeletal, cardiac and insect flight muscles. The recent demonstration by immuno-staining that each of the PEVK, N2A and B, and Ig segments extend as distinct segments support the soundness of this sequential extension/recruitment concept (Granzier et al., 2000). This model also provides the framework to integrate the single titin elasticity measurements and sarcomeric mechanics/immuno-imaging observations of striated muscle.

In the balance of this review, we focus on single titin molecule extension studies and the resulting insights into titin elasticity. In Section 2, we provide some fundamental aspects of methods for single titin molecule investigations. First, we introduce the concepts of dynamic force and entropic spring models that are centrally relevant to titin behavior under extension. Two single molecule study techniques, atomic force microscopy (AFM) and laser optical trap (LOT), are then explained. Titin molecule sample preparation techniques are also discussed. In Section 3, we give a review of major publications on recent single titin extension observations, first in a chronological order then in summary comparisons. In Section 4, we provide a more comprehensive evaluation of titin domain folding/unfolding observations in the context of protein folding and energetics. In Section 5, we summarize the elastic properties of filamentous titin motifs in relationship with cardiac and skeletal muscle elasticity in physiology/pathology, and provide brief concluding remarks on forthcoming advances.

2. Methods—single molecule extension

The last decade of the 20th century was a time of rapid advances in the study of single molecules by a variety of methods, including fluorescence, electron microscopy, laser optical trap (LOT), and scanning probe microscopy such as atomic force microscopy (AFM). One area of investigation that has been particularly fruitful is the study of the mechanical properties of single molecules such as DNA, modular proteins, and synthetic polymers. These experiments have yielded exciting and novel information on the shape, dynamics and mechanical properties of individual molecules that previously could only be inferred from the average properties of polymers in bulk. From the pioneering work of Ashkin and colleagues (Ashkin, 1980; Ashkin et al., 1986), LOT has been developed into a tool that can simultaneously manipulate multiple molecules individually while accurately measure forces on the order of 1.0 pN (Block, 1995; Visscher and Block, 1998; Mehta et al., 1999). AFM was invented a few years later (Binnig et al., 1986; Hansma et al., 1988) and is now recognized as a powerful imaging and force measurement tool with a large dynamic range, from ångstroms to microns, and with utility in high vacuum, at cryo to high temperature, and in aqueous environments. Titin is an ideal subject for mechanical studies, because it plays a mechanical role in muscle and due to its large mass (3–4 MDa) and extended structure of ca. 1 μm in length, it can be more easily manipulated than smaller molecules. In 1997, two LOT studies of titin were reported (Kellermayer et al., 1997; Tskhovrebova et al., 1997). Around the same time, the first AFM measurement of titin elasticity was reported (Rief et al., 1997a). These studies set the stage for an explosive growth of interests and insights of how external force alters the rates and pathways of folding and interactions of proteins. We will first describe the conceptual basis of the interplay between external force and the rate and energetics of protein folding and interaction. We will then critically evaluate the observations and data interpretations of titin elasticity based on these two single molecule techniques.

2.1. *Molecular forces and energy surfaces*

While there are myriad ways to apply forces to a bulk material, currently only tension and torsion (Strick et al., 2000; Tsuda et al., 1996; Yasuda et al., 1996) can be applied to single molecules. Since titin has little torsional rigidity (Fig. 3), only tension is applicable and transmitted along titin. In contrast, torsion is relevant to DNA and protein filaments with greater flexural rigidity. The application of tension to a molecule can stretch the molecule, break down intramolecular folds, rupture protein/protein or protein/ligand interactions, and weaken or rupture the attachment to the force-measuring instrument. Thus, specific application of tension to single molecules allows numerous important biological observations to be made.

The force-induced rupturing of protein/ligand interactions or protein unfolding may be conceptualized on the basis of transition-state theory that describes chemical or physical reactions as taking place on a potential energy surface with many possible reaction paths from reactants to products. The different paths that any given reaction may take will have different barrier heights and possibly local minima on the reaction surface. The most likely path, i.e. the one with the lowest barriers, will be a saddle point on the energy surface. An example of such a path on the surface with one intermediate state is represented in Fig. 4A. The bound or folded state is in the deep well and the unbound or unfolded state is where the energy is zero to the right. For simple reactions such as two atoms reacting to form a diatomic molecule, the length scale of the reaction coordinate can be directly related to bond length. For complex bonding surfaces as those found in protein–ligand complexes and in folded proteins, a direct and simple correlation between the molecular distance and the reaction coordinate does not exist.

The length of the reaction coordinate may give an indication of the nature of the transition state. For a shorter reaction width, the transition state is more like the bound state than the unbound state. Conversely, a longer reaction width indicates a transition state that is more like the unbound products.

In biological systems, the macromolecules themselves are covalently linked polymers that form complex folded conformations and interact with one another and the water molecules that surround them through a network of non-covalent interactions. These interactions are complex and dynamic, associating and dissociating at all times, with the molecular motions significantly damped by the viscous environment. Mechanical stress alters the magnitude and the rate of interactions, with a tendency toward the disassembly of the structure when an external stretching force is applied. This bias towards bond rupture by the application of stretching force was first

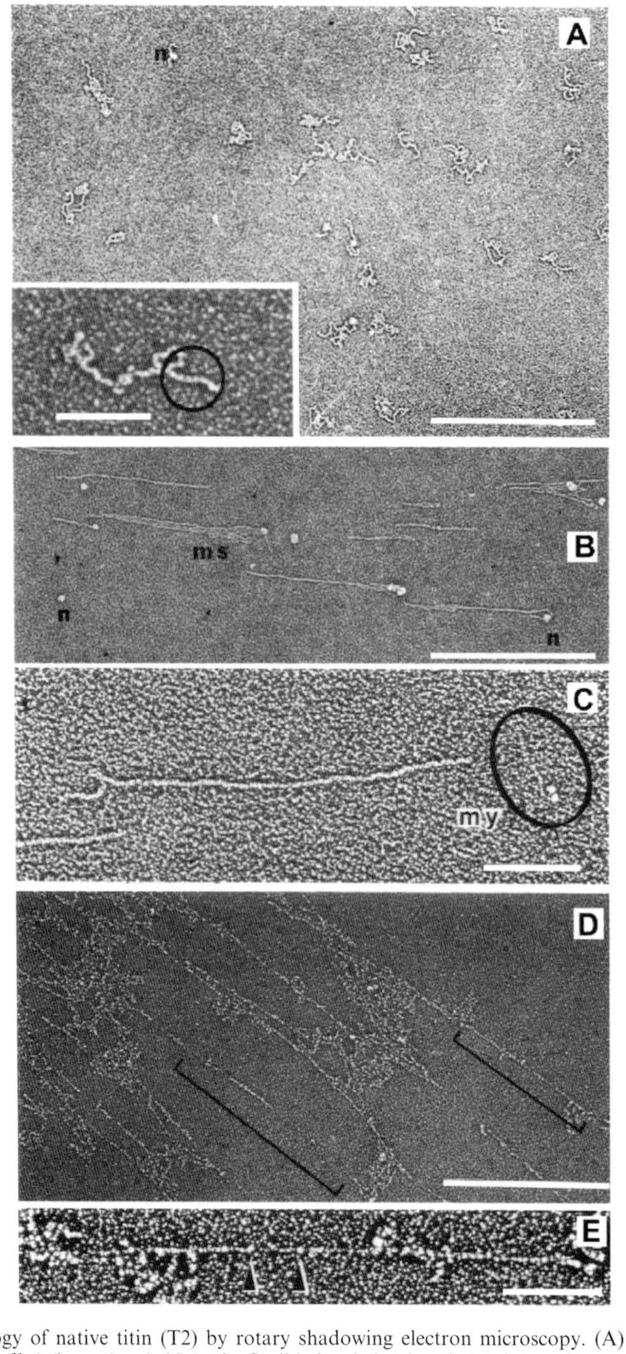

Fig. 3. Molecular morphology of native titin (T2) by rotary shadowing electron microscopy. (A) Highly convoluted titin strands (0.5 M ammonium formate buffer) (bar = 1 μm). Note the flexible beaded string that twists and turns sharply (Inset. bar = 0.2 μm). (B) Linearized titin strands on tilt-sprayed mica (0.5 M KCl buffer). Hydrodynamic shear force linearized the flexible titin strands. Note the multi-stranded regions (ms) and the nodules (n). (bar = 1 μm). (C) High-magnification view selected from (B). Note the relative dimension of titin and myosin (my). The beaded string morphology is less conspicuous in this image and the titin strands appear uniform in width (bar = 0.2 μm). (D) Reticular network and stretching of titin strands (in 0.5 M KCl). Note the thinning of many strands (brackets) (bar = 1 μm). (E) Selected high-magnification view of (D). Note the different thickness of these strands and the irregularly spaced beads (arrowheads), reflecting the unfolding of titin domains (bar = 0.2 μm). Figure after Fig. 2 in Wang et al. (1984) with permission, and more details may be found therein.

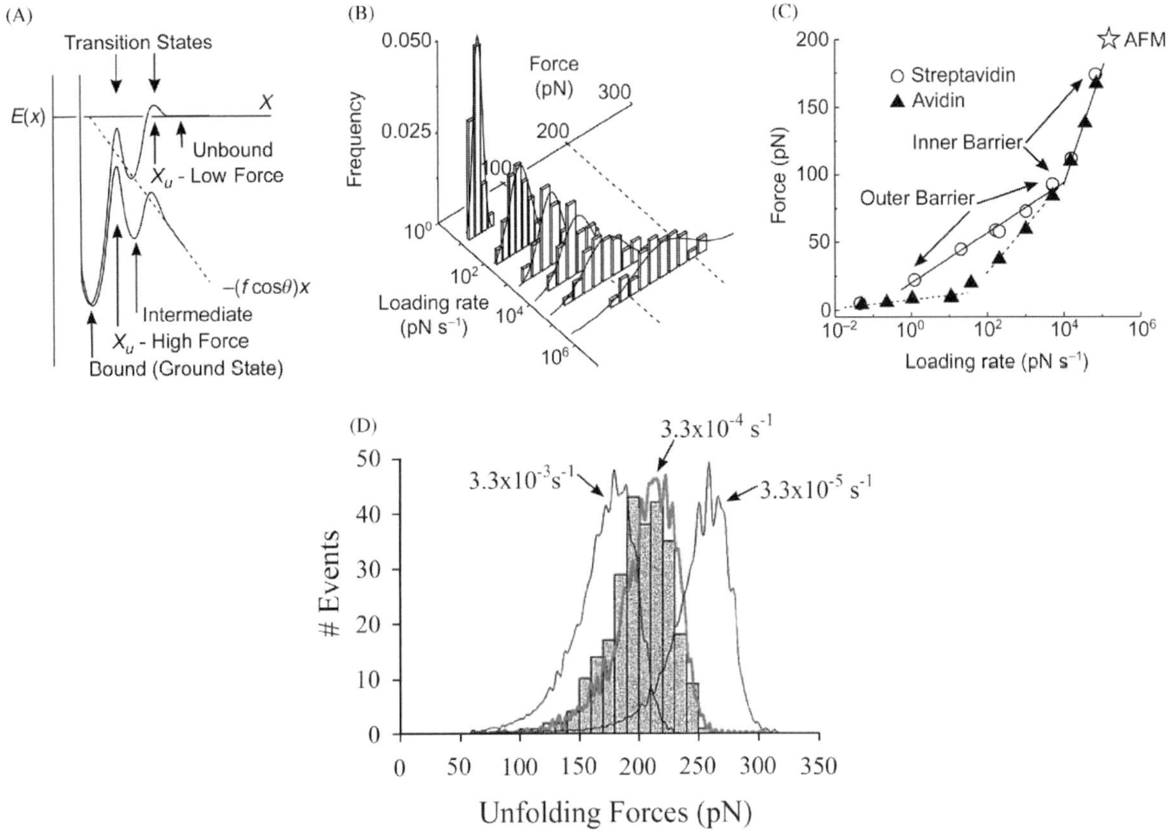

Fig. 4. Effects of applied force in weak interactions. (A) Idealised interaction potential with transition states. A force applied at an angle θ to the reaction coordinate x, adds a mechanical potential $-(f\cos\theta)x$ which tilts the reaction landscape and lowers the barrier. The energy contours near the transition states are highly curved and their shape and position are not affected significantly by the applied force. The transition states at low and high forces are indicated by x_u. (B) Force histograms from measurements on single biotin–streptavidin interactions at different loading rates. Solid curves are Gaussian fits used to determine the center of the force peak. This plot demonstrates the shift in the peak position and the increasing peak width with increasing loading rate. (C) Dynamic strength spectra for biotin–avidin and biotin–streptavidin interactions. Three distinct regions are seen for biotin–avidin, whereas only two are seen for biotin–streptavidin. At lower loading rates, the interactions are dominated by the outer barriers. See Merkel et al. (1999) for more details. Panels (A–C) were modified from Merkel et al. (1999) with permission from *Nature*. (D) Histogram of unfolding forces for an octamer of I27 Ig domains in a polymer stretched with an AFM, adapted from (Carrion-Vazquez et al., 1999b) with permission from *Proc. Natl. Acad. Sci. USA*. The solid curves are Monte-Carlo simulations of the mean unfolding force of the I27 polymer with different unfolding lifetimes, an unfolding distance of 0.25 nm and a pulling rate of 0.6 nm/ms. This plot demonstrates that the bond-strength (the peak in the histogram) is sensitive to both the loading rate and the bond lifetime.

discussed by Bell in his model based on the kinetic theory of the strength of solids (Bell, 1978). Evans and Ritchie have extended the work of Bell to the dynamic application of force that occurs frequently in biological systems (Evans, 1998; Evans, 1999). They applied Kramer's theory of chemical kinetics in a Brownian field (Kramers, 1940) to the bond strength of biological interactions in water. The viscous damping of the molecular motion leads to bond lifetimes at least 1000-fold longer than that predicted from the bond excitation frequencies used in Bell's theory. The bond lifetime in Bell's theory is written as,

$$\tau = \tau_0 e^{(E_0 - \gamma f)/k_B T)} \tag{1}$$

where E_0 is the bond energy, f is the applied force per bond, τ_0 is the reciprocal of the bond vibrational frequency (10^{-13} s), and γ is a characteristic of the solid. Bell postulated that this formula could be applied to receptor-ligand pairs if $\tau(f = 0)$ is equivalent to the inverse of the off-rate of the ligand and that γ must be approximately r_0 (the distance between the receptor and ligand) so that $\tau = \tau_0$ when $f = E_0/r_0$. During the application of tension upon the interface, the force is assumed to act along the reaction coordinate, x, from 0 at the bound or native state to reach x_u, at the transition state at the top of the energy barrier (Fig. 4A). This model implies that that the bond lifetime should decrease exponentially with increasing force:

$$\tau(f) = \tau_0 e^{(-f x_u)/k_B T}, \tag{2}$$

where $k_B T$ is the thermal energy scale. From the above equation, it is seen readily that the bond lifetime varies as a function of the applied force, f, and the width of the reaction coordinate, x_u. As the width of the reaction coordinate increases, the bond lifetime decreases more rapidly for a given applied force. In general, receptor-ligand bond lifetimes can range from milliseconds to hours.

The decrease in bond lifetime with applied force results directly from the effect the force has upon the reaction coordinate. For complex molecular systems, the attachment point for applying force to the molecules may not allow for force directly along the reaction coordinate. When a mechanical force is applied at an angle θ to the reaction coordinate, the reaction surface potential is distorted and linearly tilted by $-fx\cos\theta$, with $f\cos\theta$ being the force component along the relevant reaction coordinate (Fig. 4A). The effect of this tilting of the potential is that the energy barriers are lowered, the probability of crossing increases, and therefore the reaction rate increases. The increased probability leads to an increase in the rate of bond rupture. For a three state unbinding reaction illustrated in Fig. 4A, the outer barrier determines the crossing probability under smaller force, and the relevant width of the reaction coordinate is large. However, under larger forces the inner barrier becomes higher than the outer one and determines the crossing probability. So the width of the reaction coordinate is smaller at higher force.

In the analysis of Evans and Ritchie (1997), the activation energy for bond rupture, rather than the bond free energy, is the main determinant of the bond strength under a mechanical load. Also in biological applications, the applied force usually varies with time. Evans and Ritchie's analysis showed that the weak non-covalent interactions found in protein-ligand interactions, as well as those involved in protein folding would be sensitive to the loading rate of the applied force. The probability of bond rupture under a given load is the product of the off-rate under the load and the likelihood of bond survival under the load. As the load increases with time, the off-rate increases in time and the likelihood of bond survival decreases in time, which leads to a peak in the statistical distribution of rupture events (Fig. 4B). The peak in the distribution is regarded as the bond strength and it increases with higher loading rates. The width of the distribution also increases with higher loading rates (Fig. 4B). The bond strength varies linearly with the log of the loading rate, while the slope depends upon both the bond lifetime and the width of the reaction coordinate (Fig. 4C). Because the width is in the exponent of the bond lifetime function, small changes in the width can have a large effect. For a given bond lifetime, an increase in the width of the reaction coordinate will decrease the slope of the rupture force vs. log(loading rate) plot (Rief et al., 1998). Therefore a larger reaction coordinate width reduces the sensitivity of the interaction to the loading rate. This difference in sensitivity to applied force of homologous proteins with similar folding pattern and stability is therefore a manifestation of different dynamic force unfolding reaction widths. For a protein to withstand unfolding by the rapid application of force, it would require a short reaction width. If a protein is to unfold over a large range of loading rates, it has to have a wide reaction width so that it will unfold at high rates and low forces. The sensitivity of bond strength to the folded lifetime has been demonstrated by Monte Carlo simulations of the distribution of unfolding forces for an Ig domain (Fig. 4D). While the shape of the distribution does not change significantly from varying the folded lifetime over two orders of magnitude, the simulated bond strength changes significantly. The experiments behind Fig. 4D and their interpretation will be discussed more fully in Section 3.2.

For many biological processes, the energy landscapes can be expected to be more complex due to the large number of interactions in these biological interfaces. These barriers can have different positions along the reaction coordinate, i.e. widths and heights (Fig. 4A). At low loading rates, the unbinding kinetics is dominated by a barrier farthest from the potential well. As the loading rate is increased there is a change of the potential surface and crossover to the kinetics being dominated by the inner barriers. The force vs. loading rate plot for biotin-avidin best demonstrates this behavior (Fig. 4C), where three different slopes indicate three different reaction widths in the unbinding reaction pathway (Merkel et al., 1999). The two inner barriers (higher force loading) at 0.12 nm and 0.3 nm compare well with the position of prominent transition states in molecular dynamics simulation of the unbinding of biotin from avidin at ca. 0.1 nm and 0.4 nm (Izrailev et al., 1997). The outer barrier as determined by dynamic force spectroscopy is at 3 nm, whereas the last transition in the molecular dynamics calculations is seen at 1.4 nm. This two-fold difference in the outer barrier may result from the interaction of the biotin with peripheral, flexible loops at the mouth of the binding pocket (Chu et al., 1998). The molecular dynamics calculations are made at much shorter time scales and much higher loading rate than in real physical measurements. The loop dynamics at the picosecond time scale used in the molecular dynamics calculations are very different from those at the millisecond time scale in force spectroscopy. These results indicate that dynamic force spectroscopy of biological interactions can be a powerful tool for revealing subtleties of the interaction potential which cannot be probed by other experimental means. While a strict physical interpretation of the reaction surface may not be accessible without molecular modeling, the presence of hidden barriers can provide insights into the binding mechanisms that may be biologically relevant. Strunz et al. (2000) have recently modeled the energy landscape during force-induced dissociation of a macromolecular interaction. They found that even a single intermediate state leads to various, complex dependencies of the bond lifetime on the applied force.

2.2. Elasticity models for titin

Several evolving models of molecular elasticity have been applied to describe titin extension and sarcomere elasticity (Fig. 5). The original segmental extension model by Wang and coworkers proposed that, titin in situ extends in segments as a dual/multiple stage spring. The extension of the I-band segment titin led to the exponential rising of tension (ERT) model of sarcomere passive force (Granzier and Wang, 1993b; Wang et al., 1991) (Table 1). ERT of muscles that expressed different titin isoforms can be scaled by the I-band titin strain as, $\varepsilon \equiv (\mathrm{TL}_E - \mathrm{TL}_{EO})/\mathrm{TL}_{EO}$ where the subscripts denote the titin length (TL) at a given extension, (EO) denotes the onset of exponential extension, (O) denotes the rest length, E_0 and α are two adjustable parameters.

Despite the success of this theory in predicting the tension–length curves of various muscles, sarcomere mechanics provides only a limited view of the full range molecular elasticity of titin. Experimentally, it is difficult to apply a large force at will to titin in the sarcomere, because the applied force cannot exceed the yield point value that starts to dislodge titin from its attachment to the tip of thick filaments. This yield point force is estimated to be ca. 45 pN per titin for rabbit psoas muscle (Wang et al., 1991). It is equally difficult to impose small forces of less than a few pN

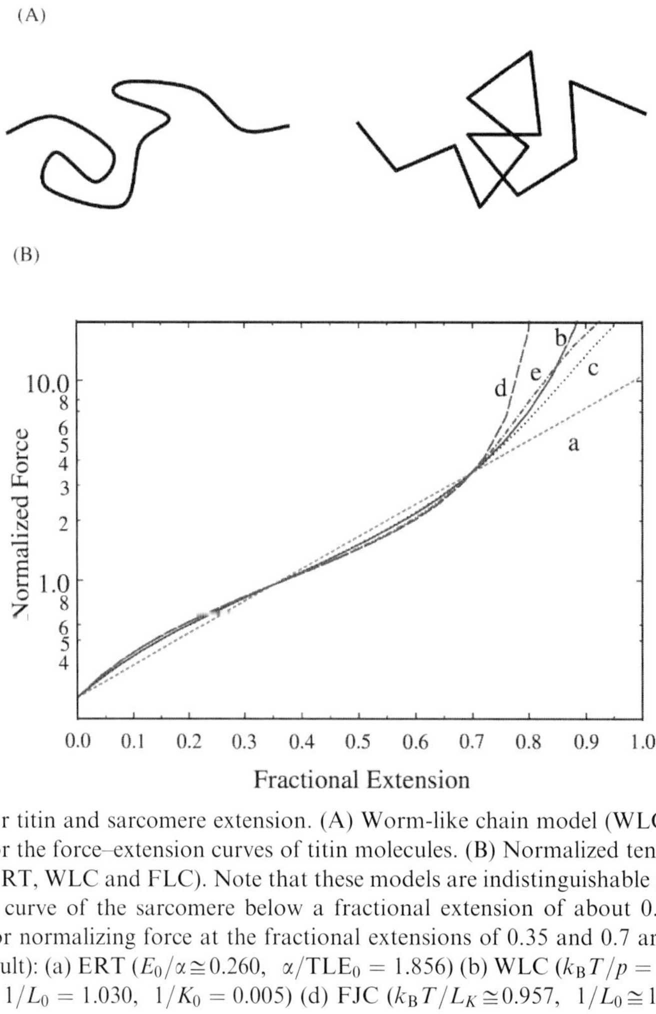

Fig. 5. Elasticity models for titin and sarcomere extension. (A) Worm-like chain model (WLC, left) and freely-jointed chain (FJC, right) model for the force–extension curves of titin molecules. (B) Normalized tension plots based on three models of titin extension (ERT, WLC and FLC). Note that these models are indistinguishable and equally applicable to explain the tension–length curve of the sarcomere below a fractional extension of about 0.8. Model equations and parameters (see Table 1) for normalizing force at the fractional extensions of 0.35 and 0.7 are as follows (Jin, Forbes and Wang, unpublished result): (a) ERT ($E_0/\alpha \cong 0.260$, $\alpha/\mathrm{TLE}_0 = 1.856$) (b) WLC ($k_B T/p = 1$, $1/L_0 = 1$, $1/K_0 = 0$) (c) WLC ($k_B T/p = 0.972$, $1/L_0 = 1.030$, $1/K_0 = 0.005$) (d) FJC ($k_B T/L_K \cong 0.957$, $1/L_0 \cong 1.217$, $1/K_0 = 0$) (e) FJC ($k_B T/L_K \cong 0.918$, $1/L_0 \cong 1.265$, $1/K_0 = 0.01$).

Table 1
Titin elasticity models and approximate behaviors for low force (a), medium force stretching (b), and high force (c)[a–c]

	Equations for titin elasticity and their behaviors at low (a), medium (b), and high (c) force stretching		
(1) Exponential resting tension (ERT)	$f = E_0\alpha^{-1}(e^{\alpha\varepsilon} - 1)$		
	(a) $f \approx E_0\varepsilon$	(b) $f \approx \frac{E_0}{\alpha}(e^{\alpha/2} - 1)$	(c) $f \approx E_0\alpha^{-1}e^{\alpha\varepsilon}$
(2) Enhanced worm-like-chain (WLC)	$f = \frac{k_B T}{p}\left(\frac{1}{4(1 - \ell/L_0 + f/K_0)^2} - \frac{1}{4} + \ell/L_0 - f/K_0\right)$		
	(a) $f \approx 3\frac{k_B T}{2p}\frac{\ell}{L_0}$	(b) $f \approx 2.5\frac{k_B T}{2p}$	(c) $f \approx K_0(\ell/L_0 - 1)$
(3) Enhanced freely-joint-chain (FJC)	$\ell/L_0 = \left(\coth\left(\frac{L_K f}{k_B T}\right) - \frac{k_B T}{L_K f}\right)(1 + f/K_0)$		
	(a) $f \approx 3\frac{k_B T}{L_K}\frac{\ell}{L_0}$	(b) $f \approx 1.8\frac{k_B T}{L_K}$	(c) $f \approx K_0(\ell/L_0 - 1)$

[a] Low force, ε, $\ell/L_0 \ll 1$.
[b] Medium force, ε, $\ell/L_0 \approx 1/2$.
[c] High ε, $\ell/L_0 > 1$. See also Fig. 5 and the text.

per titin in the sarcomere because of competing factors such as the weak crossbridges between actin and myosin in non-activating conditions (Granzier and Wang, 1993b). For these technical reasons, single molecule measurements with forces as high as several hundred pN are more versatile and greatly expand the force–extension range of titin.

Two polymer-based elasticity models (Fig. 5A), the worm-like chain (WLC) and the freely-jointed chain (FJC), have been applied with varying degrees of success to explain the force – extension behaviors of single titin molecules. The WLC model describes a semi-flexible rod undergoing entropy-driven configurations with two parameters, the persistence length, L_p, and the total chain contour length, L_o (Marko and Siggia, 1995; Ortiz and Hadziioannou, 1999) (Table 1). Recent enhancements developed in the context of DNA elasticity also include enthalpic stretch contributions via an elastic force constant, K_o (Bouchiat et al., 1999; Wang et al., 1997b). The current fitting equations, due to simplifying approximations, may deviate from exact solution by 10% in some ranges of extension. Adopting this model allows for the determination of the elastic persistence length, L_p. A larger L_p indicates that the chain is stiffer with fewer bending and orientational fluctuations. The elastic modulus, K_o, determines the degree to which the polymer chain can be extended under the applied force and represents an enthalpic contribution to the elasticity. A smaller K_0 represents a more extensible polymer chain.

The freely-jointed chain (FJC) model is based on polymer chains with a linear string of rigid rods rotating freely at the joints and with no interactions between the rods. The elastic parameters are the Kuhn segment length L_K and the total polymer length L_o. For similar filaments, $L_K = 2L_p$ gives a good first approximation (Table 1). This model has also been enhanced to include elastic stretching term, K_o (Muniz et al., 1999; Ortiz and Hadziioannou, 1999).

It is useful to note that, in this context, stiffness refers to the resistance to bending, not to extension, as it is generally used in muscle mechanics. The elasticity parameters, L_K and L_p, denote the bending configurational freedom for the filamentous chain (Fig. 5A). For an extended chain with larger end-to-end distance, the entropy of the chain from backbone bending is smaller and, hence, its free energy is higher than that of the collapsed chain. It therefore takes force to stretch the chain as work is needed to make up for the free energy increment. A smaller L_K in the FJC or L_p in the WLC allows for more bending fluctuations of an under-stretched filament, which results in a decrease in the free energy of the polymer chain and it behaves as a stiffer entropic spring. In the original WLC and FJC models, the total length of the chain backbone does not change during stretching; the inclusion of the K_o modulus gives an approximation for small backbone extension when the entropic force is transmitted along the chain. One should also note that for both of these phenomenological models, chain segments are assumed not to interact with each other, i.e. no tendency for self-organization or self-avoidance. All models predict titin as a linear spring at small extensions and as a nonlinear spring with rapidly rising tension at higher extensions. Indeed, quantitatively there is very little distinction between these models with the limited force/extension range encountered in the sarcomere. This point is illustrated in Fig. 5B, where five sets of elastic parameters from three models all led to force–extension curves that are almost completely overlapping at fractional extensions below 0.8. This is the degree of extension of titin in the normal range of sarcomere extension. Although single molecule measurements now begin to favor the WLC model, as it provides structural insights into the molecular elasticity of titin (see Section 5), distinctions among the three models are largely superficial in the context of sarcomere elasticity.

2.3. AFM and LOT techniques

Direct measurement of forces within and between molecules has remained an intractable problem for several reasons: attachment of molecules, measurement of forces in the piconewton ($pN = 10^{-12}$ N) range, and reproducible mechanical motion on the sub-nanometer scale. The last two problems were solved by the invention of the AFM and LOT. A comparison of the major features of these techniques is presented in Table 2 and they will be described in more detail below. The solution to the attachment problems follows from the materials used in the measuring techniques: glass, silicon, silicon nitride, and polymer microspheres. Many chemical methods have been successfully developed for immobilizing proteins and other types of polymeric materials on these types of surfaces.

2.3.1. Atomic force microscopy

Considering the ability of AFM to sense directly the topology of atoms, molecules and cells, its design and operation are surprisingly simple. An AFM has four basic components: a microcantilever for sensing forces; a means of detecting cantilever deflection; a piezoelectric system for moving the tip in X, Y, and Z coordinates at atomic resolution; and a computer control system (Fig. 6A). In a typical commercial system, the cantilever is microfabricated from silicon nitride or silicon using semiconductor fabrication methods. The cantilever is mounted on a tip holder that is mounted on a tubular piezoelectric element for moving the cantilever over the surface. The cantilever motion is detected by an optical lever method, where a laser is bounced off

Table 2
Comparison of atomic force microscopy (AFM) and laser optical trap (LOT)

Method	Atomic force microscopy	Laser optical trap
Force generation	Bending of cantilever obeys Hooke's law	Intensity gradient from photons from laser light source
Dynamic range	20 pN–> 20 nN with different cantilevers	0.1–400 pN[a]
Speed of extension	10 nm/s–100 μm/s	10 nm/s–100 μm/s
Force constant	Typically 0.05 N/m	Typically 0.0002 N/m
Loading rate	0.5–5000 nN/s	0.002–2.0 nN/s
Calibration	• Thermal fluctuation[b] • Added mass[d] • Cantilever dimensions and oscillation[f]	• Hydrodynamic drag[c] • Power spectrum and equipartition theorem[e] • Secondary reference, i.e. extension of DNA[g]
Strengths	• Atomic resolution images • Large dynamic force range and small contact area • Compact instrument • Reliable calibration • Fast application of force	• Gentle, small forces • Increase force by turning up laser power • Simultaneous observation of sample motion and force response • Multiple traps available
Limitations	• Cantilever must be switched to change force range	• Large contact area of bead • Potential heating of sample
Instrumentation	• Many companies offering imaging and force measuring microscopes • Z-axis force measuring instruments	• Mainly custom built systems using research grade optical microscope, IR laser, and laser optics

[a] Kellermayer et al. (1998, 1997) and Tskhovrebova et al. (1997).
[b] Hutter and Bechhoefer (1993).
[c] Bennink et al. (1999).
[d] Cleveland et al. (1993).
[e] Wang et al. (1997b).
[f] Sader et al. (1999, 1995).
[g] Wang et al. (1997b).

of the cantilever into a split photodiode. The photodiode allows bending motion of the cantilever to be detected to better than 1 Å. The cantilever behaves as a Hookian spring with a force constant, k, of the order of 0.01–1 N/m. The force on the cantilever may be determined directly from the deflection of the cantilever and the force constant of the cantilever using Hookes' law, $F = kx$. To image a sample, the tip is brought into contact with the surface at a loading of several hundreds of pN to several nN (1 nN = 1000 pN = 10^{-9} N). The cantilever is then rastered over the sample while maintaining a constant force applied to the sample. The motion of the tip holder normal to the surface that is required to maintain a constant force yields a topographic image of the surface. The AFM is capable of measuring forces in a similar manner where the tip holder is moved up and down and the deflection of the cantilever is measured.

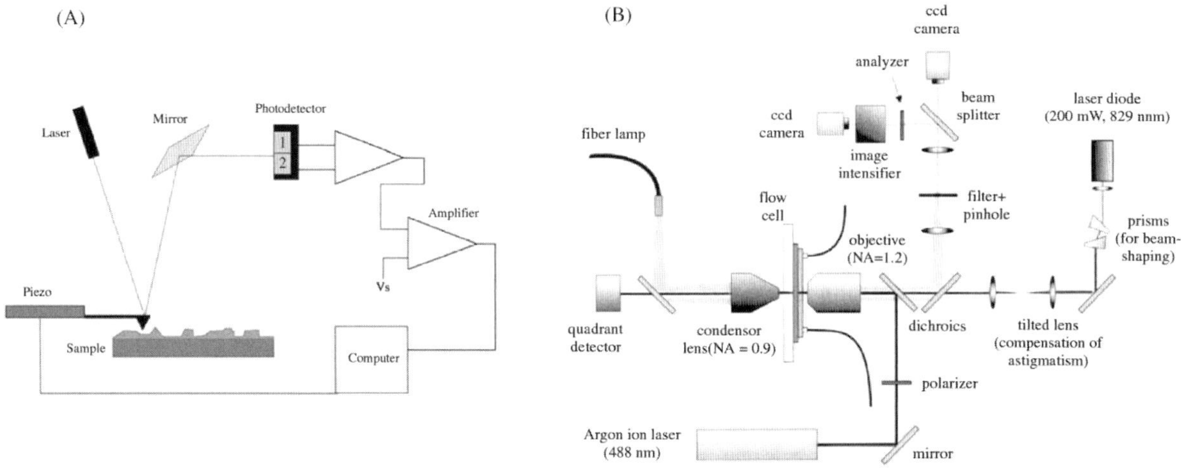

Fig. 6. Instrument design: (A) atomic force microscope; (B) laser optical trap, Bennink et al. (1999) with permission from Wiley-Liss, Inc., a Subsidiary of John Wiley & Sons, Inc. See text for the details.

Since its inception in 1986 (Binnig et al., 1986), the force measuring capabilities of the AFM have been touted as one of the most significant advantages of this new technique. It was estimated early on, that forces as small as 0.01 pN could potentially be measured with the AFM. Extensive effort for over a decade has shown that the practical limitation of this technique is on the order of 10 pN (Viani et al., 1999a). Although other techniques can be used to measure smaller forces, the AFM has the advantage of a large dynamic range in force measurement. With a single cantilever, forces ranging from tens of pN to over a nanonewton can be measured easily. Even larger forces can be measured with stiffer cantilevers. Much of the early work on force measurement with the AFM dealt with interactions between the tip and surface, and the visco-elastic properties of different material. Measurements of the interaction forces between single molecules were first reported in 1994 with the studies of the interaction between avidin and biotin (Florin et al., 1994; Lee et al., 1994), and DNA oligomers (Lee et al., 1994). Since these first unbinding force measurements, many other systems have been studied: antibodies, insulin dimers, metal ligands, among others. The first extension of a single polymer chain with an AFM was reported by Hinterdorfer et al. (1996) where they tethered an antibody and its antigen via a polyethylene glycol (PEG) linker. While the elastic property of PEG was secondary to their study of the antibody–antigen interactions, it demonstrated the potential for stretching polymers with the AFM. In 1997, Rief and coworkers reported the first study of the elastic properties of a polymer, where they stretched dextran with an AFM (Rief et al., 1997b).

In the interaction force experiments, one of the binding partners was attached to the AFM tip and the other to the surface. When the tip is brought into contact with the surface, the receptor and ligand bind. Upon retraction of the tip from the surface, the adhesion of the receptor and ligand causes the tip to adhere to the surface, which is detected as a negative deflection of the cantilever. In the polymer extension experiments, the entropic elasticity of the polymer tethering the tip and the substrate causes the cantilever to bend towards the substrate. When the force applied by the bending of the cantilever, exceeds the interaction or the tethering force, the cantilever snaps back to its equilibrium position.

The distance the cantilever bent from its equilibrium position is proportional to the force. The proportionality constant is the Hooke's constant for the cantilever. This parameter can be estimated from the dimension of the cantilever using the Young's modulus of the material. However, this is more difficult than it would seem in practice, because the silicon nitride cantilevers are formed by chemical vapor deposition, which results in a non-stoichiometric material for which there is no single Young's modulus. Several methods have been developed to work around this problem (Sader et al., 1999, 1995); however, experimental determination of the force constant is preferable. Two reliable calibration methods are based on the added mass approach (Cleveland et al., 1993) and the thermal fluctuation approach (Hutter and Bechhoefer, 1993). The latter method has been shown to yield results similar to the other methods (i.e. with less than about 10% discrepancy), as well as working in liquid and requiring no additional manipulation of the cantilever (Butt and Jaschke, 1995; Florin et al., 1995).

2.3.2. Laser optical trap

The laser optical trap (laser tweezers) technique is now recognized as a powerful, non-invasive force measurement tool that complements the AFM in its force range and its utility in aqueous environment (Svoboda and Block, 1994; Visscher and Block, 1998). The optical trap is able to apply force from near 0.1 pN to hundreds of pN to a particle by the radiation pressure from the photons traveling through the particle (Ashkin et al., 1986; Ashkin, 1992). The theory of the optical forces in laser tweezers has been well explained in a number of recent reviews (Ashkin, 1992; Block, 1992; Svoboda and Block, 1994; and other chapters in this two volume series). Briefly, an optical trap utilizes the gradient force, which draws objects with an index of refraction higher than the surrounding medium towards the high intensity region, and objects of lower index of refraction towards low intensity region (Ashkin, 1992). Thus, counting only the gradient force, the focal point of a light beam through microscope objective is a three dimensional trap for objects with their index of refraction higher than the surrounding medium. Because the photons also exert scattering force on the object in the forward direction of the beam, a true 3D trap formed by a single laser beam must have tight focus in the axial direction so that the gradient force overcomes the scattering force to prevent the object from being ejected forward (Ashkin et al., 1986; Svoboda and Block, 1994). To achieve such a tight focus, the laser beam is usually focused through a high numerical aperture microscope objective and the particle becomes trapped in the beam near the focal point. Fig. 6B shows a simplified schematic of a typical laser trap experiment. The optical microscope, being an integral part of LOT, allows the visualization of the particles being manipulated by the force trap. Usually latex or polystyrene spheres on the order of 1–3 μm in size are used for applying force to molecules. The large size of the particles needed for effective trapping can cause problems with applying forces to small molecules, but this is less of an issue for micron long polymers such as titin and DNA. The force applied to the particle is proportional to the power incident on the particle (Ashkin, 1992). While high power lasers are readily available, the practical upper limit is set by the amount of heat deposited into the sample and radiation damage. A temperature rise of 1.7°C/100 mW of laser power at the sample was estimated in one experimental setup using a 1064 nm laser (Kuo, 1998). Photon induced radiation damage can also be a major problem for many sensitive biological processes. Recent efforts have sought to reduce such problem by using selected photon wavelength including the infrared wavelength of 1064 nm produced by the common and economical continuous-wave (CW) diode-pumped Nd : YAG laser (Neuman et al., 1999). Force is measured by adjusting the trap to have a low stiffness, typically on the order of $k = 0.0002\,\mathrm{N/m}$. When the position of the trap is moved through a distance d a force proportional to kd will be exerted on the particle and the polymer attached to it.

Accurate force measurements require that the laser optical trap be carefully calibrated, preferably at each session, which can be a challenging task (Dai and Sheetz, 1998; Svoboda and Block, 1994). The force induced onto a dielectric sphere by the change of the momentum of the incident photons depends critically on photon field gradient, the index of refraction of the sphere and surrounding medium and various geometric factors; so theoretical calculations are often not precise (Ashkin, 1992). The refraction of the incident laser beam, and the concomitant shift in the position of the beam leaving the trap have been used for calibrating the trapping force of a laser trap (Smith et al., 1996). However, this method is difficult to perform in practice. In most calibration setups, the amount of force exerted on the sphere is calculated through the application of Stoke's Law, within its applicable range (Dai and Sheetz, 1998; Svoboda and Block, 1994). Stoke's Law relates the flow velocity of a fluid around a particle to the force exerted on the particle by the equation,

$$f = 6\pi \cdot r\eta \cdot v, \tag{3}$$

where r is the radius of the sphere, η is the viscosity, and v is the flow velocity. An important caveat of this method is that the fluid must flow smoothly around an object with no turbulence, i.e. a laminar flow. For particles on the order of several microns in size, the laminar flow requirement limits the force on the particle to about 10 nN, which is adequate (Gittes and Schmidt, 1998; Svoboda and Block, 1994). The LOT is calibrated by trapping the particle at a given laser power and then increasing the flow around the particle either by moving the microscope stage back and forth or pumping fluid through the experimental cell at known rates. The particle under study must also be kept several microns (i.e. more than 5–10 times the bead radius) away from the surface of the flow chamber walls, so that the surface boundary effect is minimal. Both the laser field and the laminar flow will be perturbed if measurements are made too close to the glass surface (Dai and Sheetz, 1998; Gittes and Schmidt, 1998). The flow velocity at which the particle is displaced from the trap is then used in Stoke's equation to calculate the amount of force on the particle. For higher force applications, the laser power (I) is raised to increase the stiffness of the trap. It is assumed that the slope of the F vs. I plot at low power is the same as that at high power and the trapping force is simply extrapolated. Because LOT trap is three-dimensionally non-uniform, it is crucial to keep the path of the bead in actual measurements closely similar to that during calibration, or significant errors may arise (Stout and Webb, 1998).

2.3.3. Application ranges of AFM and LOT

Both the AFM tip and trapped beads are subject to thermal motions due to collisions with the solvent molecules in the environment which determines the measurement uncertainty. Both the AFM tip and the LOT bead are under a linear restoring force of the cantilever with respect to the Hookian spring constant k. As such, one has the thermal fluctuation limiting the position of the AFM tip via the equal partition of the thermal energy as (Gittes and Schmidt, 1998)

$$E_T \cong k_B T \cong k \langle \Delta z \rangle^2 \tag{4}$$

where k_B is the Boltzman constant, T is the absolute temperature and $\langle \Delta z^2 \rangle$ is the mean square amplitude of the thermal fluctuation of the tip position in the direction of cantilever bending, z. Thus, the measurement uncertainty in the position, $\sigma_z \equiv \sqrt{\langle \Delta z^2 \rangle} \cong \sqrt{k^{-1} k_B T}$, decreases with a stiffer cantilever or trap, whereas the uncertainty in the force, $\sigma f \equiv \sqrt{\langle \Delta f^2 \rangle} \cong k \sigma z \cong \sqrt{k k_B T}$, increases with a stiffer cantilever or trap. This results in the thermal component to the signal-to-noise ratio of the force measurements being proportional to the square root of the cantilever or trap spring constant, $f/\sigma_f \propto \sqrt{k}$. The cantilever in typical AFM setups has a force constant, k, on the order of 0.01–1 N/m, this gives $\sigma_f = 10 - 100\,\text{pN}$ and $\sigma_z = 1 - 0.1\,\text{nm}$. The corresponding trap force constant is about a 1000-fold smaller and depends linearly on the applicable laser power. Thus, LOT has about 30 times lower noise in force, but simultaneously 30 times higher noise in spatial localization. The maximum forces, on the other hand, is determined by the force constant times the maximum deflection of the cantilever in AFM or the gradient trap dimension in LOT, both of the order of 500 nm. Therefore, AFM can apply force up to hundreds of nN, whereas LOT can apply forces up to a few hundred pN. LOT is more suited for low force applications in pN range where spatial resolution needs not to be higher than a few nm; AFM is more suited for higher force applications in the tens to hundreds of pN range and where a spatial resolution of better than 1 nm is needed.

Much of the recent work on the unfolding of titin domains has been done on small proteins comprised of several domains that are 4 nm in length and with an extension force of hundreds of pN. The application range consideration above shows why the AFM is a more suitable technique than the LOT for such single titin studies. More systematic considerations of the thermal fluctuations in both LOT and AFM suggest that appropriate noise filtering can improve the precision of the force measurements (Gittes and Schmidt, 1998; Liu et al., 1999; Svoboda and Block, 1994). Nevertheless, recent observations on the unfolding properties of titin domains have been made mostly with the AFM, and a small number of observations on the low-force behavior of longer titin segments with LOT techniques. As such, this review will concentrate more on the AFM results than on the LOT results. For applications in the pN force range, such as those involving molecular motors (e.g. Block, 1995; Wang et al., 1998), LOT is more suitable, and this application is more fully reviewed in the articles cited above and (Visscher and Block, 1998).

2.4. *Titin sample preparation*

Successful single molecule force measurements require that the polymer or protein under study is pure, homogeneous and monodisperse. While it is often assumed that a pure polymer is homogeneous, titin from a single tissue source exists in several different isoforms with similar molecular weights (Fig. 3A) (Freiburg et al., 2000). In cardiac titin, the tandem Ig, PEVK and N2B regions are differentially expressed and the elastic response of titin resides largely in these regions. Heterogeneity in the elastic response of a population of polymers will be difficult to model. If multiple polymers are stretched, the force that each filament experiences will be divided by the number of filaments being stretched, if all are being stretched at the same rate. The problem is that not all of the polymers in an aggregate may even be under tension, and by interacting with those filaments under tension the overall elasticity will differ from the isolated polymer. Additionally, efforts must be taken to ensure that the measurements are not obscured by extraneous interactions. These extraneous interactions can arise from contamination of the surfaces used for immobilizing the polymer or protein, as well as non-specific interactions between the protein and substrate. A synopsis of the methods used in the titin extension experiments is given in Table 3.

2.4.1. Protein preparation

Verifying the purity of proteins and polymers is rarely as accurate or straightforward as with small molecules. The extremely large size of titin exacerbates this problem and special methods must be used for characterizing the size distribution and purity of the protein preparations (Wang, 1982). The AFM measurements on native titin (Rief et al., 1997a) used bovine cardiac titin isolated as described by Pan et al. (1994). The laser optical trap measurements on native titin used rabbit back muscle (*Longissimus dorsi*) isolated as described by Soteriou et al. (1993). These procedures involve extraction of titin and myosin with a high salt buffer, followed by removal of the myosin by ion exchange or hydroxyapatite chromatography. Aggregated titin is then removed by gel filtration to yield monodisperse titin. The final preparation is mainly truncated titin, T2, (Fig. 1) where a portion of the amino terminus has been cleaved by proteolysis. In most cases T2 is also contaminated with titin binding proteins such as myomesin, M protein, attached at the carboxy-terminal or M-line end of titin. Wang et al. (1984) first demonstrated that hydroxyapatite chromatography can be used for removal of these M protein contaminations. Intact titin, T1, (Fig. 1) is apparently also present in minor amounts in T2 preparations, because a T1 specific monoclonal antibody, T12, was used successfuly to immobilize the amino-terminal (Z-line) end of titin in the LOT experiments, as will be described below. The titin T2 preparation following gel filtration in high salt remains monomeric. However, it tends to aggregate at physiological and lower salt concentrations. Interestingly, T2 also tends to aggregate at the air/water interface of droplets (Wang et al., 1984). This tendency of titin to aggregate renders some uncertainty in achieving monomeric attachment of titin for single molecule extension experiments.

Most of the AFM measurements on titin have been made using recombinant fragments (see Table 3 and Section 3). These fragments were engineered with two cysteines on the carboxy-terminus for immobilizing on a gold surface, and the amino terminus carries a His_6 tag for purifying on a nickel–NTA column (Carrion-Vazquez et al., 1999b; Rief et al., 1997a, 1998). The histidine tags allows for an easy one-column purification. These recombinant proteins have primarily been comprised of immunoglobulin domains. However, fibronectin domains have also been studied (Rief et al., 1998).

2.4.2. Attachment of proteins in AFM

To study the extension of a single polymer with the AFM, the polymer must be fixed to the cantilever tip and a flat substrate (Fig. 7A). Ideally, the attachment would be at a known position on the polymer, which could be the ends. As is often the case, the ideal situation is not the easiest to attain, and it is possible to study the extension of polymers without knowing the attachment position. One successful method is to attach polymers or proteins to a flat substrate and then "grab" them with the tip. The mechanism of "grabbing" can be either covalent attachment or physisorption—i.e. adsorption via physical forces. The desired result is to grab a single protein rather than a large number of filaments. The sharp apex of the AFM tip limits the number of molecules than can attach to the tip. The attachment of single polymers to the sharp AFM tip is

Table 3
Main observations of single molecule measurements of titin and titin motifs using AFM and LOT

	Atomic force microscopy	Laser optical trap
Titin	Bovine cardiac[a]	Rabbit *Longissimus dorsi*[b,c]
Recombinant proteins	• Human cardiac Ig domains[a,d–g] • Human skeletal Ig and Fn3 domains[h]	N/A
Assay buffer	• Phosphate buffered saline (pH 7.4) • 10 mM imidazole, 10 mM K-EGTA, 2.5 mM $MgCl_2$, 2 mM NaN_3, pH 7.0, variable $CaCl_2$ for pCa = 3–8, variable KCl for ionic strength = 35–560 mM[n]	• 25 mM imidazole, 25 mM KCl, 4 mM $MgCl_2$, 1 mM EGTA, 1 mM DTT, pH 7.4, 0.2% TWEEN, 0.2 mg/ml T4 lysozyme, 42 µM leupeptin, 10 µM E-64[b] • 19 mM histidine, 0.3 M KCl, 0.1 mM NaN_3, 1 mM EGTA 0.1 mM DTT, pH 7.5[c]
Protein attachment	• Native titin 10–100 µg/ml in PBS. Deposited onto freshly evaporated gold surface, incubated for 10 min[a] • Recombinant fragments are cysteine labeled for covalent attachment to gold surface. Deposited from a 10–100 µg/ml solution in PBS. Protein film was 20–50 nm thick[e] • Tip was pressed into the surface and the protein adsorbs onto the tip at random position	• Native titin Z-line end attached to a 3 µm polystyrene sphere via the T12 antibody[i,j]. Specific attachment • M-line end attached to a glass bead via myomesin on a 2 µm glass bead[b,j,k]. Frequently bound nonspecifically to various positions on molecule • Attached to a glass surface via the AB5 antibody[c,l]
Observations	*Extension of native titin* • Variable length spacer region followed by a saw-tooth pattern. • Peaks were spaced 25–38 nm apart with forces between 150 and 300 pN. • Saw-tooth pattern results from unfolding of IG and Fn3 domains. • Refolding observed in repeated extension–retraction cycles. • Force required for unfolding was speed dependent[a]	• Dynamic force–extension[c] • Monotonic extension curve reflect worm-like chains.
	Extension of recombinant titin segments • Saw-tooth patterns observed with the number of teeth up to the number of domains in fragment • Different domain types have different mechanical stabilities and unfolded at different forces[h]	• Static force–extension[b,c,i] • Stretch-release hysteresis loops—increased in width with degree of stretch and repeated stretching at constant length

Table 3 (*continued*)

	Atomic force microscopy	Laser optical trap
	• Unfolding folding rate at zero force is $3.3 \times 10^{-4}\,s^{-1}$ • Length of polymer contracted by 23 Å upon folding • Unfolding energy barrier height for mechanical and chemical unfolding were similar • Refolding was slower for mechanically unfolded proteins due to entropy loss and increased folding barrier[m]	• Titin refolding force plateau at low-force of 2–3 pN[b]
	• Mutations in load bearing portion of protein were identified by forced unfolding[e] • Ig domains underwent a 6.6 Å transitional extension at 100 pN. Resulting from the rupture of hydrogen bonds between the A and B β-strands of the Ig domain • Extension disappears in a mutant where the H-bond cannot form[f] • Repeated unfolding and folding resulted in the misfolding of two domains together and 'skips' in the unfolding pattern[g]	• Step-wise relaxation of stress when length change was held constant indicated domain unfolding[c]

[a] Rief et al. (1997a).
[b] Kellermayer et al. (1997).
[c] Tskhovrebova et al. (1997).
[d] Carrion-Vazquez et al. (1999b).
[e] Carrion-Vazquez et al. (1999a).
[f] Marszalek et al. (1999).
[g] Oberhauser et al. (1999).
[h] Rief et al. (1998).
[i] Furst et al. (1988).
[j] Kellermayer et al. (1998).
[k] Obermann et al. (1995).
[l] Whiting et al. (1989).
[m] Carrion-Vazquez et al. (1999b).
[n] Jin et al. (2001).

one of the fundamental advantages of the AFM in measuring the force properties of single polymers.

One way of affixing the polymer to the tip is to chemically cross-link it. A covalent bond will resist a force much larger than is required to stretch a polymer or unfold a small globular protein domain (Grandbois et al., 1999). While most AFM cantilevers are made from silicon nitride, the substoichiometric nitrogen composition results in a partial silicon characteristic of the surface. The silicon on the surface can be oxidized to silica, which may then be readily functionalized with silanes. A large number of organosilanes with different functionalities are commercially available

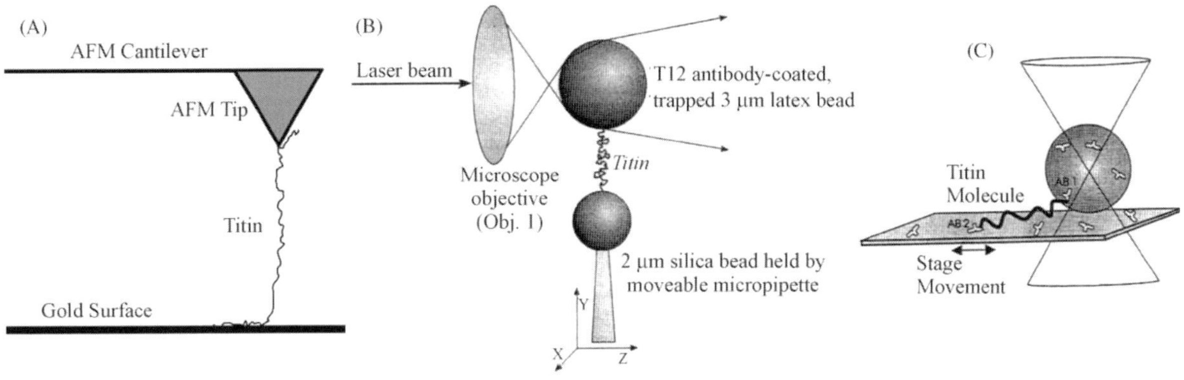

Fig. 7. Attachment methods for the extension of native titin. (A) AFM. Native titin is adsorbed onto gold surface and then picked up by pressing the tip onto the surface. A portion of titin becomes adsorbed onto the tip. This attachment gives rise to an extension curve as the tip is lifted from the surface. (B) LOT: attachment method used by Kellermayer and colleagues for stretching native titin. The Z-line end is attached to a 3 μm bead via the T12 antibody. The M-line end is attached to a silica bead coated with myomesin. Figure from Kellermayer et al. (1998) with permission from Academic Press. (C) LOT: attachment method used by Tskhovrebova and colleagues (Tskhovrebova et al., 1997). The Z-line end was attached via the T12 antibody and the M-line end via the AB5 antibody. Figure from Tskhovrebova et al. (1997) with permission from *Nature*.

from several suppliers. Typically a trimethoxy or triethoxy silane with a functional group on a long linker is used. The silane may be adsorbed to the reactive silica surface either from a solution or the gas phase. Proteins may then be attached to the silated surface via traditional cross-linking methods, such as carbodiimides or glutaraldehyde. These methods can be used to covalently attach antibodies to a substrate and tip for specific attachment of a protein for extension studies (Forbes et al., 2000).

Many proteins adhere readily to glass, silica and mica surfaces. Bovine serum albumin (BSA) has been shown to bind irreversibly to glass and silica surfaces. This characteristic has been used in a number of force measuring studies with biotinylated BSA (Florin et al., 1994; Rief et al., 1997b). The biotinylated BSA is adsorbed to the surface and then streptavidin is bound to the immobilized BSA. This immobilized streptavidin may then be used for binding biotinylated proteins. The streptavidin/biotin interaction can withstand about 200 pN (Chilkoti et al., 1995; Moy et al., 1994) with multiple interactions increasing the force required to detach the biotinylated polymer. Many other proteins also adhere strongly to these surfaces. This robust adsorption of proteins to an AFM tip has been the primary method of fixing titin to the AFM tip for extension studies. In all of the AFM studies we will describe below, the titin or titin fragment was first adsorbed onto a surface and then the AFM tip was pressed into the surface to pick up the protein for extension. While the exact mechanism of this adsorption of protein to an AFM tip is not known, it appears that the amino terminus of the protein can be picked up, allowing for a known attachment position of the protein under study. However, random positions along the polypeptide are equally likely to be adsorbed.

Mica is an ideal substrate for imaging biomolecules, because it is atomically flat and many proteins adhere well to it. In fact, many of these proteins adhere so well to mica that they behave like an adhesive when an AFM tip is pressed into them. Rather than observing the extension of the polypeptide, the tip sticks more tightly to the surface than it does in the absence of the protein. This characteristic must arise from multiple, strong interactions between the mica and protein and between the tip and the protein. While these multiple interactions may be welcome in an imaging substrate, they are undesirable for polymer extension work. A surface that allows for a specific attachment or at least relatively few attachment points is needed.

Gold films have been shown to be suitable substrates for both imaging of proteins and for force measurement. When properly evaporated onto mica, gold forms atomically flat areas with the Au (1 1 1) structure (close packed) (Derose et al., 1991; Zheng et al., 1995). Gold can also be evaporated onto glass following a thin coating of chromium. Without the chromium, the gold will not adhere and will peel away from the glass. Gold evaporated onto glass will not be atomically flat, but will be dominated by small ($\sim$1 nm) globules whose size depends upon a number of factors. A perfectly flat surface is not necessary for force measurements and gold on glass is more readily prepared than gold on mica. Gold has the additional advantage that it can be readily modified with alkane thiols, which form self-assembled monolayers that can have different functionalities. Thiols may also be used for passivating the gold surface to reduce further interaction between the protein and surface. Clean gold surfaces quickly adsorb compounds from the air, which renders the surface partially hydrophobic. All of the AFM studies of titin and titin fragment extension described below used a gold surface for immobilizing the protein. In the case of native titin, it may bind to the gold through surface cysteines or some physisorption process. In the work by Rief et al. (1997a), the native titin was stored at 0.5 mg/ml in phosphate buffered saline (PBS) with 0.5 mM dithiothreitol (DTT) and adsorbed onto the gold surface in PBS at a concentration of 10–100 μg/ml. Even with a dilution factor of 50, the DTT will be at a sufficiently high concentration that it will react with most of the available thiol reaction sites available on the gold surface. This indicates that native titin must interact with the gold surface through mechanisms other than just surface cysteines. All of the AFM studies of recombinant titin fragments have engineered the protein with two cysteines on the carboxy terminus of the protein (Carrion-Vazquez et al., 1999b; Rief et al., 1998). These recombinant proteins must be purified and stored in the presence of DTT or some other thiol reducing agent to ensure that the protein does not aggregate or form disulfide bonds. These thiol reducing agents must be removed from the buffer solutions just prior to reacting the protein with the gold surface.

2.4.3. Attachment of proteins in laser optical trap

As with the AFM, to stretch molecules with an optical trap, a single molecule must first be affixed to a stationary fixture at one end and a moveable fixture at the other. With the optical trap, the moveable fixture can be either a bead or a slide (Fig. 7B and C). The ideal "handles" are those specific to designated area of titin and strong enough to withstand the developed force. Kellermayer and colleagues (Kellermayer et al., 1997, 1998) used the T12 antibody (Furst et al., 1988) to affix native rabbit muscle titin to a 3 μm latex sphere. The T12 antibody binds about 100 nm from the Z-line end of (titin, it therefore binds only to the contaminating T1 in the T2 preparation). Kellermayer and colleagues used several different methods to immobilize the M-line end of titin. A polystyrene bead was coated with either the T51 antibody to titin or myomesin (Kellermayer et al., 1997; Obermann et al., 1995). They also used 2 μm silica spheres, which had been coated with myomesin (Kellermayer et al., 1998). The M-binding sphere was held in a micropipette (Fig. 7B). The myomesin–silica spheres increased the chance of tethering the titin between the two spheres, but the beads often bound non-specifically to various positions along the molecule. Tskhovrebova et al. (1997) have taken a different approach by immobilizing one end of the titin molecule on a glass slide that moved on a stage and using a latex microsphere in the optical trap as the stationary anchor, as well as for force sensing. They also used the T12 antibody to immobilize the Z-line end of titin and the AB5 antibody to immobilize the M-line end of the protein (Whiting et al., 1989). Wang et al. (1997a) used stage affixed, streptavidin derivatized, latex beads to immobilize one end of biotinylated titin, and another streptavidin coated latex bead bound to the distal end of titin for the trap. The attachment sites were distributed randomly along the length of titin due to the biotinylation of titin. The high affinity and strong unbinding force of biotin/streptavidin, even in denaturing medium, allowed for specific attachment and made it possible to stretch titin in 8 M urea (Wang et al., 1997a).

3. Measurements—single titin molecule extension

3.1. Chronological review

Atomic force microscopy imaging of titin was first reported in 1996 (Hallett et al., 1996). Their work used the imaging ability of AFM and confirmed the filamentous conformation of titin molecules as seen in earlier EM studies (see Section 1 and Fig. 3). As in the EM investigations, titin sample preparation appears to be an important factor in the outcome of the experiments.

For single titin elasticity, LOT investigations were first reported in meeting presentations in 1996 and early 1997 (Kellermayer et al., 1997; Tskhovrebova et al., 1996; Wang et al., 1997a). Later in the same year three full publications appeared simultaneously, one in Nature (Tskhovrebova et al., 1997) and the other two in Science (Kellermayer et al., 1997; Rief et al., 1997a), accompanied by commentaries (Erickson, 1997; Keller, 1997). These three papers mark the beginning of an exciting period of single titin molecule stretching efforts and a number of significant results have followed (Carrion-Vazquez et al., 1999a, 1999b; Kellermayer et al., 1998; Marszalek et al., 1999; Oberhauser et al., 1999, 1998; Rief et al., 1998). Additional results have also been reported in recent scientific meetings and are rapidly forthcoming. Reviews on these single molecule studies are appearing (e.g. Kellermayer et al., 2000).

In the first full-length publication, Tskhovrebova and coworkers (Tskhovrebova et al., 1997) performed mechanical experiments on single molecules of titin to determine their viscoelastic properties using the LOT technique. They found that titin is elastic on a fast (0.1 s) time scale and the force–extension data can be fitted with random-coil polymer models having two different segments of entropic springs. One segment was attributed to the entropy of straightening folded Ig and Fn3 modules of titin; the other was attributed to the extension of the polypeptide chain in the PEVK region. Both FJC and WLC models (see Section 2 for model details) fitted the data similarly well with L_K (segment length) $\cong 2L_p$ (persistence length) $\cong 0.35$ nm for the PEVK region and the over 10-fold greater elastic parameters for the string of folded Ig/Fn3 domains (Fig. 8A). On a slower time scale of seconds and above a certain force threshold of a few tens to hundreds of pN, the stretched titin molecule displayed stress-relaxation in rapid steps of 3–10 pN. This drop of

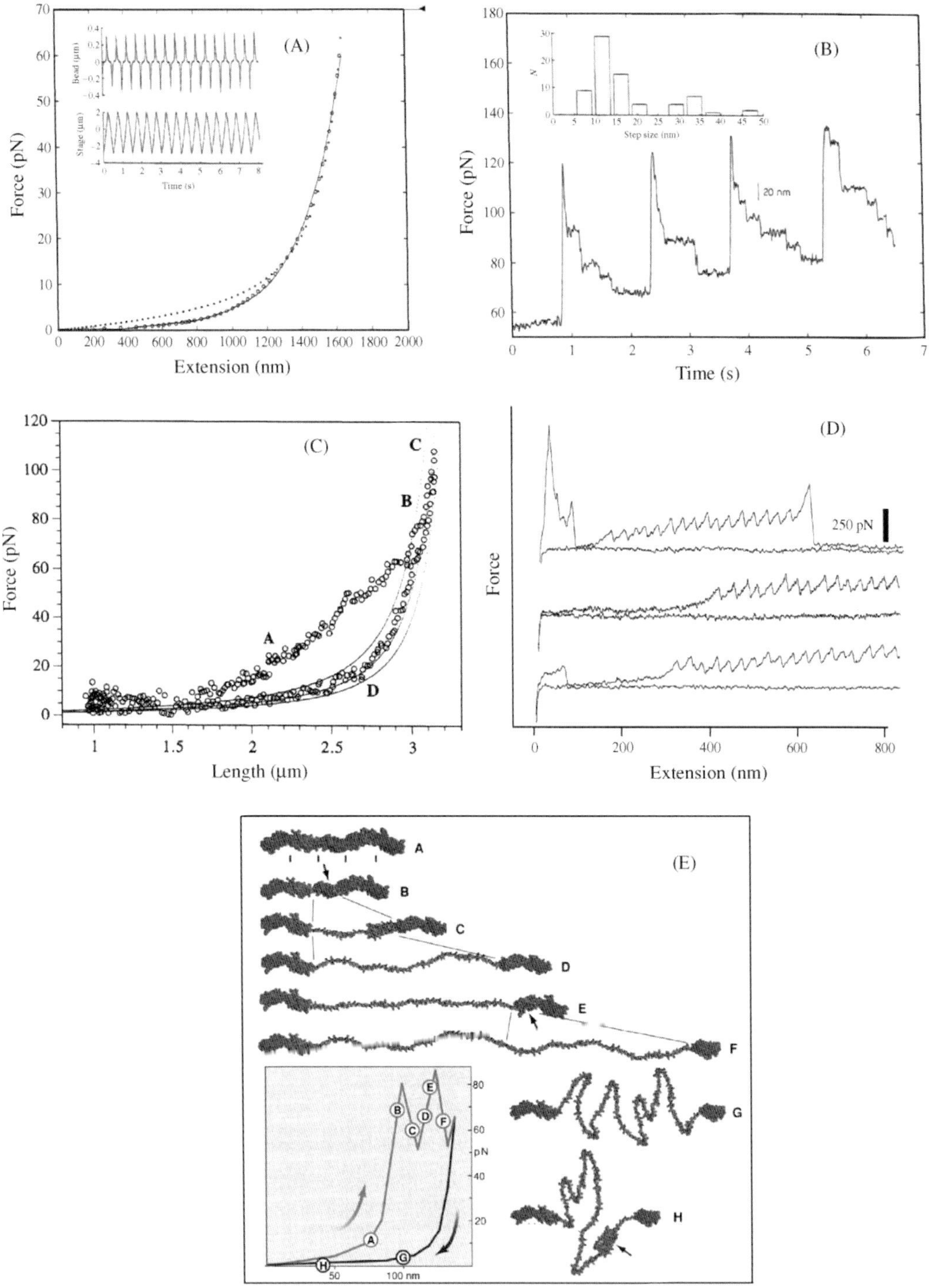
(A)
Force (pN)
Extension (nm)
0
10
20
30
40
50
60
70
0 200 400 600 800 1000 1200 1400 1600 1800 2000
Bead (μm)
Stage (μm)
Time (s)
(B)
Force (pN)
Time (s)
60
80
100
120
140
160
180
0 1 2 3 4 5 6 7
N
Step size (nm)
20 nm
(C)
Force (pN)
Length (μm)
0
20
40
60
80
100
120
1 1.5 2 2.5 3
A
B
C
D
(D)
Force
Extension (nm)
250 pN
0 200 400 600 800
(E)
A
B
C
D
E
F
G
H
pN
20
40
60
80
50
100 nm

tension was correlated to an increase of the entropic chain length of 19 ± 11 nm from yielding of internal structures, probably the unfolding of Ig and Fn3 domains (Fig. 8B). It is clear that the observed unfolding relaxation behavior deserves further quantification.

Kellermayer and coworkers (Kellermayer et al., 1997) performed similar LOT measurements of the force required to stretch a single titin and similarly revealed that titin behaves as a highly nonlinear entropic spring. The transition from a rapid to a more gradual tension increase around 20–30 pN per titin molecule was attributed to Ig and Fn3 domain unfolding. Refolding in a low-force transition below 2.5 pN was also observed. The authors fitted their initial tension vs. extension data and the force release data to entropic spring models, and observed that a persistence length of 2.0 nm most likely describes single denatured titin behavior. Only one-segment WLC fits were performed. The stretch–relaxation force hysteresis was attributed to a difference between the unfolding and refolding kinetics of titin folded domains. The authors also concluded that scaling the molecular data up to sarcomeric dimensions reproduced many features of the passive force vs. extension curve of muscle fibers. Although it was noted that a fraction of the molecule (5–40 percent) remained permanently unfolded, the authors did not resolve the two sequential WLCs (c.f. Tskhovrebova et al. (1997)). In their LOT setup apparently one to three T1 titins were stretched at a time.

Rief and coworkers (Rief et al., 1997a) used atomic force microscopy (AFM) to investigate the mechanical properties of titin. Individual titin molecules were repeatedly stretched between a gold substrate and an AFM tip, and the force–extension behavior was recorded and analyzed. At large extensions, the restoring force exhibited a striking saw-tooth-like pattern between 150 and 300 pN, with a periodicity that varied between 25 and 28 nm (see Fig. 8D). They proposed that these saw-tooth-like patterns rose from the unfolding of titin Ig domains. They confirmed this by stretching engineered recombinant titin Ig segments of two different lengths and observed the same pattern. The unfolding forces increased linearly from 150 to 300 pN when pulling speed was varied exponentially from 0.01 μm/s to over 10 μm/s. Upon relaxation, refolding of Ig domains was observed and repeat stretching was demonstrated.

Fig. 8. Representative data from the different methods of investigating titin elasticity. (A) Force–extension of native rabbit skeletal muscle titin by an optical trap (Tskhovrebova et al., 1997) with permission from *Nature*. The experiment data (solid line) are fitted with either a single component WLC model (dotted line) or a two-component WLC model (open circles). (B) Stepwise stress–relaxation of native rabbit skeletal muscle titin by an optical trap (Tskhovrebova et al., 1997). The titin was stretched in 250 nm steps and then allowed to relax for over 1 s. When the force reached 120 pN, domains began to unfold to give rise to the staircase pattern (Tskhovrebova et al., 1997) with permission from *Nature*. (C) Extension of native rabbit skeletal muscle titin by an optical trap (figure from Kellermayer et al. (1998) with permission from Academic press). In the trap, titin molecule straightens and PEVK extends in A and titin Ig/Fn3 domains start to unfold between A and B. Complete unfolding of titin domains from B to C gives rise to a curve that is well-fit with the WLC model of chain dynamics. Upon release, the completely unfolded titin polypeptide chain behaves as a worm like chain. (D) Extension of native human cardiac titin (figure with permission from Rief et al. (1997a) © [1997] The American Association for Advancement of Science). The striking saw-tooth patterns at high force reflect the sequential unfolding of titin Ig/Fn3 domains under stress. Notice the long extensions in the lower two extension curves, which indicate the extension of a portion of the protein before domains unfold. (E) Cartoon of the relationship between the structure of the extended titin domains and positions on the force curve (figure with permission from Erickson (1997) © [1997] The American Association for Advancement of Science). Lettering on the force curve in the inset corresponds with possible structures of the titin filament at these extensions.

Kellermayer and coworkers (Kellermayer et al., 1998) extended their single-molecule mechanical works to focus on Ig and Fn3 complete unfolding at high forces (Fig. 8C). The force exerted on single titin molecules by the LOT was claimed to be above 400 pN. The high external forces resulted in complete mechanical unfolding of the molecule, characterized by the disappearance of force hysteresis at high forces. This also allowed WLC fitting to the unfolded chain to a greater chain length. Titin refolded following complete denaturation, as the hysteresis at low forces reappeared in subsequent stretch–release cycles. The broad force range of the unfolding force was attributed to variations between Ig and Fn3 modules in the native titin having different activation energies for their unfolding. However, the LOT force calibration in this work is complicated (see Section 2).

Rief and coworkers (Rief et al., 1998) also followed up their AFM work on titin and compared the unfolding forces of defined regions of different titin isoforms. Constructs comprising six to eight Ig domains located in the mechanically active I-band segment of titin were compared to those containing Fn3 and Ig domains from the A-band part. They improved the spatial resolution of their AFM to detect differences in length as low as a few amino acids per titin folded domain. The authors showed that the unfolding forces ranged between 150 and 300 pN and differed systematically between the constructs. Fn3 domains in titin exhibited 20% lower unfolding forces than Ig domains and Fn3 domains from tenascin unfolded at forces only half those of titin Fn3 domains (see below). The authors suggest that the tightly folded titin domains are designed to maintain their structural integrity and, hence, unfolding is unlikely under physiological sarcomere forces.

Oberhauser and coworkers (Oberhauser et al., 1998) also studied the extracellular matrix protein tenascin using AFM techniques and compared the results with titin. They found that tenascin, which has many Fn3 modules, is also an elastic protein and could be stretched to several times its resting length. The investigation closely paralleled the earlier AFM studies on titin unfolding. Force–extension curves showed a saw-tooth pattern, with peaks of force around 100 pN that depended linearly on log (pulling speed). The leading edges of the force peaks were well fitted by WLC models with $L_p \cong 0.4$ nm and each was approximately 25 nm apart in the chain contour length. The authors found similar stretch-induced unfolding of its Fn3 domains in recombinant tenascin fragments encompassing the 15 Fn3 domains of tenascin. Refolding of tenascin after stretching was observed when the force was reduced to near zero and showed a double-exponential recovery with time constants of 42 domains refolded per second and 0.5 domains per second.

In yet another extension of the AFM work on titin, Carrion-Vazquez and coworkers (Carrion-Vazquez et al., 1999a) demonstrated that AFM techniques could detect contour length changes due to the insertion of a few amino acids in an Ig domain. The authors amplified the mechanical features of a single domain by engineering polyproteins composed of up to 12 identical repeats. They then showed that upon mechanical unfolding, mutant polyproteins containing five extra glycine residues extended 2.0 nm per module farther than the wild-type polyproteins. When similar insertions were near the N or C termini, the 4 nm per amino acid increase did not show up in the unfolding saw-tooth pattern.

In a more comprehensive report, Carrion-Vazquez and coworkers (Carrion-Vazquez et al., 1999b) also investigated the detailed folding/unfolding pathway of a titin Ig module using AFM and found insightful correspondence with chemical induced folding/unfolding pathway. By engineering a protein made of tandem repeats of identical Ig modules, the authors were able to get explicit pulling-rate dependent data on the unfolding rate of a single protein domain that can be accurately extrapolated to zero force. When this was compared with chemical unfolding rates for untethered modules extrapolated to 0 M denaturant, the two results were in good agreement. Their evidence also supported that the transition state for unfolding appears at the same position in terms of the reaction coordinate on the folding pathway when assessed by either the AFM force or chemical stress.

In the next work, Marszalek and coworkers (Marszalek et al., 1999) resolved a mechanical unfolding intermediate in a titin Ig domain, and supported this finding with careful computer simulations and mutation investigations. Applying AFM extension to a string of multiple copies of one Ig domain, the authors found an abrupt extension of each domain by about 0.7 nm before the first unfolding event. This fast initial extension before a full unfolding event was thought to produce a reversible unfolding intermediate that is about 15% longer. Steered molecular dynamics simulations on the unfolding of the same Ig module showed that a pair of hydrogen bonds near the amino terminus of the domain may rupture first under force extension and cause an extension of about 0.6 nm, in good agreement with the AFM number of 0.7 nm. Furthermore, it was shown that disruption of these hydrogen bonds by site-directed mutagenesis eliminated the unfolding intermediate. It was noted that the existence of this unfolding intermediate might also extend titin domains up to 15% of their slack length and contribute to titin elasticity in sarcomere.

In the latest report of their series of AFM observations on titin constructs, Oberhauser and coworkers (Oberhauser et al., 1999) demonstrated that, after repeated mechanical extension/relaxation cycles, tandem modular proteins can misfold into a structure formed by two neighboring modules. The misfolding altered the mechanical topology of the modules and was shown to be fully reversible and about as stable as the original fold. This showed that if titin domains unfold under sarcomere stress, it could refold into unusual conformations. The work demonstrated further that AFM has the power to capture, in real time, rare misfolding events at the level of a single protein.

3.2. Comparison of AFM and LOT unfolding data

The data on titin stretching obtained from the AFM and LOT at first glance appear quantitatively and qualitatively different, especially the absolute force at which titin domains unfold and the lack of saw-teeth in the LOT studies from Kellermayer et al. (1997). Many factors may contribute to these discrepancies. The difference in instrumentation design, force calibration, protein attachment, geometry of stretching and data acquisition has been discussed above. The large force and loading rate of AFM and the lower force and slower loading rate of LOT are also major factors here.

One possibility arises from the fact that AFM was performed on a film of titin or titin fragments adsorbed onto a substrate. It is conceivable that the protein adheres to the substrate at multiple points and, upon stretch, the protein–substrate interaction ruptures sequentially, without unfolding of domains. Several facts argue against the sequential desorption of the protein from the surface. The rupture forces seen for native titin and titin fragments are widely dispersed, whereas those for the I27 homopolymer form a tight distribution which can be fit by the bond lifetime model (Fig. 4D). The increase of contour length following one of the ruptures is exactly what one would expect for a protein domain with 89 amino acids. Each domain would have to adhere to the surface at the same position to exhibit this behavior and still have the ruptures result from surface interactions rather than domain unfolding. The best evidence for domain unfolding comes from the study by Carrion-Vazquez et al. (1999a) where five glycine residues were inserted into the core of the immunoglobulin domain. Upon extending this polyprotein with the AFM, they found that the increase in the contour length is what would be expected for the addition of the five amino acids to the sequence. Adhesion to the substrate cannot explain this difference between the wild-type I27 domain and that with an insert in the core of the domain. These results argue strongly that the saw-tooth pattern results from domain unfolding and not rupture of protein–substrate adhesion.

The apparent lack of the saw-tooth pattern in LOT is likely due to several reasons. First, the loading rates on titin are significantly different. Kellermayer et al. (1977) stretched the titin at a constant rate from 1–6 pN/s (90 nm/s) until a maximum force was reached and then the polymer was relaxed. This loading rate is much less than that of AFM measurements and unfolding dynamics of the titin domain can be significantly different (see Section 2.1). Second, the AFM cantilever is much stiffer than LOT trap. The stretch-induced unfolding of each titin domain should increase the contour length by an amount of about 25 nm. In AFM, the stiff cantilever recoils very little and the force (which depends sharply on the ratio of the end-to-end distance to the contour length) is reduced greatly. The force has to increase gradually to cause another domain to unfold, leading to a regular saw-tooth pattern. In LOT, the softer trap stiffness means the large backward movement of the bead causes the end-to-end distance of the titin segment to also increase, so the force on the molecule drops only slightly after a domain unfolding event. This contributes to the absence of any clear saw-tooth pattern in LOT experiments even as domains unfold (Kellermayer et al., 1997, 2000; Tskhovrebova et al., 1997).

Kellermayer et al. (1998, 1997) observed a striking hysteresis by comparing the extension and relaxation curves for native titin. The extension curve followed the response of the WLC model up to about 20 pN where upon a "stretch transition" occurred, which results in less elastic force for a given extension. Kellermayer and colleagues have suggested that this transition may be a structural change. Evans and Ritchie (1999) showed that this change in the slope of the F vs. d curve could be attributed to the steady unfolding of domains connected by a flexible linker. The low force loading of LOT at ca. 13 pN/s results in the domains unfolding at a much lower force without the saw-tooth discontinuities seen in the AFM experiments. In the extension portion of the curve, the low force data was fit with the WLC model with a persistence length of 4 Å and a contour length of 2.2 μm. The release curve of the same segment fits the WLC model with a persistence length of 9 Å and a contour length of 3.0 μm. The hysteresis was caused by the mechanical unfolding of the Ig and Fn3 domains in titin. After the stretched titin was allowed to relax, the domains refolded and upon restretching the hysteresis curve was recovered.

Tskhovrebova et al. (1997) used two different methods to extend titin. For the dynamic extension of native titin, the bead was held in the optical trap and the stage was driven with a triangle wave with a frequency of 2 Hz. The amplitude of the stage motion was ca. 4 μm peak to peak, with a maximum force of about 70 pN. The extension/retraction rates were held constant, however, the loading force increased nearly exponentially. The maximal loading rate was of the order of 1 nN/s. For the static extension method, titin was stretched in steps of about 250 nm and then held at each stage until the stress relaxation was complete on the order of seconds. The force jumped to over 120 pN after each step extension. The effective loading rate was much higher than that in the dynamic extension case, favoring domain unfolding during the waiting periods, rather than during the stretching period. The dynamic force–extension of native titin yielded a curve that could be readily fit by the WLC model. In Fig. 8A, the solid line is the data and the dots represent the fit to a single component WLC, with a persistence length of 0.76 nm and a contour length of 1932 nm. The open circles are for a two component WLC model, with persistence lengths of 4.15 and 0.19 nm and contour lengths of 1174 and 791 nm, respectively. The dynamic extension was accomplished by moving the stage at a rate of 16 μm/s. At this rapid rate of extension and low force, the Ig and Fn3 domains will not unfold and all of the extension was attributed to the straightening of the molecule and the extension of the PEVK segment.

The static extension of native titin by Tskhovrebova et al. (1997) (Fig. 8B), in contrast, shows the unfolding of domains. The titin was stretched by 250 nm and then was held at that position until the stress–relaxation was complete. As is clearly seen in Fig. 8B, the force increases rapidly upon the initial stretch and then the force decreases in a stepwise manner, with an average step of 20 nm. The steps are less than the fully extended length of 38 nm for 100-residue domains, because the forces are not great enough to fully extend the molecule. This stepwise reduction in force when titin is stretched and held at a certain length was modeled by Rief et al. (1998) using the lifetime and unfolding reaction width determined from their AFM experiments. Their Monte-Carlo simulation of the static extension of titin yields the same stepwise unfolding behavior seen experimentally with the optical trap system.

3.3. Molecular elasticity results

The different results obtained for the different methods of extension with optical traps and AFM exemplify the importance of the loading rate or time that the load is applied to the polymer in determining the outcome of single molecule mechanics. While much of these differences are attributable to Ig and Fn3 domain unfolding/refolding, the PEVK and other titin motifs must also contribute to the overall molecular elasticity. As described in Section 2.2, several models have been used for chain molecule extension behavior and are approximations at various levels of simplifications. Clear limitations of these models include the lack of direct loading rate and history-dependence of the extension force. Nevertheless, these models have offered useful parameter for comparing molecular elasticity of titin motifs.

Single molecule measurements generally favor the use of the WLC model as the basis for parameterizing the full range of titin elasticity. The results suggest that the elasticity of whole molecule titin results from the extension of two components: a stiff entropic spring for titin motifs including PEVK, N2-A and N2-B unique insertions and the unfolded peptides from the Ig and Fn3 modules, and a soft entropic spring consisting of folded beads of Ig and Fn3 domains (Fig. 3 and Freiburg et al., 2000). The persistence lengths deduced from the single titin measurements are summarized in Table 4. The stiff entropic string has a reported persistence length range of 0.15–0.8 nm (Jin et al., 2000; Kellermayer et al., 1998, 1997; Rief et al., 1997a; Tskhovrebova et al., 1997). The soft entropic spring has a corresponding persistence length range of 4.6–15 nm

Table 4
Summary of the persistence lengths for titin and titin motifs

Proteins	Persistence length (WLC, nm)	Experimental conditions
Titin	0.15 ± 0.1[a] (PEVK/N2)	• LOT. Two segment fits. Force range 0 to 60 pN. Room Temperature. In 0.3 M KCl.
	4.6 ± 0.9[a] (Ig/Fn3)	• FJC fits gave a corresponding Kuhn length values of 0.35 ± 0.03 nm (PEVK) and 5.3 ± 0.08 nm (Ig/Fn3 domains).
	1.5–2.0[b,c]	• LOT. Attributed to the "fully unfolded T1." Single-segment WLC fits only and contour length was only 4 μm. 1–3 titin stretched each time and the *Lp* value distributed accordingly. About 60 mM buffer ionic strength, low Ca^{2+}.
	15[f]	• Dynamic light scattering. T2 with folded Ig/Fn3 in 0.3 M NaCl and 0.25 M potassium Phosphate buffer (pH = 7.0).
Poly-Ig recombinant	0.4[d]	• AFM. Leading edges of titin Ig8 sawtooth-like stretch curves. In 200 mM PBS over Au surface at room temperature, low Ca^{2+}.
PEVK recombinant	0.3–0.7[e]	• AFM. 51 kDa PEVK cloned segment. Ca^{2+} and ionic strength dependency.

[a] Tskhovrebova et al. (1997).
[b] Kellermayer et al. (1998).
[c] Kellermayer et al. (1997).
[d] Rief et al. (1997a).
[e] Jin et al. (2001).
[f] Higuchi et al. (1993).

(Higuchi et al., 1993; Kellermayer et al., 1997; Tskhovrebova et al., 1997). The whole titin acts as a serial spring and lengthens first its soft spring under low force of less about 5 pN/titin and then its stiff spring under high force of larger about 5 pN.

Tskhovrebova et al. (1997) made an effort to characterize both the soft and stiff springs by two segment fits to the LOT force–extension data. Most other studies did not attempt to extract beyond single segment fits (Higuchi et al., 1993, Kellermayer et al., 1998, 1997; Rief et al., 1997a). Furthermore, the persistence parameter for each titin motif appears to be dependent significantly on solution ionic conditions (Jin et al., 2000), which are different for each research groups (Table 4). Nevertheless, it is a consensus that the soft spring segment, with persistence length of the order of its 4 nm-long Ig and Fn3 domains, is about 10 times more compliant than that of the stiff spring segment, with the persistence length of the order of the 0.38 nm peptide bond. Additional support for such a broad distribution of stiffness within titin also comes from recent immunoelectron microscopy studies (Freiburg et al., 2000; Granzier et al., 1996; Linke et al., 1998, 1996, 1999; Trombitas et al., 1998), using site-specific antibody markers to resolve details of the extension of various titin motifs (Freiburg et al., 2000). However, a fuller discussion of these non-single titin results is beyond the scope of this review.

4. Dynamic force spectroscopy and mechanical unfolding/folding

The forced unfolding of Fn3 and Ig domains has provided valuable insight into how titin domains might behave under more stressful conditions in vivo. Additionally, these experiments have provided new information on how proteins respond to mechanical stress. The power of dynamic force spectroscopy in revealing subtle behavioral changes in protein folding is illustrated by the work of Oberhauser et al. (1999) who demonstrated that the Ig modules in a homopolymer of I27 can misfold to a dimeric form after a few mechanical unfolding–refolding cycles. This approach opens the door to investigate the means of reversing undesirable misfolding of proteins, a topic of major physiological and biotechnology interests. It also demonstrates the efficiency of refolding of titin domains after repeated mechanical unfolding. Many eukaryotic proteins have evolved as modular constructs that must fold co-translationally, which is facilitated by chaperonins (Netzer and Hartl, 1997). In prokaryotes, proteins are believed to fold post-translationally, which may explain, in part, why many modular proteins are not expressed well or appear as inclusion bodies of misfolded proteins in *E. coli*. However, if a protein misfolds, intracellular repair mechanisms will correct the misfolding, e.g. GroEL/GroES allows *E. coli* proteins to refold by first unfolding them (Shtilerman et al., 1999). It is conceivable that both mechanical and chemical unfolding of proteins may be an integral part of the cellular repair mechanisms of misfolded proteins.

Repeated extension and relaxation of native titin results in the loss of the saw-tooth pattern and the hysteresis in the force–distance curves. One possible cause may be the slow and rate limiting proline *cis–trans* isomerization as first postulated by Kellermayer et al. (1997). Recent studies by Plaxco et al. (1997) showed that two Fn3 domains, with similar, characteristic beta folds and 30% sequence identity, had refolding rates different by two orders of magnitude. The folding of the slow folding module was limited by proline isomerization, which can be catalyzed by the addition of prolyl peptidyl *cis–trans* isomerase (PPI), giving a three-fold reduction in the folding time. If the repeated extension of native titin results in an increase in the amount of *cis*-proline isomers, this could account for the effects observed during the repeated mechanical unfolding of the protein. However, other mechanisms that can lead to a kinetic folding trap cannot be discounted. The possibility that prolyl isomerase or chaperonins are involved in accelerating titin folding in solution and in situ is well worth exploring.

The folded lifetime at zero force and the width of the unfolding potential (or length along the reaction coordinate, see Fig. 4A) are the two defining parameters of how the protein will respond to load. Rief et al. (1998) showed that for native titin, $\tau_0 = 30,000 \pm 10,000$ s and $x_u = 0.3 \pm 0.1$ nm based on the average unfolding force at different speeds. However, for human tenascin Fn3 domains, $x_u = 0.55 \pm 0.1$ nm, which indicates that tenascin Fn3 domains are more sensitive to an applied force than is native titin. From the I27 polyprotein, Carrion-Vazquez et al. (1999b) were able to demonstrate that the I27 Ig domain was even more stable to applied unfolding forces with $\tau_0 = 3000$ s and $x_u = 0.25$ nm. One can infer that the domains of titin have evolved to be stable to high forces when they are applied for a short period of time. Any applied force will bias a protein to unfold if it is applied for a long enough time. If a protein is stable to high forces applied over a short period of time, it follows from Eq. (2) that it will also be more stable to small forces applied over a longer period of time. This results from the shorter reaction width and the concomitant lower reduction in lifetime for any given force applied to the protein.

The notion of mechanical unfolding of proteins has captured the imagination of the protein folding community, which has given rise to several theoretical analyses of forced unfolding of proteins (Gao et al., 2000; Klimov and Thirumalai, 1999, 2000; Krammer et al., 1999; Lu et al., 1998; Lu and Schulten, 1999a, b, 2000; Paci and Karplus, 1999, 2000; Rohs et al., 1999; Socci et al., 1999; Zhang et al., 1999). The thorough understanding of two-state folding proteins (proteins which have no intermediates between the folded and unfolded states) is crucial for understanding how proteins with more complex folding processes fold in cells (Eaton et al., 2000; Levitt et al., 1997; Onuchic et al., 1997). Because many load-bearing protein structures are aggregate systems, a thorough understanding of how individual protein molecules respond to force will provide fundamental insights into how supramolecular assemblies perform their functions in cells and tissues.

Since β-barrels are fundamental and major folding motifs in nature, the mechanical unfolding of titin Ig domains are likely of general significance and application. It is therefore worthwhile to summarize the current state of knowledge of the unfolding properties of the titin Ig, Fn3 and tenascin domains as characterized by AFM. An important observation was that the average unfolding force for the modules of native titin was dependent upon the log of the extension speed (Rief et al., 1997a). While this behavior was predicted by Evans and Ritchie (1997), the results from Rief and colleagues represents the first experimental evidence for the predicted speed dependence of forced rupture of biomolecular interactions (Rief et al., 1998). Comparison of the speed dependence of native titin with that of a recombinant tenascin fragment (primarily Fn3 domains) demonstrate the importance of the loading rate in the characterization of the mechanical properties of a protein. The tenascin domains have a folded lifetime similar to that of the domains in titin. However, the unfolding reaction width was larger than the average width of the titin domains. This increased width leads to a lower slope in the plot of rupture force vs. log (extension rate). This decreased sensitivity to the loading rate implies that tenascin cannot withstand rapid extensions as well as titin. By using recombinant fragments of different parts of skeletal titin, Rief et al. (1998) showed that these different domains have different average rupture forces, even though the rupture of individual domains ranged between 150 and 300 pN. Whether this heterogeneity in rupture forces plays a physiological role or is merely the consequence of different amino acid sequences needed for the protein to fulfill its other roles is yet to be determined. At this point it would be pure speculation as to whether these differences in the stability of the titin domains to applied force has evolved to fulfill some function or simply adventitious.

These early experiments on native proteins and fragments of the proteins suggested that the AFM could be a powerful tool for studying protein folding. However, the heterogeneity of the native titin and the recombinant fragments makes a detailed analysis a difficult task. Carrion-Vazquez et al. (1999b) have made a "cleaner" system for studying the forced unfolding properties of titin Ig domains by preparing a homopolymer of Ig domains. Recombinant methods were used to prepare the polymer with 12 identical domains of module 27 from the I-band of human cardiac titin. With a polymer of identical Ig domains, any variation in the unfolding properties is due to random processes and not to differences in the intrinsic properties of the domain. This is demonstrated well by the histogram of unfolding forces for an octamer of the I27 domain (Fig. 4D). The histogram of the unfolding forces of the eight I27 domains can be fit readily by a Monte-Carlo simulation of the mean unfolding forces with an unfolding distance of 0.25 nm, a folded lifetime of 3000 s and a pulling rate of 0.6 nm/ms. This folded lifetime is an order of magnitude shorter than that determined by Rief et al. (1998) for native titin. The identical domains of the recombinant homopolymers allow experimentally determined parameters to be assigned as properties of the domains. This ability has allowed Carrion-Vazquez et al. (1999b) to compare directly chemical unfolding of isolated I27 domains with the force unfolding of polymers of the same domain. Extrapolating the chemical unfolding data to a zero concentration of guanidinium hydrochloride gives a folded lifetime of 2000 s and a folding rate of 32 s^{-1}. Forced unfolding of the polymer of I27 gives a folded lifetime of 3000 s and a folding rate of 1.2 s^{-1}. The kinetic rates for unfolding and folding can be used to estimate the activation free energy of unfolding and folding. The activation free energy for unfolding the I27 domain was found to be 22.0 kcal/mol for chemical unfolding and 22.2 kcal/mol for forced unfolding. The activation free energy for folding of I27 was found to be 15.4 kcal/mol for chemically unfolded protein and 17.3 kcal/mol for force unfolded protein. The barrier heights to unfolding by the two methods are almost identical which suggests that the unfolding pathways are similar. The barriers to refolding by the two methods are significantly different. The 1.9 kcal/mol difference can be attributed to the entropic cost of tethering. As was discussed above, even a small force applied to a polypeptide chain will greatly decrease the folded lifetime. The extension of the protein decreases the conformational entropy of the polypeptide chain, which will slow down the refolding process. The similarity of the reaction coordinates between chemical and forced unfolding is at first surprising. However, considering the β-barrel structure of the Ig domains and the fact that the amino and carboxy termini are about 180° apart this seems quite reasonable. There are now both computational and experimental evidence that the first step of unfolding of the Ig domains is the rupture of hydrogen bonds between the A and B β-strands (Marszalek et al., 1999). The rupturing of these hydrogen bonds leads to a 6.6 Å extension of the Ig domain at a force of about 100 pN. Molecular dynamics modeling of the extension of Ig domains indicates that the next step is the rupture of the patch of hydrogen bonds between the A′ and G β-strands (Lu et al., 1998) followed by the extension of the polypeptide. The effect of chemical denaturants on protein structure is to both weaken hydrogen bonds and weaken hydrophobic interactions. The folding of small, single domain proteins arises largely from the cooperative formation of hydrogen bonds. If strain placed on hydrogen bonds causes them to first weaken and then rupture, a sequence similar to that induced by chemical denaturants, then the unfolding pathways taken by the two unfolding methods would be expected to be similar. However, a small change in protein conformation could have dramatic effects upon the forced unfolding pathway, while the chemical unfolding pathway could remain the same. If the amino and carboxy termini are closer together, then the forced unfolding pathway will be very different from what it is for the Ig domain with the termini opposite from one another (Klimov and Thirumalai, 1999). It very well may be that the Ig domain is a special case where the unfolding pathways are similar for chemical and forced unfolding.

While other arrangements of the termini of the domains forming a large polyprotein such as titin are possible, having them 180° apart is clearly an advantage for a protein designed to sustain a force along its length. If the termini were close to one another, the strain would not be distributed throughout the molecule as it is with the Ig domains. This could lead to sequential unfolding of the secondary structural folds rather than the two-state unfolding observed with the Ig domains. While a polymer of T4 lysozyme has been studied, it was not stretched between its termini and Yang et al. (2000) did not investigate the kinetics of the folding and unfolding.

5. Summary and discussion

5.1. Elasticity: from single molecules to tissues

Mechanically, titin is a modular polymer that acts as a multistage spring that adjusts both its length and its apparent extension stiffness in response to axial stress. Without external force, the necklace of beads acts as an entropic spring. Each folded domain, by its characteristic folding and idiosyncratic resistance to stretch, acts as a latent spring. Under external force, each domain "yields" and unfolds at a threshold values of force and loading rate. The disordered polypeptides are then "drafted" to participate in the entropic elasticity of the newly adjusted spring that now has a longer contour length and a different spring constant. Upon relaxation, the "draftees" fold individually and are "discharged" in an order that is determined by the rate and pathway of folding. Titin therefore undergoes the same "yield" phenomenon as the sarcomere and, indeed, as any composite biological material.

The behavior of this multistage spring therefore bears striking resemblance to the segmental extension of the sarcomere subjected to stretch. In the sarcomere, the tension rises mildly first and then exponentially until the threshold value at the yield point, where it levels off or drops. The force rise can be resolved into two mechanical phases by plotting log tension vs. sarcomere length (Wang et al., 1991, 1993), or three phases by plotting stiffness vs. tension (Granzier and Wang, 1993a, b). Each mechanical phase varies with the type of titin isoforms expressed in the muscle. It is conceivable that these mechanical phases correspond structurally to the straightening of Ig domains and the extension of PEVK or even the rupture of interfilament interactions between actin–nebulin co-filaments and myosin thick filaments. However, the major yield point of the sarcomere is controlled by the strength of interaction of titin with the thick filaments (Wang et al. 1991, 1993). Such "external force" imposed by thick filaments on titin may prevent the unfolding of Ig and Fn3 domains in the A band region of titin.

The fact that Ig and Fn3 modules do not unfold/refold significantly during physiological cycles is important, because it preserves the efficiency of striated muscles, as the extension/relaxation hysteresis is kept to a minimum. The unfolding of domains provides a large reserve of titin chain length and can act as a safety mechanism against sarcomere damage or fracturing under extreme pathological extension. The refolding ability provides a pathway to regenerate the original multistage elastic structure titin.

5.2. Emerging trends

We can expect more concerted efforts in detailed examination of other titin motifs using a new generation of single molecule research tools and new techniques of expressing engineered domains and polyproteins. Recently, we have undertaken an AFM study of the elasticity of a 52 KDa PEVK motif of human fetal skeletal muscle titin (Jin et al., 2000). The worm-like chain (WLC) models adequately describe its force–extension curves. The persistence length varies with Ca^{2+} concentration and ionic strength, suggesting that elasticity may be modulated by ionic conditions in muscle. Other constructs of specific titin regions are sure to follow.

The molecular insights reviewed here were made possible by single molecule techniques that are evolving constantly and rapidly. In biological AFM, current advances include the wider adaptation of the acoustic and magnetic tapping modes used under physiological conditions, development of faster AFM, smaller cantilevers (Viani et al., 1999a, b), functionalized nanotubes (Cheung et al., 2000; Nagao et al., 2000; Woolley et al., 2000), better characterization of substrate effects (Forbes, 2001; Glazier et al., 2000; Rao et al., 1997), continued advance in high resolution cryo-AFM (Czajkowsky and Shao, 1998; Shao et al., 1996, 1995) and instrumentation for force/volume/quantitative phase/other modes of data acquisition (see, e.g., another chapter in this series, Zlatanova et al., 2000). LOT technology has also been improved with better calibration methods (Bennink et al., 1999; Wang et al., 1997b), force-feedback modes (Wang et al., 1997b), multiple optical traps (Svoboda and Block, 1994; Visscher and Block, 1998, 1996), high-force setups (Kellermayer et al., 1998), and reduced sample heating and damage (Kuo, 1998; Neuman et al., 1999). For both LOT and AFM, more systematic considerations of the thermal noise behaviour and filtering may improve the measurement signal-to-noise ratio.

Other techniques for force measurements in biological systems, including membrane micropipette aspiration/bioforce transducer (Evans et al., 1995) and the more traditional surface force apparatus (Helm et al., 1991), are also being improved to compliment these two core single molecule techniques, AFM and LOT (see, e.g., a review by Leckband (2000)). Methods that allow simultaneous measurements, e.g. AFM and fluorescence, and the application of energy transfer to study protein folding/unfolding at the single molecule level, are likely to provide the next quantum leap in our pursuit of single molecule mechanics and single molecule biology (Fernandez et al., 2001; Mehta et al., 1999; Weiss, 1999; Zhuang et al., 2000).

Note added in proof

Recent structural and mechanical studies of the PEVK segment shed new light upon the structural basis of elasticity in this region of titin. Previously, the PEVK region of titin was assumed to be made up of random coils or unfolded polymer as in the mechanical analysis discussed above. New studies have now shown that the PEVK segment is modular and contains tandem repeats of modules containing 28 amino acids (Freiburg et al., 2000; Greaser, 2001; Gutierrez-Cruz et al., 2001). The existence of folded structure, in the form of polyproline-II helices (PPII), is demonstrated by NMR and circular dichroism of synthetic peptides and cloned PEVK protein (Ma et al., 2001). Measurement of the molecular elasticity of cloned PEVK domain using AFM force spectroscopy is found to contain both entropic and enthalpic contributions, and its elasticity is sensitive to the total ionic strength of the environment (Jin et al., 2001). These direct measurements corroborate the deduced behavior of PEVK from mechanical measurements of single myofibrils as studied by Linke and colleagues (Linke, 2000; Linke et al., 1998; Minajeva et al., 2001).

The elasticity and organizational roles of titin have also found relevance in tissues other than striated sarcomeres. Recent titin localization and mechanical studies of chromosomes clearly indicate important roles for titin in chromosome structure and mechanics (Machado and Andrew, 2000). The reported presence of smooth muscle titin, and its organizational roles in myosin assembly, raises the possibility that titin may yet be a new class of elastic cytoskeletal molecules that are widespread in muscle as well as nonmuscle cells (Keller et al., 2000). Defects in titin have now been linked to several types of muscular dystrophy (Haravuori et al., 2001; Sorimachi et al., 2000; van der Ven et al., 2000).

Acknowledgements

We thank Dr. M. Kellermayer and the other two expert referees for their insightful comments on the original version of this review and detailed suggestions, which we have gratefully adopted. We are also grateful for the helpful comments from Dr. Mike Lewis and other members of the Wang research group.

References

Ashkin, A., 1980. Applications of laser-radiation pressure. Science 210, 1081–1088.

Ashkin, A., 1992. Forces of a single-beam gradient laser trap on a dielectric sphere in the ray optics regime. Biophys. J. 61, 569–582.

Ashkin, A., Dziedzic, J.M., Bjorkholm, J.E., Chu, S., 1986. Obervation of a single-beam gradient force optical trap for dielectric particles. Opt. Lett. 11, 288–290.

Bell, G.I., 1978. Models for the specific adhesion of cells to cells. Science 200, 618–627.

Bennink, M.L., Scharer, O.D., Kanaar, R., Sakata-Sogawa, K., Schins, J.M., Kanger, J.S., de Grooth, B.G., Greve, J., 1999. Single-molecule manipulation of double-stranded DNA using optical tweezers: interaction studies of DNA with RecA and YOYO-1. Cytometry 36, 200–208.

Binnig, G., Quate, C.F., Gerber, C., 1986. Atomic force microscope. Phys. Rev. Lett. 56, 930–933.

Block, S.M., 1992. Making light work with optical tweezers. Nature 360, 493–495.

Block, S.M., 1995. Nanometers and piconewtons: the macromolecular mechanics of kinesin. Trends Cell Biol. 5, 169–175.

Bouchiat, C., Wang, M.D., Allemand, J.F., Strick, T., Block, S.M., Croquette, V., 1999. Estimating the persistence length of a worm-like chain molecule from force–extension measurements. Biophys. J. 76, 409–413.

Butt, H.-J., Jaschke, M., 1995. Calculation of thermal noise in atomic force microscopy. Nanotechnology 6, 1–7.

Carrion-Vazquez, M., Marszalek, P.E., Oberhauser, A.F., Fernandez, J.M., 1999a. Atomic force microscopy captures length phenotypes in single proteins. Proc. Natl. Acad. Sci. USA 96, 11288–11292.

Carrion-Vazquez, M., Oberhauser, A.F., Fowler, S.B., Marszalek, P.E., Broedel, S.E., Clarke, J., Fernandez, J.M., 1999b. Mechanical and chemical unfolding of a single protein: a comparison. Proc. Natl. Acad. Sci. USA 96, 3694–3699.

Cheung, C.L., Hafner, J.H., Lieber, C.M., 2000. Carbon nanotube atomic force microscopy tips: direct growth by chemical vapor deposition and application to high-resolution imaging. Proc. Natl. Acad. Sci. USA 97, 3809–3813.

Chilkoti, A., Boland, T., Ratner, B.D., Stayton, P.S., 1995. The relationship between ligand-binding thermodynamics and protein–ligand interaction forces measured by atomic force microscopy. Biophys. J. 69, 2125–2130.

Chu, V., Freitag, S., Le Trong, I., Stenkamp, R.E., Stayton, P.S., 1998. Thermodynamic and structural consequences of flexible loop deletion by circular permutation in the streptavidin–biotin system. Protein Sci. 7, 848–859.

Cleveland, J.P., Manne, S., Bocek, D., Hansma, P.K., 1993. A nondestructive method for determining the spring constant of cantilevers for scanning force microscopy. Rev. Sci. Instr. 64, 403–405.

Czajkowsky, D.M., Shao, Z.F., 1998. Submolecular resolution of single macromolecules with atomic force microscopy. FEBS Lett. 430, 51–54.

Dai, J.W., Sheetz, M.P., 1998. Cell membrane mechanics. In: Sheetz, M.P. (Ed.), Methods in Cell Biology, Vol. 55. Academic Press, San Diego, pp. 157–171.

Derose, J.A., Thundat, T., Nagahara, L.A., Lindsay, S.M., 1991. Gold grown epitaxially on mica—conditions for large area flat faces. Surf. Sci. 256, 102–108.

Eaton, W.A., Munoz, V., Hagen, S.J., Jas, G.S., Lapidus, L.J., Henry, E.R., Hofrichter, J., 2000. Fast kinetics and mechanisms in protein folding. Annu. Rev. Biophys. Biomol. Struct. 29, 327–359.

Erickson, H.P., 1997. Protein biophysics—stretching single protein molecules: titin is a weird spring. Science 276, 1090–1092.

Evans, E., 1998. Energy landscapes of biomolecular adhesion and receptor anchoring at interfaces explored with dynamic force spectroscopy. Faraday Discuss. 111, 1–16.

Evans, E., 1999. Looking inside molecular bonds at biological interfaces with dynamic force spectroscopy. Biophys. Chem. 82, 83–97.

Evans, E., Ritchie, K., 1997. Dynamic strength of molecular adhesion bonds. Biophys. J. 72, 1541–1555.

Evans, E., Ritchie, K., 1999. Strength of a weak bond connecting flexible polymer chains. Biophys. J. 76, 2439–2447.

Evans, E., Ritchie, K., Merkel, R., 1995. Sensitive force technique to probe molecular adhesion and structural linkages at biological interfaces. Biophys. J. 68, 2580–2587.

Fernandez, J.M., Chu, S., Oberhauser, A.F., 2001. RNA structure—pulling on hair(pins). Science 292, 653–654.

Florin, E.L., Moy, V.T., Gaub, H.E., 1994. Adhesion forces between individual ligand–receptor pairs. Science 264, 415–417.

Florin, E.L., Rief, M., Lehmann, H., Ludwig, M., Dornmair, C., Moy, V.T., Gaub, H.E., 1995. Sensing specific molecular-interactions with the atomic-force microscope. Biosens. Bioelectron. 10, 895–901.

Forbes, J.G., Jin, A.J., Wang, K., 2000 AFM investigation of native nebulin. Biophys. J. 78, 380A.

Forbes, J.G., Jin, A.J., Wang, K., 2001. Atomic force microscope study of the effect of the immobilization substrate on the structure and force-extension curves of a multimeric protein. Langmuir 17, 3067–3075.

Freiburg, A., Trombitas, K., Hell, W., Cazorla, O., Fougerousse, F., Centner, T., Kolmerer, B., Witt, C., Beckmann, J.S., Gregorio, C.C., Granzier, H., Labeit, S., 2000. Series of exon-skipping events in the elastic spring region of titin as the structural basis for myofibrillar elastic diversity. Circ. Res. 86, 1114–1121.

Funatsu, T., Higuchi, H., Ishiwata, S., 1990. Elastic filaments in skeletal muscle revealed by selective removal of thin filaments with plasma gelsolin. J. Cell Biol. 110, 53–62.

Furst, D., Osborn, M., Nave, R., Weber, K., 1988. The organization of titin filaments in the half sarcomere revealed by monoclonal antibodies in immunoelectron microscopy: a map of ten nonrepetitive epitopes starting at the z line extends close to the m line. J. Cell Biol. 106, 1563–1572.

Gao, M., Isralewitz, B., Lu, H., Schulten, K., 2000. Steered molecular dynamics simulation of the unfolding of multiple titin immunoglobulin domains. Biophys. J. 78, 28A.

Gittes, F., Schmidt, C.F., 1998. Signals and noise in micromechanical measurements. In Methods in cell biology, Vol. 55, (ed. M.P. Sheetz), pp. 129–156.

Glazier, S.A., Vanderah, D.J., Plant, A.L., Bayley, H., Valincius, G., Kasianowicz, J.J., 2000. Reconstitution of the pore-forming toxin alpha-hemolysin in phospholipid/18-octadecyl-1-thiahexa(ethylene oxide) and phospholipid/n-octadecanethiol supported bilayer membranes. Langmuir 16, 10428–10435.

Grandbois, M., Beyer, M., Rief, M., Clausen-Schaumann, H., Gaub, H.E., 1999. How strong is a covalent bond? Science 283, 1727–1730.

Granzier, H., Helmes, M., Cazorla, O., McNabb, M., Labeit, D., Wu, Y.M., Yamasaki, R., Redkar, A., Kellermayer, M., Labeit, S., Trombitas, K., 2000. Mechanical properties of titin isoforms. Adv. Exp. Med. Biol. 481, 283–304.

Granzier, H., Helmes, M., Trombitas, K., 1996. Nonuniform elasticity of titin in cardiac myocytes: a study using immunoelectron microscopy and cellular mechanics. Biophys. J. 70, 430–442.

Granzier, H.L., Irving, T.C., 1995. Passive tension in cardiac muscle: contribution of collagen, titin, microtubules, and intermediate filaments. Biophys. J. 68, 1027–1044.

Granzier, H.L.M., Wang, K., 1993a. Interplay between passive tension and strong and weak binding cross-bridges in insect indirect flight-muscle—a functional dissection by gelsolin-mediated thin filament removal. J. Gen. Physiol. 101, 235–270.

Granzier, H.L.M., Wang, K., 1993b. Passive tension and stiffness of vertebrate skeletal and insect flight muscles: the contribution of weak cross-bridges and elastic filaments. Biophys. J. 65, 2141–2159.

Greaser, M., 2001. Identification of new repeating motifs in titin. Proteins 43, 145–149.

Gutierrez-Cruz, G., Van Heerden, A., Wang, K., 2001. Modular motif, structural folds and affinity profiles of PEVK segment of human fetal skeletal muscle titin. J. Biol. Chem 276, 7442–7449.

Hallett, P., Tskhovrebova, L., Trinick, J., Offer, G., Miles, M.J., 1996. Improvements in atomic force microscopy protocols for imaging fibrous proteins. J. Vac. Sci. Technol. B 14, 1444–1448.

Hansma, P.K., Elings, V.B., Marti, O., Bracker, C.E., 1988. Scanning tunneling microscopy and atomic force microscopy: Application to biology and technology. Science 242, 209–242.

Haravuori, H., Vihola, A., Straub, V., Auranen, M., Richard, I., Marchand, S., Voit, T., Labeit, S., Somer, H., Peltonen, L., Beckmann, J.S., Udd, B., 2001. Secondary calpain3 deficiency in 2q-linked muscular dystrophy—titin is the candidate gene. Neurology 56, 869–877.

Helm, C.A., Knoll, W., Israelachvili, J.N., 1991. Measurement of ligand receptor interactions. Proc. Natl. Acad. Sci. USA 88, 8169–8173.

Higuchi, H., Nakauchi, Y., Maruyama, K., Fujime, S., 1993. Characterization of beta-connectin (titin 2) from striated muscle by dynamic light scattering. Biophys. J. 65, 1906–1915.

Hinterdorfer, P., Baumgartner, W., Gruber, H.J., Schilcher, K., Schindler, H., 1996. Detection and localization of individual antibody-antigen recognition events by atomic force microscopy. Proc. Natl. Acad. Sci. USA 93, 3477–3481.

Horowits, R., 1999. The physiological role of titin in striated muscle. Rev. Physiol. Biochem. Pharmacol. 138, 57–96.

Horowits, R., Kempner, E.S., Bisher, M.E., Podolsky, R.J., 1986. A physiological role for titin and nebulin in skeletal muscle. Nature 323, 160–164.

Horowits, R., Maruyama, K., Podolsky, R.J., 1989. Elastic behavior of connectin filaments during thick filament movement in activated skeletal muscle. J. Cell Biol. 109, 2169–2176.

Hutter, J.L., Bechhoefer, J., 1993. Calibration of atomic-force microscope tips. Rev. Sci. Instr. 64, 1868–1873.

Huxley, H.E., 1990. Sliding filaments and molecular motile systems. J. Biol. Chem. 265, 8347–8350.

Izrailev, S., Stepaniants, S., Balsera, M., Oono, Y., Schulten, K., 1997. Molecular dynamics study of unbinding of the avidin–biotin complex. Biophys. J. 72, 1568–1581.

Jin, A.J., Forbes, J.G., Wang, K., 2000. Atomic force microscopy studies of a titin PEVK segment. Biophys. J. 78, 381A.

Jin, A.J., Forbes, J.G., Wang, K., 2001. Molecular elasticity of the PEVK segment of titin: Entropic and enthalpic contributions. Submitted.

Keller, T.C.S., 1997. Muscle structure—molecular bungees. Nature 387, 233–235.

Keller, T.C.S., Eilertsen, K., Higginbotham, M., Kazmierski, S., Kim, K.T., Velichkova, M., 2000. Role of titin in nonmuscle and smooth muscle cells. In Elastic filaments of the cell, Vol. 481, pp. 265–281.

Kellermayer, M.S.Z., Smith, S.B., Granzier, H.L., Bustamante, C., 1997. Folding-unfolding transitions in single titin molecules characterized with laser tweezers. Science 276, 1112–1116.

Kellermayer, M.S.Z., Smith, S.B., Bustamante, C., Granzier, H.L., 1998. Complete unfolding of the titin molecule under external force. J. Struct. Biol. 122, 197–205.

Kellermayer, M.S.Z., Smith, S.B., Bustamante, C., Granzier, H.L., 2000. Mechanical manipulation of single titin molecules with laser tweezers. Adv. Exp. Med. Biol. 481, 111–126.

Klimov, D.K., Thirumalai, D., 1999. Stretching single-domain proteins: phase diagram and kinetics of force-induced unfolding. Proc. Natl. Acad. Sci. USA 96, 6166–6170.

Klimov, D.K., Thirumalai, D., 2000. Native topology determines force-induced unfolding pathways in globular proteins. Proc. Natl. Acad. Sci. USA 97, 7254–7259.

Kramers, H.A., 1940. Brownian motion in a field of force and the diffusion model of chemical kinetics. Physica 7, 284–304.

Krammer, A., Lu, H., Isralewitz, B., Schulten, K., Vogel, V., 1999. Forced unfolding of the fibronectin type III module reveals a tensile molecular recognition switch. Proc. Natl. Acad. Sci. USA 96, 1351–1356.

Kuo, S.C., 1998. A simple assay for local heating by optical tweezers. In: Sheetz, M.P. (Ed.), Methods in Cell Biology, Vol. 55. Academic Press, San Diego, pp. 43–45.

Labeit, S., Kolmerer, B., 1995. Titins—giant proteins in charge of muscle ultrastructure and elasticity. Science 270, 293–296.

Leckband, D., 2000. Measuring the forces that control protein interactions. Annu. Rev. Biophys. Biomol. Struct. 29, 1–26.

Lee, G.U., Chrisey, L.A., Colton, R.J., 1994. Direct measurement of the forces between complementary strands of DNA. Science 266, 771–773.

Levitt, M., Gerstein, M., Huang, E., Subbiah, S., Tsai, J., 1997. Protein folding: the endgame. Annu. Rev. Biochem. 66, 549–579.

Linke, W.A., 2000. Titin elasticity in the context of the sarcomere: Force and extensibility measurements on single myofibrils. In Elastic filaments of the cell, Vol. 481, pp. 179–206.

Linke, W.A., Ivemeyer, M., Mundel, P., Stockmeier, M.R., Kolmerer, B., 1998. Nature of PEVK-titin elasticity in skeletal muscle. Proc. Natl. Acad. Sci. USA 95, 8052–8057.

Linke, W.A., Ivemeyer, M., Olivieri, N., Kolmerer, B., Ruegg, J.C., Labeit, S., 1996. Towards a molecular understanding of the elasticity of titin. J. Mol. Biol. 261, 62–71.

Linke, W.A., Rudy, D.E., Centner, T., Gautel, M., Witt, C., Labeit, S., Gregorio, C.C., 1999. I-band titin in cardiac muscle is a three-element molecular spring and is critical for maintaining thin filament structure. J. Cell Biol. 146, 631–644.

Liu, Y.Z., Leuba, S.H., Lindsay, S.M., 1999. Relationship between stiffness and force in single molecule pulling experiments. Langmuir 15, 8547–8548.

Lu, H., Isralewitz, B., Krammer, A., Vogel, V., Schulten, K., 1998. Unfolding of titin immunoglobulin domains by steered molecular dynamics simulation. Biophys. J. 75, 662–671.

Lu, H., Schulten, K., 1999a. Steered molecular dynamics simulation of conformational changes of immunoglobulin domain i27 interpret atomic force microscopy observations. Chem. Phys. 247, 141–153.

Lu, H., Schulten, K., 1999b. Steered molecular dynamics simulations of force-induced protein domain unfolding. Proteins-Struct. Function Genet. 35, 453–463.

Lu, H., Schulten, K., 2000a. The key event in force-induced unfolding of titin's immunoglobulin domains. Biophys. J. 79, 51–65.

Ma, K., Kan, L.S., Wang, K., 2001. Polyproline ii helix is a key structural motif of the elastic PEVK segment of titin. Biochemistry 40, 3427–3438.

Machado, C., Andrew, D.J., 2000. Titin as a chromosomal protein. In Elastic filaments of the cell, Vol. 481, pp. 221–236.

Marko, J.F., Siggia, E.D., 1995. Stretching DNA. Macromolecules 28, 8759–8770.

Marszalek, P.E., Lu, H., Li, H.B., Carrion-Vazquez, M., Oberhauser, A.F., Schulten, K., Fernandez, J.M., 1999. Mechanical unfolding intermediates in titin modules. Nature 402, 100–103.

Maruyama, K., 1994. Connectin, an elastic protein of striated muscle. Biophys. Chem. 50, 73–85.

Maruyama, K., 1997. Connectin/titin, giant elastic protein of muscle. FASEB J. 11, 341–345.

Maruyama, K., Kimura, S., Ohashi, K., Kuwano, Y., 1981. Connectin, an elastic protein of muscle. Identification of titin with connectin. J. Biochem. (Tokyo) 89, 701–709.

Mehta, A.D., Rief, M., Spudich, J.A., Smith, D.A., Simmons, R.M., 1999. Single-molecule biomechanics with optical methods. Science 283, 1689–1695.

Merkel, R., Nassoy, P., Leung, A., Ritchie, K., Evans, E., 1999. Energy landscapes of receptor–ligand bonds explored with dynamic force spectroscopy. Nature 397, 50–53.

Minajeva, A., Kulke, M., Fernandez, J.M., Linke, W.A., 2001. Unfolding of titin domains explains the viscoelastic behavior of skeletal myofibrils. Biophys. J. 80, 1442–1451.

Moy, V.T., Florin, E.L., Gaub, H.E., 1994. Intermolecular forces and energies between ligands and receptors. Science 266, 257–259.

Muniz, J., Marin, J.L., Huerta, M., Del Rio, J., Larios, A., Trujillo, X., 1999. Passive forces in mammalian skeletal muscle: a freely-jointed and worm-like chain. Gen. Physiol. Biophys. 18, 155–163.

Nagao, E., Nishijima, H., Akita, S., Nakayama, Y., Dvorak, J.A., 2000. The cell biological application of carbon nanotube probes for atomic force microscopy: comparative studies of malaria-infected erythrocytes. J. Electron Microsc. 49, 453–458.

Netzer, W.J., Hartl, F.U., 1997. Recombination of protein domains facilitated by co-translational folding in eukaryotes. Nature 388, 343–349.

Oberhauser, A.F., Marszalek, P.E., Carrion-Vazquez, M., Fernandez, J.M., 1999. Single protein misfolding events captured by atomic force microscopy. Nat. Struct. Biol. 6, 1025–1028.

Oberhauser, A.F., Marszalek, P.E., Erickson, H.P., Fernandez, J.M., 1998. The molecular elasticity of the extracellular matrix protein tenascin. Nature 393, 181–185.

Obermann, W.M., Plessmann, U., Weber, K., Furst, D.O., 1995. Purification and biochemical characterization of myomesin, a myosin-binding and titin-binding protein, from bovine skeletal muscle. Eur. J. Biochem. 233, 110–115.

Onuchic, J.N., LutheySchulten, Z., Wolynes, P.G., 1997. Theory of protein folding: the energy landscape perspective. Annu. Rev. Phys. Chem. 48, 545–600.

Ortiz, C., Hadziioannou, G., 1999. Entropic elasticity of single polymer chains of poly(methacrylic acid) measured by atomic force microscopy. Macromolecules 32, 780–787.

Paci, E., Karplus, M., 1999. Forced unfolding of fibronectin type 3 modules: an analysis by biased molecular dynamics simulations. J. Mol. Biol. 288, 441–459.

Paci, E., Karplus, M., 2000. Unfolding proteins by external forces and temperature: the importance of topology and energetics. Proc. Natl. Acad. Sci. USA 97, 6521–6526.

Pan, K.-M., Damodaran, S., Greaser, M.L., 1994. Isolation and characterization of titin t1 from bovine cardiac muscle. Biochemistry 33, 8255–8261.

Plaxco, K.W., Spitzfaden, C., Campbell, I.D., Dobson, C.M., 1997. A comparison of the folding kinetics and thermodynamics of two homologous fibronectin type III modules. J. Mol. Biol. 270, 763–770.

Rao, N.M., Plant, A.L., Silin, V., Wight, S., Hui, S.W., 1997. Characterization of biomimetic surfaces formed from cell membranes. Biophys. J. 73, 3066–3077.

Rief, M., Gautel, M., Oesterhelt, F., Fernandez, J.M., Gaub, H.E., 1997a. Reversible unfolding of individual titin immunoglobulin domains by afm. Science 276, 1109–1112.

Rief, M., Oesterhelt, F., Heymann, B., Gaub, H.E., 1997b. Single molecule force spectroscopy on polysaccharides by atomic force microscopy. Science 275, 1295–1297.

Rief, M., Gautel, M., Schemmel, A., Gaub, H.E., 1998. The mechanical stability of immunoglobulin and fibronectin III domains in the muscle protein titin measured by atomic force microscopy. Biophys. J. 75, 3008–3014.

Rohs, R., Etchebest, C., Lavery, R., 1999. Unraveling proteins: a molecular mechanics study. Biophys. J. 76, 2760–2768.

Sader, J.E., Chon, J.W.M., Mulvaney, P., 1999. Calibration of rectangular atomic force microscope cantilevers. Rev. Sci. Instr. 70, 3967–3969.

Sader, J.E., Larson, I., Mulvaney, P., White, L.R., 1995. Method for the calibration of atomic-force microscope cantilevers. Rev. Sci. Instr. 66, 3789–3798.

Shao, Z.F., Mou, J., Czajkowsky, D.M., Yang, J., Yuan, J.Y., 1996. Biological atomic force microscopy: what is achieved and what is needed. Adv. Phys. 45, 1–86.

Shao, Z.F., Yang, J., Somlyo, A.P., 1995. Biological atomic force microscopy: from microns to nanometers and beyond. Annu. Rev. Cell Dev. Biol. 11, 241–265.

Shtilerman, M., Lorimer, G.H., Englander, S.W., 1999. Chaperonin function: folding by forced unfolding. Science 284, 822–825.

Smith, S.B., Cui, Y.J., Bustamante, C., 1996. Overstretching b-DNA: the elastic response of individual double-stranded and single-stranded DNA molecules. Science 271, 795–799.

Socci, N.D., Onuchic, J.N., Wolynes, P.G., 1999. Stretching lattice models of protein folding. Proc. Natl. Acad. Sci. USA 96, 2031–2035.

Sorimachi, H., Ono, Y., Suzuki, K., 2000. Skeletal muscle-specific calpain, p94, and connectin/titin: Their physiological functions and relationship to limb-girdle muscular dystrophy type 2a. In Elastic filaments of the cell, Vol. 481, pp. 383–397.

Soteriou, A., Gamage, M., Trinick, J., 1993. A survey of interactions made by the giant protein titin. J. Cell Sci. 104, 119–123.

Strick, T.R., Allemand, J.F., Bensimon, D., Croquette, V., 2000. Stress-induced structural transitions in DNA and proteins. Annu. Rev. Biophys. Biomolec. Struct 29, 523–543.

Strunz, T., Oroszlan, K., Schumakovitch, I., Guntherodt, H.J., Hegner, M., 2000. Model energy landscapes and the force-induced dissociation of ligand–receptor bonds. Biophys. J. 79, 1206–1212.

Svoboda, K., Block, S.M., 1994. Biological applications of optical forces. In: Stroud, R.M., Cantor, C.R., Pollard, T.D. (Eds.), Annual Reviews of Biophysics and Biomolecular Structures, Vol. 23. Annual Reviews, Inc., Palo Alto, CA, pp. 247–285.

Trinick, J., Tskhovrebova, L., 1999. Titin: a molecular control freak. Trends Cell Biol. 9, 377–380.

Trombitas, K., Greaser, M., Labeit, S., Jin, J.P., Kellermayer, M., Helmes, M., Granzier, H., 1998. Titin extensibility in situ: entropic elasticity of permanently folded and permanently unfolded molecular segments. J. Cell Biol. 140, 853–859.

Trombitas, K., Pollack, G.H., 1993. Elastic properties of connecting filaments along the sarcomere. Adv. Exp. Med. Biol. 332, 71–79.

Tskhovrebova, L., Trinick, J., Sleep, J., Simmons, R., 1996. Optical tweezer measurements of force and extension in individual titin molecules. J. Muscle Res. Cell Motil. 17, 113.

Tskhovrebova, L., Trinick, J., Sleep, J.A., Simmons, R.M., 1997. Elasticity and unfolding of single molecules of the giant muscle protein titin. Nature 387, 308–312.

Tsuda, Y., Yasutake, H., Ishijima, A., Yanagida, T., 1996. Torsional rigidity of single actin filaments and actin–actin bond breaking force under torsion measured directly by in vitro micromanipulation. Proc. Natl. Acad. Sci. USA 93, 12937–12942.

van der Ven, P.F.M., Wiesner, S., Salmikangas, P., Auerbach, D., Himmel, M., Kempa, S., Hayess, K., Pacholsky, D., Taivainen, A., Schroder, R., Carpen, O., Furst, D.O., 2000. Indications for a novel muscular dystrophy pathway: Gamma-filamin, the muscle-specific filamin isoform, interacts with myotilin. J. Cell Biol. 151, 235–247.

Viani, M.B., Schaffer, T.E., Chand, A., Rief, M., Gaub, H.E., Hansma, P.K., 1999a. Small cantilevers for force spectroscopy of single molecules. J. Appl. Phys. 86, 2258–2262.

Viani, M.B., Schaffer, T.E., Paloczi, G.T., Pietrasanta, L.I., Smith, B.L., Thompson, J.B., Richter, M., Rief, M., Gaub, H.E., Plaxco, K.W., Cleland, A.N., Hansma, H.G., Hansma, P.K., 1999b. Fast imaging and fast force spectroscopy of single biopolymers with a new atomic force microscope designed for small cantilevers. Rev. Sci. Instr. 70, 4300–4303.

Visscher, K., Block, S.M., 1998. Versatile optical traps with feedback control. Methods Enzymol. 298, 460–489.

Visscher, K., Gross, S.P., Block, S.M., 1996. Construction of multiple-beam optical traps with nanometer-resolution position sensing. IEEE J. Sel. Top. Quantum Electron. 2, 1066–1076.

Wang, K., 1982. Purification of titin and nebulin. Methods Enzymol. 85, 264–274.

Wang, K., 1996. Titin/connectin and nebulin: giant protein rulers of muscle structure and function. Adv. Biophys. 33, 123–134.

Wang, K., 1999. Titin. In: Kries, T., Vale, R. (Eds.), Guidebook to the Cytoskeletal and Motor Proteins. Oxford University Press, London, pp. 481–486.

Wang, K., McCarter, R., Wright, J., Beverly, J., Ramirez-Mitchell, R., 1991. Regulation of skeletal muscle stiffness and elasticity by titin isoforms: a test of the segmental extension model of resting tension. Proc. Natl. Acad. Sci. USA 88, 7101–7105.

Wang, K., McCarter, R., Wright, J., Beverly, J., Ramirez-Mitchell, R., 1993. Viscoelasticity of the sarcomere matrix of skeletal muscles: the titin–myosin composite filament is a dual-stage molecular spring. Biophys. J. 64, 1161–1177.

Wang, K., McClure, J., Tu, A., 1979. Titin: major myofibrillar components of striated muscle. Proc. Natl. Acad. Sci. USA 76, 3698–3702.

Wang, K., Ramirez-Mitchell, R., Palter, D., 1984. Titin is an extraordinarily long, flexible, and slender myofibrillar protein. Proc. Natl. Acad. Sci. USA 81, 3685–3689.

Wang, K., Wright, J., Gutierrez, G., Brady, A., Sheetz, M., 1997a. Titin elasticity: laser trap measurement of force–extension curves of isolated titin molecules. Biophys. J. 72, MAMA6.

Wang, M.D., Schnitzer, M.J., Yin, H., Landick, R., Gelles, J., Block, S.M., 1998. Force and velocity measured for single molecules of RNA polymerase. Science 282, 902–907.

Wang, M.D., Yin, H., Landick, R., Gelles, J., Block, S.M., 1997b. Stretching DNA with optical tweezers. Biophys. J. 72, 1335–1346.

Weiss, S., 1999. Fluorescence spectroscopy of single biomolecules. Science 283, 1676–1683.

Whiting, A., Wardale, J., Trinick, J., 1989. Does titin regulate the length of muscle thick filaments? J. Mol. Biol. 205, 263–268.

Woolley, A.T., Guillemette, C., Cheung, C.L., Housman, D.E., Lieber, C.M., 2000. Direct haplotyping of kilobase-size DNA using carbon nanotube probes. Nat. Biotechnol. 18, 760–763.

Yang, G.L., Cecconi, C., Baase, W.A., Vetter, I.R., Breyer, W.A., Haack, J.A., Matthews, B.W., Dahlquist, F.W., Bustamante, C., 2000. Solid-state synthesis and mechanical unfolding of polymers of T4 lysozyme. Proc. Natl. Acad. Sci. USA 97, 139–144.

Yasuda, R., Miyata, H., Kinosita, K., 1996. Direct measurement of the torsional rigidity of single actin filaments. J. Mol. Biol. 263, 227–236.

Zhang, B., Xu, G.Z., Evans, J.S., 1999. A kinetic molecular model of the reversible unfolding and refolding of titin under force extension. Biophys. J. 77, 1306–1315.

Zheng, X.Y., Ding, Y.Z., Bottomley, L.A., Allison, D.P., Warmack, R.J., 1995. Discontinuous gold island films on mica. J. Vac. Sci. Technol. B 13, 1320–1324.

Zhuang, X.W., Ha, T., Kim, H.D., Centner, T., Labeit, S., Chu, S., 2000. Fluorescence quenching: A tool for single-molecule protein-folding study. Proc. Natl. Acad. Sci. U.S.A. 97, 14241–14244.

Zlatanova, J., Lindsay, S.M., Leuba, S.H., 2000. Single molecule force spectroscopy in biology using the atomic force microscope. Prog. Biophys. Mol. Biol. 74, 37–61.

Biology at the Single Molecule Level
S.H. Leuba an J. Zlatanova (Eds)

Review

Analysis of single-molecule mechanical recordings: application to acto-myosin interactions

Alex E. Knight[a,*], Claudia Veigel[a], Charles Chambers[b], Justin E. Molloy[a]

[a] *Biology Department, University of York, P.O. Box 373, York YO10 5YW, UK*
[b] *Quotient Communications Ltd., 1 Langford Arch, London Road, Pampisford, Cambridge CB2 4EE, UK*

Abstract

Several laboratories have now developed methods to make single-molecule mechanical recordings from interacting pairs of biological molecules. The mechanical work done (product of force and distance) by a single biomolecular interaction is usually of the same order as thermal energy. Recordings made from non-processive, intermittently interacting, molecular motors such as acto-myosin therefore contain a large background of thermal noise. We have applied Page's test to analyse mechanical interactions between muscle myosin II's and F-actin recorded using an optical tweezers based single-molecule mechanical transducer. We compare Page's test with other variance-based methods and find it to be a robust method for analysing both simulated and real data sets. We discuss some of the problems associated with automatic detection of transient mechanical events in noisy data signals, and show that if the start and end points of individual events are known accurately then the events may be synchronised and combined to give more detailed information about different mechanical states. © 2001 Elsevier Science Ltd. All rights reserved.

Contents

*Corresponding author. Tel.: +44-1904-432829; fax: +44-1904-432860. http://www.motility.york.ac.uk.
E-mail addresses: aek4@york.ac.uk (A.E. Knight).

1. Introduction

Recent years have seen the development of a number of technologies that allow us to observe the behaviour of individual molecules, including scanning probe microscopies, optical tweezers and single-molecule fluorescence (e.g. papers in Vol. 74/1 and this issue and Uppenbrink and Clery, 1999). Single-molecule experiments can give dramatic visual confirmation of molecular mechanisms that have to some extent already been deduced from macroscopic studies; for example, the direct observation of F1 ATPase rotation (Noji et al., 1997). Such experiments can also give novel information about details of the molecular mechanisms which may enable us to distinguish between model-dependent interpretations of bulk measurement data. Finally, single-molecule experiments are sometimes the only way possible to study the mechanochemistry of motor proteins such as RNA polymerase (Wang et al., 1998; Davenport et al., 2000), kinesins (Visscher et al., 1999; Vale et al., 1996; Kojima et al., 1997), dyneins (Shingyoji et al., 1998; Sakakibara et al., 1999) and unconventional myosins (Veigel et al., 1999; Mehta et al., 1999b). To date, molecular motors have been the best studied "prototype" system for single-molecule techniques in biology, and it is very likely that the lessons learned in experimental methods, data analysis and interpretation will be important for future studies of a wide range of biological molecules.

Muscle myosin II's are the most extensively studied class of molecular motor: The proteins (myosin and actin) are easily purified in large quantities for biochemistry and these experiments, combined with mechanical studies of single intact muscle fibres, have told us much about the chemical and mechanical behaviours of the acto-myosin ATPase cycle (e.g. Cooke, 1998). It is clear from such studies that understanding the effects of mechanical load on the biochemical kinetics is key to a full understanding of the mechanochemical coupling process. Unfortunately, the highly ordered structure of the muscle sarcomeric thick and thin filament lattice means that different myosin molecules are held in different local environments and therefore experience different loads (e.g. White and Thorson, 1973). Dealing with the effects of mechanical load both prior and subsequent to acto-myosin binding has been both a theoretical and experimental challenge (e.g. Huxley, 1957; Hill, 1974). The unknown load dependence of individual biochemical steps has therefore lead to some degree of ambiguity in interpretation of the macroscopic data.

The current view of the mechanism of force production by acto-myosin is that myosin binds to actin with the products of ATP hydrolysis (ADP and phosphate) bound in the catalytic site (cross-bridge attachment). Then, as the products are released, myosin changes conformation to produce a movement or "power-stroke" (Huxley, 1969). In the absence of nucleotide, actin and myosin form a tightly bound "rigor" complex. Binding of a new ATP molecule to myosin causes the rigor complex to dissociate (cross-bridge detachment) and subsequent ATP hydrolysis resets the original myosin conformation so that the cycle can be repeated (Lymn and Taylor, 1971). During the cycle, part of the cross-bridge becomes distorted by the working-stroke and mechanical work is stored in this elastic deformation. In this way, the acto-myosin cross-bridge captures the sudden changes in chemical potential associated with steps in the biochemical cycle and is able to do external work on a much slower time scale; e.g. as muscle shortens or vesicles are transported (Huxley and Simmons, 1971). There are three key features to this mechanism: (1) one ATP molecule is broken down per mechanical cycle; (2) the size of the working-stroke is determined by the conformation of myosin at the start and end of the attached period; (3) the cross-bridge is elastic (see review by Cooke, 1998). The biochemical and mechanical cycles are "coupled" because different myosin biochemical states have different affinities for actin. Also, product release is hugely accelerated when myosin is tightly bound to actin. Because product release occurs rapidly after myosin binds to actin the force generating steps are not easily resolved by optical tweezers measurements (but see Section 5.2). So, most of the published single-molecule mechanical data for muscle myosins are dominated by the tightly bound, nucleotide free, "rigor" state. The mechanical "Holy Grail" will be to identify the properties of acto-myosin products states occurring at the start of the myosin "power stroke".

What questions can single-molecule mechanical studies of acto-myosin hope to resolve? First, we need to confirm that the basic mechanical parameters of the acto-myosin interaction namely, the displacement and force produced, and the mechanical stiffness of the complex, are the same as measured in muscle fibres. Then we need to know the time course of the events that occur during a single interaction, for example, the duration of different mechanical and biochemical states, and how they relate to bulk biochemical kinetics measurements. Finally, we need to know the strain dependence of each of the various biochemical and mechanical steps that we identify.

In this paper, we briefly describe how single-molecule acto-myosin interactions may be measured; we review some of the methods of data analysis and introduce a modified version of Page's test that we have adapted to identify intermittent single-molecule binding events within noisy data signals; finally, we discuss why improved event detection techniques might lead to a better understanding of the mechanism of force production by acto-myosin.

2. In vitro assays and optical tweezers

In vitro motility assays using purified molecules (the "motor" and its "track") have enabled the action of molecular motors to be observed directly by light microscopy in a buffer containing the appropriate substrates (Scholey, 1993). For example, myosin motors will move along actin tracks in a buffer containing Mg · ATP (Kron and Spudich, 1986), as will kinesin motors on microtubules

(Vale et al., 1985); RNA polymerase will move along template DNA molecules in the presence of the four NTPs (Schafer et al., 1991). One of the components (track or motor) is immobilised on the microscope cover slip, while the other is visualised by a suitable form of light microscopy. (For example, a fluorescent label or a plastic bead may be attached to the protein.) Using such assays, it has been possible to measure a variety of different properties of motor function under unloaded conditions. (The method can also be extended to loaded conditions by using a centrifuge microscope (Oiwa et al., 1990; Hall et al., 1993), or by imposing viscous drag (Hunt et al., 1994)).

One example is an important motor property known as the mechanical *processivity*. This is a measure of the number of mechanical interactions that are made, on average, for each diffusional encounter between motor and track. Processivity is determined by the ratio of the probability for making a forward step versus that for detachment of the motor from the track (Veigel et al., in prep.). Hence, RNA polymerase is extremely processive, as it takes many thousands of steps to transcribe a gene. Conventional kinesin takes an average of 100 steps for each encounter it makes with a microtubule (Block et al., 1990), myosin V probably makes about 30 steps for each contact it makes with F-actin (Mehta et al., 1999b). In contrast, myosin IIs are referred to as "non-processive" motors because, on average, they take only one step. However, it is noteworthy that the two heads of myosin may interact with F-actin sequentially (Tyska et al., 1999) which implies that even skeletal muscle myosin II might be (very) weakly processive. So, it is important to emphasise the fact that processivity is a continuum. The concept of processivity is important both because it helps us understand the functional role of a particular motor system in the living cell and because the methods used to study weakly and strongly processive motors are somewhat different.

Because highly processive motors take multiple steps before diffusing away from their track, the translation produced by a single motor molecule is directly observable. This makes processive motors in some ways more tractable to study at the single-molecule level. In contrast, weakly or non-processive motors, like the muscle myosin II's, require large numbers of active motors to be present in order to produce smooth sliding motion of F-actin. Because muscle myosin IIs interact rapidly and intermittently with F-actin the individual events are difficult to detect. The breakthrough came in 1994 when Finer et al. (1994) devised an optical tweezers based method to record single acto-myosin interactions.

2.1. Optical tweezers based mechanotransducers

Optical tweezers (also known as optical traps) are a versatile tool for manipulating microscopic objects and making mechanical measurements (Mehta et al., 1999a). A laser beam is focused by a microscope objective into an experimental chamber, producing a steep gradient of light intensity close to the focal point. Micron-sized objects, of higher refractive index than the surrounding medium, become "trapped" in this region, which acts like a potential well (Ashkin et al., 1986; Svoboda and Block, 1994). The optical trap has the useful property that it behaves like a spring. Over a certain range, an optical trap generates a restoring force proportional to displacement of the object from the trap centre. The constant of proportionality is the trap stiffness, κ_{trap}, that may be controlled by adjusting the laser intensity. To increase the variety of experiments that may be performed, multiple optical tweezers may be produced, either by splitting the laser beam using a polarising beamsplitter or by rapidly scanning it with acousto-optic deflectors (Visscher et al., 1996; Molloy, 1998). If the position of the trapped object is measured with sub-nanometre resolution the optical tweezer and its associated trapped particle can be used as a transducer capable of measuring the force, movement and stiffness produced by interacting biological molecules.

In the following sections, we will discuss the methodology used to study acto-myosin-II interactions, including the methods of analysis that we have developed which enable single interactions to be picked out from the background of thermal noise.

2.2. The "three bead" geometry

The experimental arrangement used by Finer et al. (1994) is illustrated in Fig. 1A. Most workers since have adopted a very similar configuration. Actin and myosin cannot be manipulated directly with the optical tweezers and so must be attached to "handles"—in this case, 1 μm diameter plastic beads. Two beads are held in twin optical traps; the beads are attached to the ends of an actin filament, which is held taut between them. The myosin to be studied is attached to a third bead that is adhered to the microscope cover slip surface using nitrocellulose. The third bead acts as a fixed "pedestal" to raise the molecule above the surface. The actin filament is then held in the vicinity of the third bead and subsequent interactions are observed by monitoring the bead positions using four-quadrant photodiode detectors. Conceptually, three basic types of mechanical experiment may be performed: *Displacement mode*, in which optical tweezer stiffness is made low so that myosin can produce its maximum movement in an unhindered fashion; *Force mode*, in which the optical tweezer is made much stiffer than the myosin molecule (often by using a feedback loop) so that the acto-myosin complex is held isometric and allowed to develop its maximum force; and *Perturbation mode*, in which force or movement is externally applied to the proteins in order to measure the distortion or position dependence of the interaction (e.g. the immediate stiffness, or the strain dependence of binding or unbinding kinetics). Each of these types of experimental modes has some associated complications, which are discussed in the following sections.

2.3. Single-molecule recordings

Single-molecule mechanical data usually consist of digital time series information about the position of the beads held in the optical tweezers. Ideally, they should be recorded with sufficient bandwidth that the motion of the beads can be oversampled with an overall measurement noise well below ($< -20\,\mathrm{dB}$) the expected signal level.

Fig. 2 shows acto-myosin interactions recorded using the "three bead" method. The position of both of the optically trapped beads was recorded, in the x axis (defined as the direction parallel to the actin filament). In the unusual, but illustrative, record shown here, the actin filament breaks towards the end of the recording. This record demonstrates several features:

(1) *Thermal noise dominates the record.* It is obvious that the data is very noisy. This is due to the thermal or Brownian motion of the actin—bead complex. This noise is unavoidable at room temperature, but serves as a useful probe of the system properties. The variance of the displacement signal $\langle x^2 \rangle$ is inversely proportional to the stiffness of the system in the x direction, κ_x:

$$\langle x^2 \rangle = \frac{kT}{\kappa_x}, \tag{1}$$

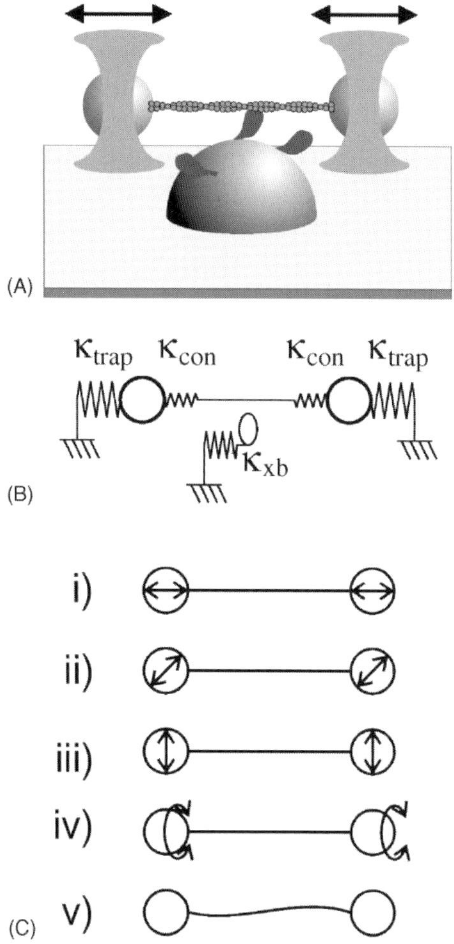

Fig. 1. Model of acto-myosin interactions in the optical tweezers; (A) A cartoon showing the experimental arrangement used. Two plastic beads (grey spheres) are held in twin optical tweezers (hourglass shapes). An actin filament is held stretched between them. The whole assembly is brought close to a third bead attached to a microscope cover slip. This allows the attached myosin heads (black blobs) to interact with the actin filament. The position of the beads (black arrows) is monitored using a quadrant photodiode. (B) A mechanical model of the arrangement shown in A. The compliance of the actin filament is negligible for the purposes of these experiments and is therefore not shown. The important compliances can be considered to be at the attachment point of the filament to the beads (corresponding to κ_{con}), in the traps themselves (κ_{trap}) and in the myosin cross-bridge (κ_{xb}). (C) Motions of the actin–bead assembly between interactions with myosin. These thermally driven motions mean that successive acto-myosin interactions start at different x positions. (i)–(iii) Translations in the x, y and z directions; (iv) rotation of the assembly about the filament axis; (v) vibrational modes of the actin filament. The x translational motion is the primary factor in the randomisation of the starting positions of acto-myosin interactions. The other translations and the vibrational modes of the actin filament allow it to sample different areas of the surface, thereby potentially interacting with different myosin heads. The rotational motion changes the position of the most favourable myosin binding sites on the actin filament (because of its helical symmetry) (see footnote 1).

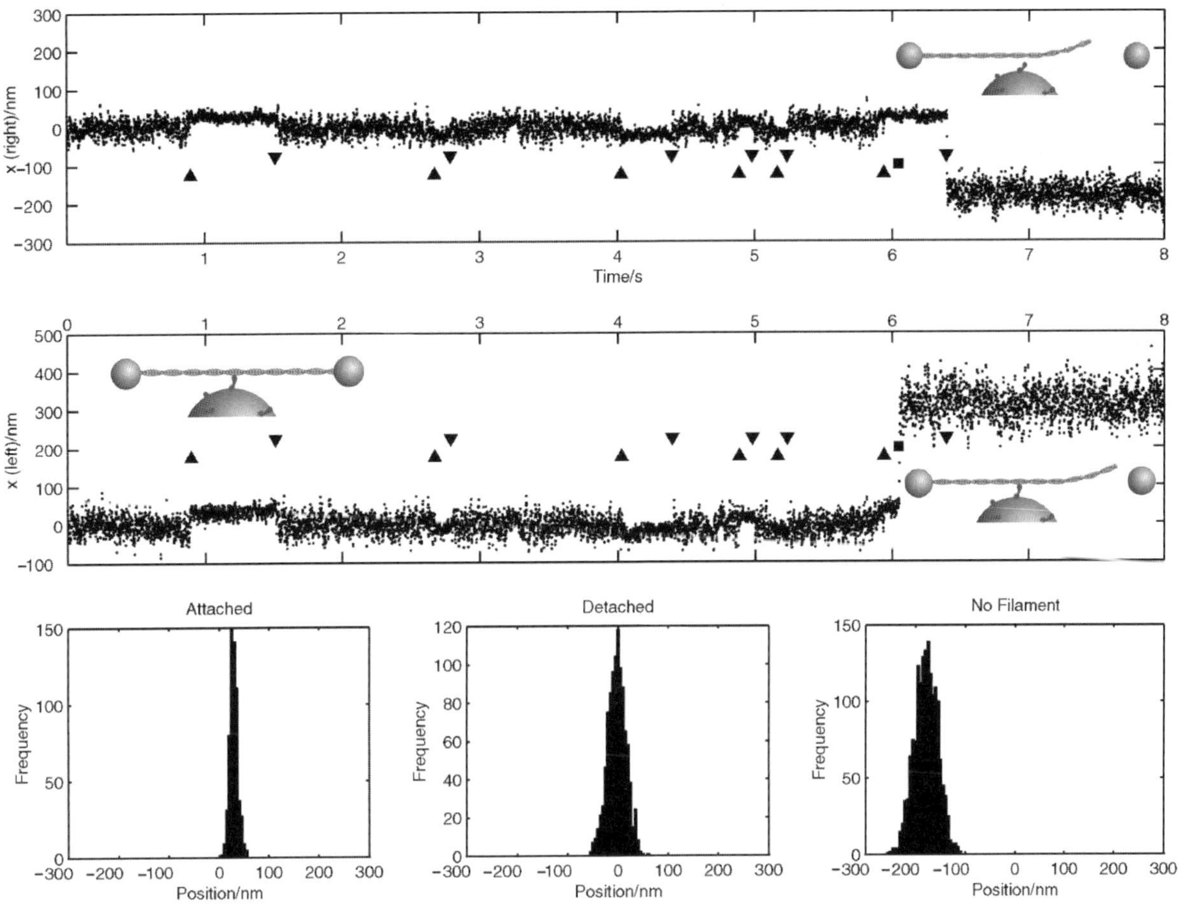

Fig. 2. Optical trap data. This figure illustrates some of the features of optical trap recordings of acto-myosin interactions. (Rabbit skeletal muscle myosin II (heavy meromyosin subfragment) and actin; 3 μM Mg·ATP; other conditions as Veigel et al., 1998.) Upper and middle panels show the position traces from the right and left beads, respectively. The three lower panels show position histograms from the left bead during an attached interval (0.9–1.5 s), a detached interval (1.6–2.6 s) and with no actin filament connecting the beads (6.4–8.0 s). The fortuitous actin filament breakage occurred during an attachment event, resulting in the time difference between the changes in bead position. Note that the two beads move in opposite directions, because the filament was under tension. The width of each histogram corresponds approximately to the system stiffness under each condition: *attached*: SD = 8.3 nm corresponding to $\kappa_x = 0.058\,\mathrm{pN\,nm^{-1}}$; *detached:* SD = 18.3, $\kappa_x = 0.012\,\mathrm{pN\,nm^{-1}} \approx 2\kappa_{\mathrm{trap}}$; *no filament:* SD = 23.4 nm, $\kappa_x = 0.007\,\mathrm{pN\,nm^{-1}} \approx \kappa_{\mathrm{trap}}$ (See Section 2 for more detail).

where kT is thermal energy, given by Boltzmann's constant k and the absolute temperature, T (at room temperature, kT is about 4 pN nm). The frequency spectrum of this thermal noise has a single pole "roll-off" frequency, f_c, which can be calculated using Stoke's law knowing the effective radius, r, of the beads (either one or two beads, corrected for proximity to the boundary layer), the viscosity, η, of water and the system stiffness, κ_x:

$$f_c = \frac{1}{2\pi}\frac{\kappa_x}{6\pi\eta r}. \tag{2}$$

In the detached state, the system stiffness (κ_x) is dominated by the combined trap stiffness, $2\kappa_{\text{trap}}$ because the actin filament connection between the traps combines their stiffnesses in parallel.

(2) *Myosin binding to actin increases the system stiffness.* During periods when myosin binds to the actin filament, system stiffness increases and data variance falls (the signal becomes quieter). These intervals correspond to individual acto-myosin interactions or binding "events" and the system stiffness is then a composite of trap stiffness, acto-myosin cross-bridge stiffness, κ_{xb}, and series compliance, κ_{con} (see e.g. Veigel et al., 1998, Appendix) which, essentially reduces to

$$\kappa_x \approx \frac{\kappa_{\text{con}}\,\kappa_{\text{xb}}}{\kappa_{\text{con}} + \kappa_{\text{xb}}}. \tag{3}$$

Note that if κ_{con} is large compared to κ_{trap} then κ_x will increase during acto-myosin binding and the amplitude of thermal motion will be significantly reduced. This property can then be exploited to identify the start and end of individual events within the noisy signals.

(3) *Stiffness experienced by a detached bead pair is twice the individual trap stiffness.* Note that towards the end of this particular record, the actin filament breaks. Because the actin filament was held under tension (in order to keep it taut) the beads now return to the centre of the optical traps (left and right beads move in opposite directions). Also, because the traps are then no longer connected by the actin filament, the stiffnesses no longer summate, so κ_x for each bead falls from $2\kappa_{\text{trap}}$ to κ_{trap}.

(4) *Estimation of cross-bridge stiffness.* Careful inspection of Fig. 2 shows that the actin filament connection happens to break at a time while myosin is bound. This means that a sudden force of about 1.8 pN (equal to the tension in the actin filament) is applied to the myosin molecule. We found for this particular example that the myosin cross-bridge is deflected by about 0.8 nm by this force giving $\kappa_{\text{xb}} \approx 2.2\,\text{pN}\,\text{nm}^{-1}$.

The variances of different sections of the data (see histograms, Fig. 2, lower panels) enable κ_x to be determined during attached and detached intervals, and the single trap stiffness κ_{trap} to be determined when the filament breaks. Also, this particular record, gives an estimate of κ_{xb}. These parameters are important both to enable automated data analysis (see below) and to enable correction for extension of the series compliance during the myosin power-stroke (typically a correction of about 10%). However, in practice these parameters are usually measured more accurately by other methods (see Section 2.6).

2.4. Displacement mode

Low trap stiffness means that thermal motion of the trapped beads is large (Eq. (1)) and the viscous damping of the bead held in water is high. Bead positions are therefore correlated in time and have a "roll-off" frequency, f_c in the 100 Hz range (Eq. (2)). During a mechanical "event", myosin binds to F-actin and the system stiffness is increased; also (on average) the mean position of the actin filament should be shifted. The rapid kinetics of skeletal myosin II, and the relative slowness of the optical tweezers mechanical transducer (see Section 2.6), make it difficult to observe the intermittent acto-myosin-II binding events. To make them stand out from the thermal noise they must be of sufficient duration ($> 10/f_c$ s) so that the mechanical properties of bound versus free states of F-actin may be reliably distinguished. Experimentally, this is achieved by manipulating the event duration by adjusting the ATP concentration until the average bound lifetime ($\approx 20/f_c$) enables about 75% of all events to be detected reliably. For this reason most of the published acto-myosin event lifetimes are dominated by the process of ATP binding to the acto-myosin rigor state.

Most workers now agree that the underlying thermal motion of the system randomises the starting position at which the myosin initially binds to actin (see Fig. 1 and Molloy et al., 1995). Therefore in order to estimate the size of the displacement produced by myosin, many events must be measured and their average displacement determined. The spread of the estimates is given by the thermal motion of the bead–actin–bead assembly (see Section 2.3, Eq. (1)) in the absence of myosin attachment. Therefore the standard error of the mean, SEM, of a sample size, n, is given by Eq. (4). For example, if the detached stiffness = 0.04 pN nm^{-1} then 100 events must be measured to obtain 1 nm precision.

$$\mathrm{SEM} = \sqrt{\frac{kT/\kappa_{x(\text{detached})}}{n}}. \tag{4}$$

2.5. Feedback mode

To measure the maximum force produced by a motor protein, negative feedback is applied between the position signal from one of the beads and the device controlling the laser tweezer position. The movement of the trap required to hold the bead in fixed position then gives the amount of force being exerted (because the trap is a linear spring of known stiffness). This approach was used by several workers to measure a force for myosin of 2–4 pN (e.g. Finer et al., 1994; Molloy et al., 1995). However, it was later realised that there was an unforeseen problem with these measurements. The connection between the actin filament and the bead is relatively compliant (Dupuis et al., 1997; Mehta et al., 1997; Veigel et al., 1998) and, worse still, this compliance is variable between preparations. Since this compliance is in series with the motor, the true force produced will be underestimated. Series compliance is also an issue in displacement mode measurements (above), but because κ_{trap} is much lower during these measurements a much smaller correction factor is required.

A technique which gets around this problem altogether and allows the motor's properties to be characterised under different load regimes is a different kind of feedback known as a *force clamp* (Visscher and Block, 1998). Instead of keeping the bead position constant, this form of feedback keeps the force on the bead constant, i.e. it maintains an *isotonic* condition. This is achieved by keeping the bead at a constant distance from the centre of the trap. Because the force is constant, the extension of the series compliance is also constant and can be ignored. However, this technique only works with processive motors which cause slow and continuous translations. The use of feedback is not discussed further in this article.

2.6. Perturbation mode

Use of "forcing functions" allows a variety of additional mechanical experiments and calibrations to be performed. For example, oscillating the optical tweezer (and thereby a trapped bead) with a waveform of known amplitude is one method of calibrating the position sensor. Monitoring the displacement of the bead from the centre of the trap while moving it at a constant velocity (e.g. using a triangle wave) can be used to measure the stiffness of the optical trap if we know the viscous drag on the bead (Svoboda and Block, 1994).

High frequency oscillations can also be used to increase the time resolution of event detection. Using Brownian motion to detect events is limited in time resolution because of the roll-off frequency of the thermal noise (Eq. (2)), typically around 100 Hz. Resolution is limited not by the bandwidth of the detector system or methods of analysis but by the lack of signal in this frequency range. By oscillating one optical tweezer at low amplitude and high frequency (e.g. 1 kHz) and measuring the amplitude or phase of the 1 kHz signal at the other bead we can increase the time resolution to ~ 1 ms. This is very useful for motors with relatively fast kinetics, such as muscle myosins.

In order to measure myosin and connection stiffnesses much larger forces must be applied so that the acto-myosin complex is measurably distorted. To do this a low frequency, high amplitude, forcing function (e.g. 100 Hz, 300 nm) is used and the position of both beads recorded (see Veigel et al., 1998).

3. Simulation of optical tweezers experiments

We have modelled the mechanical system as described above to generate simulated data sets on which we can test out various event detection algorithms. We find that the theoretical description fits well with our experimental data as we measure it, except that real data sets sometimes contain other sources of noise, drift, variation in connection stiffness, additional measurement noise due to rotation of asymmetric beads and so on. In the model, we assume two states, *attached* (myosin bound to actin) and *detached* (bead–actin–bead system moves freely within the optical tweezers). The states differ in their stiffness (κ_x) and in their distribution of lifetimes (given by a time constant in each case). These parameters are given in Table 1. In each state the system undergoes random thermal motion with the correct variance and power density spectrum given by Eqs. (1) and (2). In detached states the mean position is defined as zero. When the motor attaches, the

Table 1
Parameters used in optical tweezers simulation[a]

State	State parameters	
	Stiffness (pN nm^{-1})	Time constant (s)
Attached	0.25	0.05
Detached	0.03	0.50
General parameters		
Bead radius (nm)	550	
Sampling frequency (Hz)	2000	
Myosin power stroke (nm)	5	
Temperature (K)	296	

[a] The stiffness values given are for HMM (attached) and the trap stiffness (detached) (Veigel et al., 1998). The time constants are derived from the rate constant for detachment under typical conditions. The value for the myosin power stroke is that obtained in our laboratory previously (Molloy et al., 1995; Veigel et al., 1998).

mean position is instantaneously displaced by a distance, d (assumed to be $+5\,\text{nm}$) from the position *at the moment of binding*. This new position is then the mean position for the subsequent attached period. When the motor detaches the mean position returns to zero. This model is based on observations of the behaviour of skeletal myosin in optical trapping experiments. Other types of myosin, such as myosin Is, could be simulated by a relatively straightforward elaboration of the method used here.

We have implemented a "Monte Carlo" simulation based on this model in the MATLAB environment (The MathWorks, Inc., Natick, MA) using custom written routines and Signal Processing and Statistics Toolboxes. The simulation program outputs an x data trace and information about the underlying state of the system. Thermal noise for each state is simulated by selecting random numbers from a Gaussian distribution with an appropriate mean and variance. This data is then passed through a single pole energy-normalised Butterworth filter, to give the correct frequency spectrum for over-damped thermally driven motion. The cut-off frequency, f_c, of the filter is calculated using Eq. (2). Examples of simulated data can be seen in Figs. 6 and 7.

We used these simulated data sets to test the various automatic methods of data analysis (event detection algorithms) described below.

4. Analysis of single-molecule displacement recordings

A wide variety of methods have been applied to the analysis of single-molecule mechanical recordings, ranging from classical signal-processing methods such as the Fourier transform, through statistical methods borrowed from ion-channel studies, to specialised methods which aim to identify the acto-myosin attachment events within the data. The aim of all these methods, however, is to extract the biologically relevant information, such as the motor's working stroke, or binding and unbinding rate constants, from the data.

4.1. Spectral analysis

Several approaches to the analysis of single-molecule recordings rely upon transformation of the time-series signal into the frequency domain. The most familiar method is probably the use of the discrete Fourier transform or DFT. This approach was important in analyses of early mechanical recordings made from low numbers of myosin molecules (Ishijima et al., 1991). However, it is less useful for single-molecule data recordings in which the signals contain local changes in spectral properties with abrupt discontinuities. It is still useful for analysing signals where the system remains in one state; for example, the corner frequency of the power spectrum of a trapped bead can be used to estimate trap stiffness (Eq. (2)) (Svoboda and Block, 1994).

Spectral changes could in principle be used to identify interactions between actin and myosin. This could be done by generating a time–frequency plot using either a windowed Fourier transform, or preferably some form of wavelet transform (e.g. Graps, 1995). As far as we are aware, this approach has not been tried.

A related technique, autocorrelation analysis, is useful for extracting rate constants from noisy single-molecule data, and has been used for the analysis of single-molecule fluorescence data (Dickson et al., 1997).

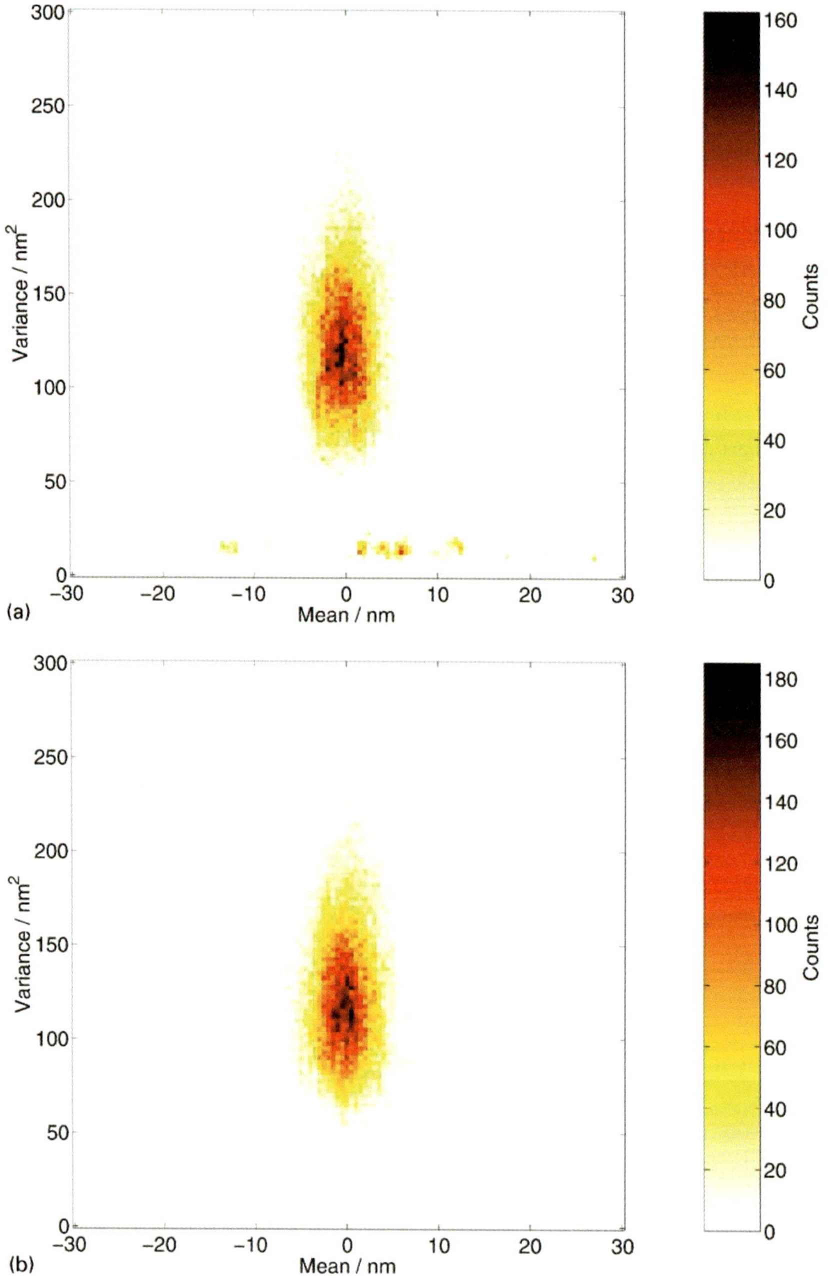

Fig. 3ab.

4.2. Mean–variance analysis

Mean–variance analysis (Guilford et al., 1997) was originally developed to study ion channel data. The method works by first calculating the local mean and variance for a short sliding

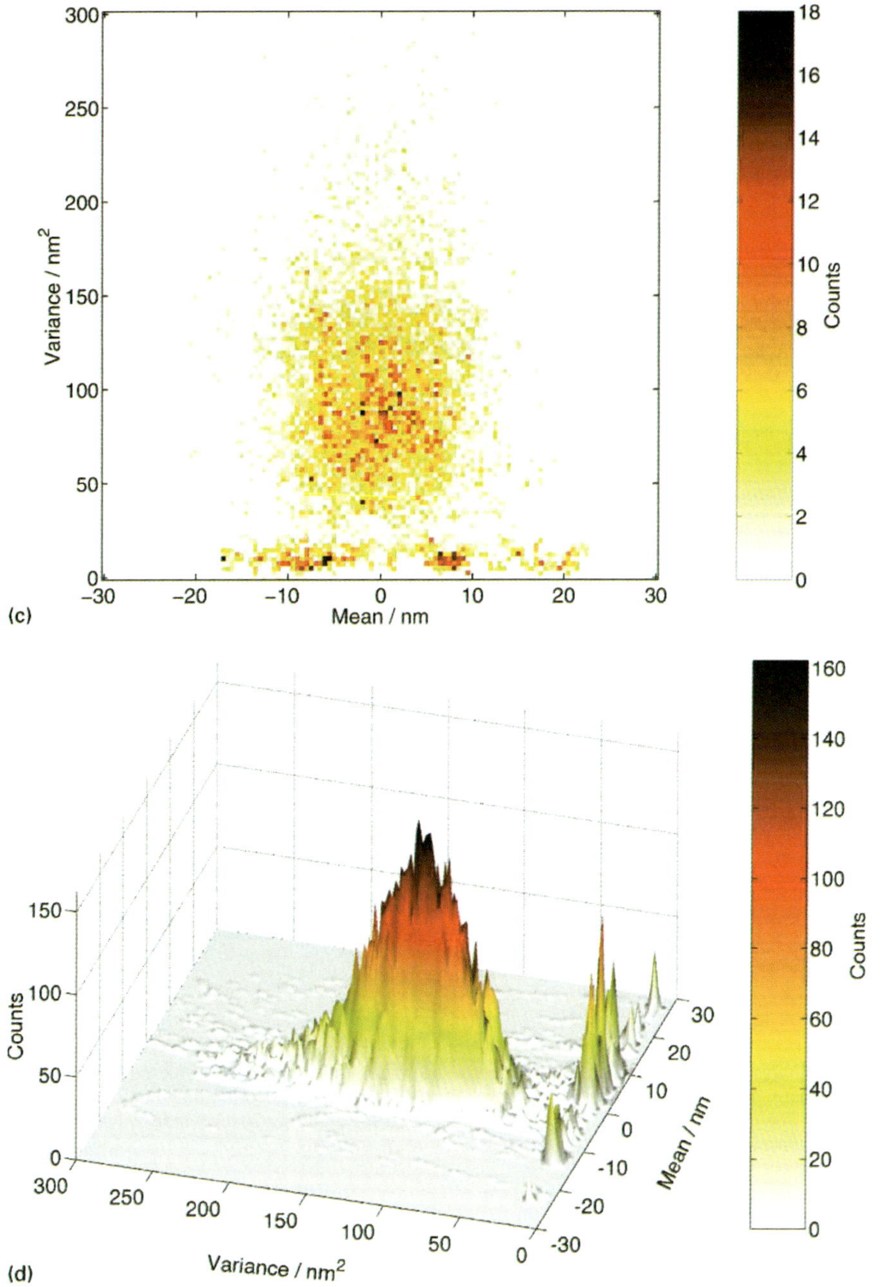

Fig. 3cd.

window which is passed through the data. The resulting mean and variance values are then binned and plotted as a three-dimensional histogram (see Fig. 3). This is a very useful method for visualising a large data set and assessing its quality. For example, if there is an increase in stiffness when myosin binds to actin, the detached and attached data points should form two separate

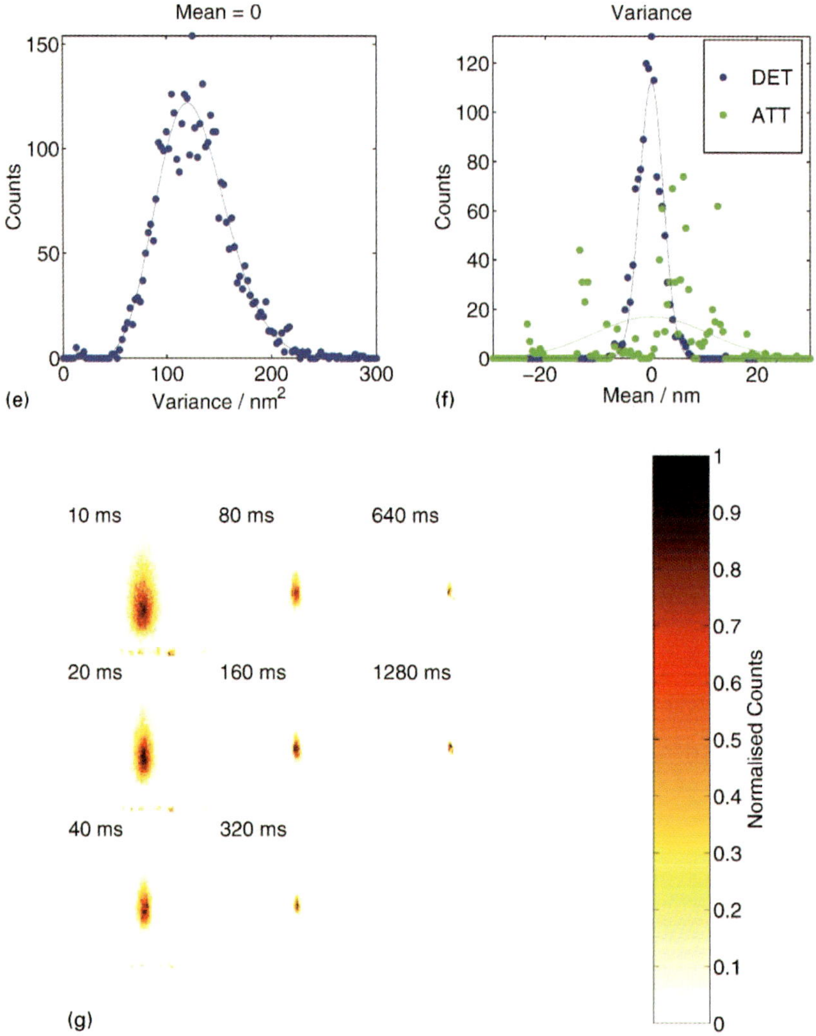

Fig. 3. Mean–variance analysis. This figure illustrates the method of mean–variance analysis (see Guilford et al., 1997) applied to the analysis of optical trap data. (a) A mean–variance histogram of a simulated data run containing 50 attachment events (see Table 1 for simulation parameters). The mean and variance values were calculated for a sliding 20 ms window and summarised as a histogram. Here each histogram bin is coloured according to the number of points it contains, according to the colour scale at right. Note that the histogram axes used here are both linear, whereas Guilford et al. used a log scale for the variance and counts. The data points fall into two distinct populations; the oval area centred about a mean of zero corresponds to the detached data points, whereas the elongated region at the bottom of the plot corresponds to the attached data points. (d) The same histogram rendered as a surface plot. Note that the attached points occur as a series of peaks. Each peak corresponds to one, or a few, events. (e) Cross-section through the histogram at mean = 0. The curve is a χ^2 fit to the data. If points with a low variance are excluded, then the χ^2 curve is a good fit to the data. (f) Cross sections through the histogram at variance values corresponding to the detached and attached stiffnesses. The curves are normal fits to the data. The data corresponding to detached intervals are fit well to a normal distribution, but the attached intervals are much more 'spiky' and are not well described by a normal distribution. The mean value calculated from the histogram bins in this example was 1.8 nm, whereas the mean position of the model events was 5.6 nm. (b) Simulation as in (a) but with attached stiffness 0.0429 pN nm^{-1} (i.e., variance of signal only reduced by 30% from attached level). Note that only one population is visible, in agreement with (Guilford et al., 1997, Fig. 8). (c) A mean–variance histogram generated from an experimental data file using skeletal muscle myosin II (heavy meromyosin subfragment). The distribution is similar to (a), except that the detached stiffness is higher, and the amount of data accumulated is much lower. (g) The effect of changing window length on the mean variance histogram. Note that the width of the detached population decreases in both mean and variances axes with increased window length. Furthermore, the attached population gradually disappears as window length becomes greater than event length. This can be used to obtain an estimate of attached lifetimes.

(although possibly overlapping) distributions, centred about different variances. Furthermore, the distribution of data points assigned to the attached state should (for a motor protein) have a mean displacement that is shifted away from the arbitrarily defined zero mean of the detached data pool. Fig. 3a and d show a mean–variance histogram produced from a long run of simulated data. In this example, the data points from intervals where myosin is detached from actin fall within a roughly oval area of the plot. The centroid of the distribution has a variance value determined by a combination of κ_{trap}, window size and f_c. A cross-section through the variance plane shows a χ^2 distribution (Fig. 3e). On the mean axis the centroid is defined, arbitrarily as zero (the average resting position) and the mean plane cross-section is Gaussian, with half-width again determined by κ_{trap}, window size and f_c (Fig. 3f). The data points corresponding to myosin attachment events form a separate pool that lies below the detached data pool. The higher system stiffness during attached intervals means that the variance values are lower and form a narrower distribution. In this simulated data set, the attached data pool is quite distinct from the detached data points. In the mean dimension the data points are spread out and shifted in the positive direction. This can be explained by the variation in starting position and the myosin's working stroke, respectively. According to the Molloy et al. (1995) idea, the mean plane cross-section through the attached pool should be Gaussian with a half-width determined by $2\kappa_{trap}$ (and should be independent of chosen window size and f_c).

Using this method, attachment events can be identified by imposing a variance threshold. This would then be similar to the method described in Section 4.5 (below). However, it is not straightforward to highlight the attached data points in the raw, time series, data and therefore to assess the reliability of the method by inspection.

Fig. 3f shows that the attached data points are in fact *not* normally distributed. This is because each event has a different mean position and will therefore form a separate peak, with a height determined by the *duration* and therefore number of data points making up the individual event. In other words the overall mean position is arbitrarily and randomly weighted by the duration of the separate events. To reliably measure the working stroke by this approach, it would be necessary to sample the full range of both displacement and event duration, which would require an extremely large data set.

If the change in noise during attachment intervals is smaller than the simulated data set just discussed (e.g. this is the case with many real data sets) the mean variance plots no longer have discrete pools (Fig. 3b). In this case we cannot distinguish attached and detached states using a simple variance thresholding method. Guilford et al. (1997) overcame this problem by curve-fitting to a detached data set and later subtraction of the detached distribution. However, it is quite difficult to estimate the correct amplitude for the subtraction process since it requires a priori knowledge of the ratio of attached to detached durations in the data series. Fig. 3c shows that the two populations can be clearly distinguished without recourse to such measures if the experimental data is of sufficiently high quality.

Because the mean–variance method does not directly identify the start and end of individual binding events, it does not easily enable ensemble averaging (Section 5.2) of the data to be performed. An indirect method of analysis is also necessary to determine attached lifetimes (Fig. 3g).

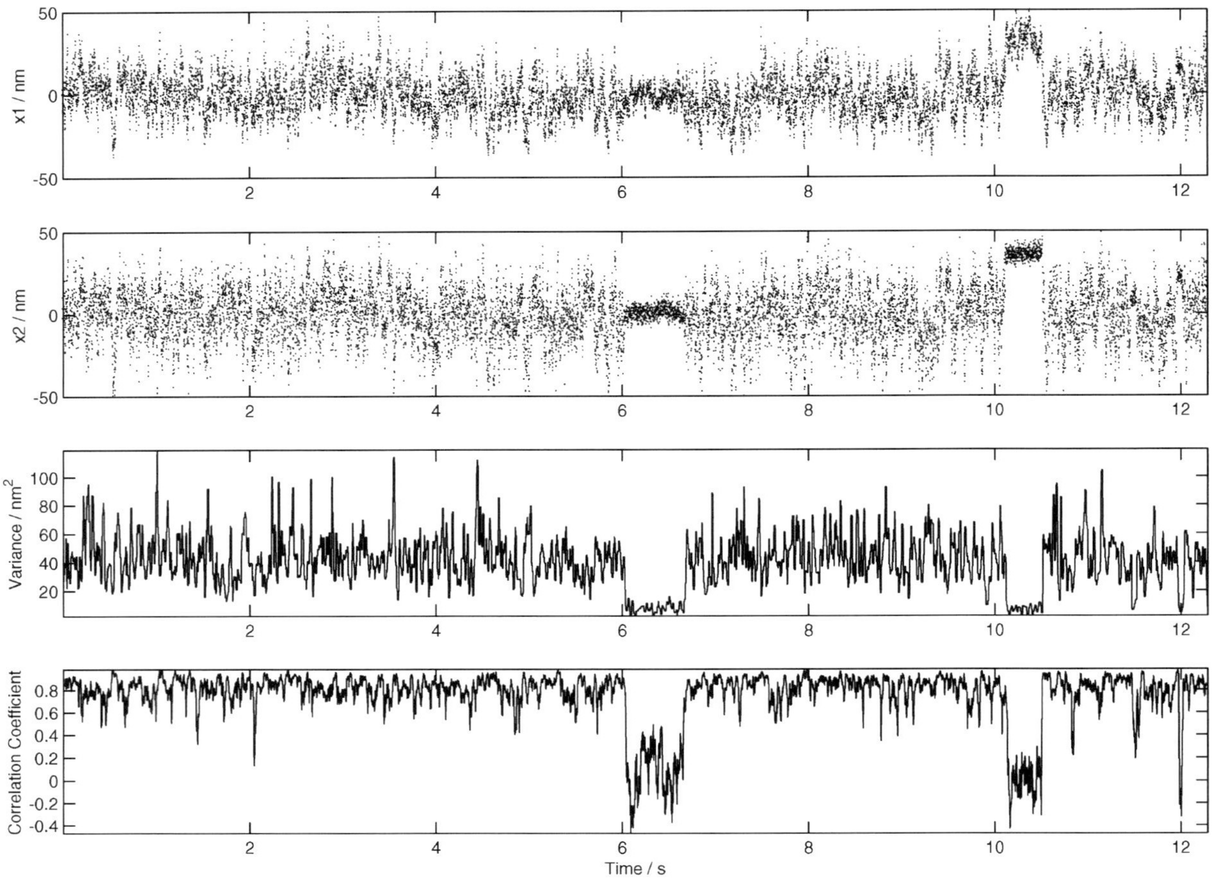

Fig. 4. Detecting events by correlated thermal diffusion. This figure shows the use of the correlation coefficient to detect acto-myosin attachments. The upper two panels show real data traces obtained from the left and right bead detectors during an experiment using rabbit skeletal muscle myosin II (conditions as for Fig. 2). The third panel shows the running variance calculated for the right bead trace (as described in Section 4.5). The lower panel shows the correlation coefficient between the traces calculated for a 35 point (equivalent to 35 ms) sliding window (see Section 4.4). During longer attachments there is a clear drop in the correlation coefficient, although the signal during attachments is quite noisy. Shorter attachments are not so easily detected (see Mehta et al., 1997 for further details of this method).

4.3. Identification of individual events

In contrast to the methods discussed above, which analyse the record as a single entity, we favour a two-stage approach

1. Identify time periods in the record where myosin is attached to actin ("events").
2. Perform statistical analysis of individual events to use in later analysis.

This approach has many advantages. Each event becomes a discrete datum so that events may be weighted either equally or with known bias in the measurement of parameters such as rate constants and displacements. Events from many records can be collected together to give large sample sizes for statistical analysis. All of the characteristics of the individual events are retained so that we can analyse the lifetime of attachments or look for changes occurring during attachment. Furthermore, we retain the ability to go back to the original data set and evaluate the quality of the data or look for patterns in the properties of successive events. For example, we might want to know if event amplitudes are correlated in time; alternatively we might want to discard data from periods where there is a large amount of drift or a low attached stiffness (i.e. high series compliance).

Analysis of individual events is also crucial when correlated measurements are made with other techniques, such as single-molecule fluorescence (Ishijima et al., 1998). The following sections discuss alternative methods for the identification of events in single-molecule recordings.

4.4. *Correlated diffusion*

This method, devised by Mehta et al. (1997) detects myosin binding by monitoring the drop in correlation between motion of the beads held at either end of the actin filament. When the bead–actin–bead assembly is held taut, motions of the two beads will be strongly correlated, i.e., they will tend to move synchronously (if $\kappa_{trap} < \kappa_{con}$). However, when myosin binds in the central region of the filament, series compliance at either end causes bead motion to become uncorrelated (see Fig. 4) (if $\kappa_{con} < \kappa_{xb}$). So, this method of event detection works well if $\kappa_{trap} < \kappa_{con} < \kappa_{xb}$. In a way the method is complementary to that of the running variance method which simply requires that both κ_{con} and κ_{xb} are greater than κ_{trap} (see below). A visual form of this analysis can be created during an experiment by plotting the x-positions of the beads against each another using an oscilloscope in "$x-y$" plotting mode. During detached intervals the plot forms a 45° ellipse (because the two bead positions are correlated). During attached intervals the data forms a much smaller circular pattern, because the increase in stiffness reduces the variance of both signals, and loss of correlation makes it circular.

4.5. *Running variance*

We find that monitoring changes in variance of the position signal with time is one of the most reliable and easily checked methods for monitoring acto-myosin binding events. This can be done in real-time using an analogue, "rms-to-dc converter" circuit to monitor signal variance over a defined bandwidth. However, we usually perform this analysis off-line on the digital data. We can then apply filters based upon the roll-off, f_c, (Eq. (2)) of the thermal noise present in the data. (We determine f_c by spectral analysis of data series in which no myosin interactions are present.)

1. A running mean is calculated using a window size of $10/f_c$ s (usually about 50 ms). This running mean is then subtracted from the data to remove changes in the mean position. This removes discontinuities caused by event steps.
2. The running variance is then calculated over a window size of $2/f_c$ s (10 ms). This data is then low-pass filtered by calculating its running median value, using a window size of $5/f_c$ s (usually about 25 ms).
3. This filtered variance data is then thresholded to identify events in the original data series. The threshold level is set automatically for the simulated data, because the stiffnesses for the two states are known. For experimental data, the threshold is either set manually, based on the

observed bimodal distribution of variances, or automatically, based on measured attached and detached stiffnesses (see above). The threshold level may need to be adjusted during long (>30 s) data runs, as the actin–bead connection stiffness may change over time.
4. "Myosin attached" event amplitudes are calculated from the average of the event data relative to the local mean position calculated from the adjacent regions of "unattached" data series.

The disadvantages of the method lie with the heavy filtering necessary and the use of a threshold. The former tends to smooth out the edges of variance events leading to timing errors; the latter can cause problems where the threshold value is not correct or the data noisy for some uncharacterised reason (tumbling of non-spherical beads, large series compliance). This leads to events being split into multiple "sub-events" and to "false event" detection. These problems are ameliorated by including some form of hysteresis in the thresholding algorithm analogous to the use of a Schmitt trigger circuit in electronics.

4.6. *Page's test*

Page's test (Han et al., 1999; Page, 1954) is a statistical method for detecting changes in the underlying probability density function (PDF) of noisy signals. We have implemented a modified version of this method to detect changes in signal variance. For the method to work the optical tweezers data is pre-processed by subtracting a running mean from the signal to remove displacement shifts but leave variance changes. We then expect the distributions of attached and detached positions to be Gaussians with means of zero but with different variances (Fig. 5). The variances of these distributions can be calculated from the known, expected or observed stiffness of each state (Eq. (1)). For each data point, with position x the probability density $f(x)$ can then be calculated for each distribution

$$f(x) = \frac{e^{-(x-\bar{x})^2/2v}}{\sqrt{2\pi v}}, \tag{5}$$

where v is the variance and $\bar{x}$ the mean of each distribution (in this case, 0). The log odds ratio g is then calculated

$$g(x) = \log_{10}\left(\frac{f_{\mathrm{ATT}}(x)}{f_{\mathrm{DET}}(x)}\right), \tag{6}$$

Fig. 5. Page's test: probability density functions. The upper panel shows the theoretical distributions (strictly speaking, the probability density function or PDF) of positions for a trapped bead when a myosin is attached (solid line) or detached (dashed line). These Gaussian PDFs were calculated using a mean of zero and a variance corresponding to kT/κ_x, where κ_x is the system stiffness according to Table 1. The next step is to determine whether a given position value comes from the attached or the detached distribution; i.e. because of the greater spread of the detached distribution, a data point near the zero position is more likely to belong to the attached distribution; whereas a point at, let us say, $+20$ nm is more likely to belong to the detached distribution. We express this mathematically by finding $g(x)$, the log odds ratio of the two distributions at a given x position (Eq. (6)). The central panel plots the ratio of the two PDFs (attached/detached). The shaded area corresponds to the values of x where the attached probability is greater than the detached. The logarithm of this ratio is plotted in the bottom panel. Again the shaded area indicates values of x where the attached probability is greater than the detached. Note that a value in this central region will give a small positive $g(x)$; values outside it will give a large negative value. This property is exploited in the calculation of the cusum statistic.

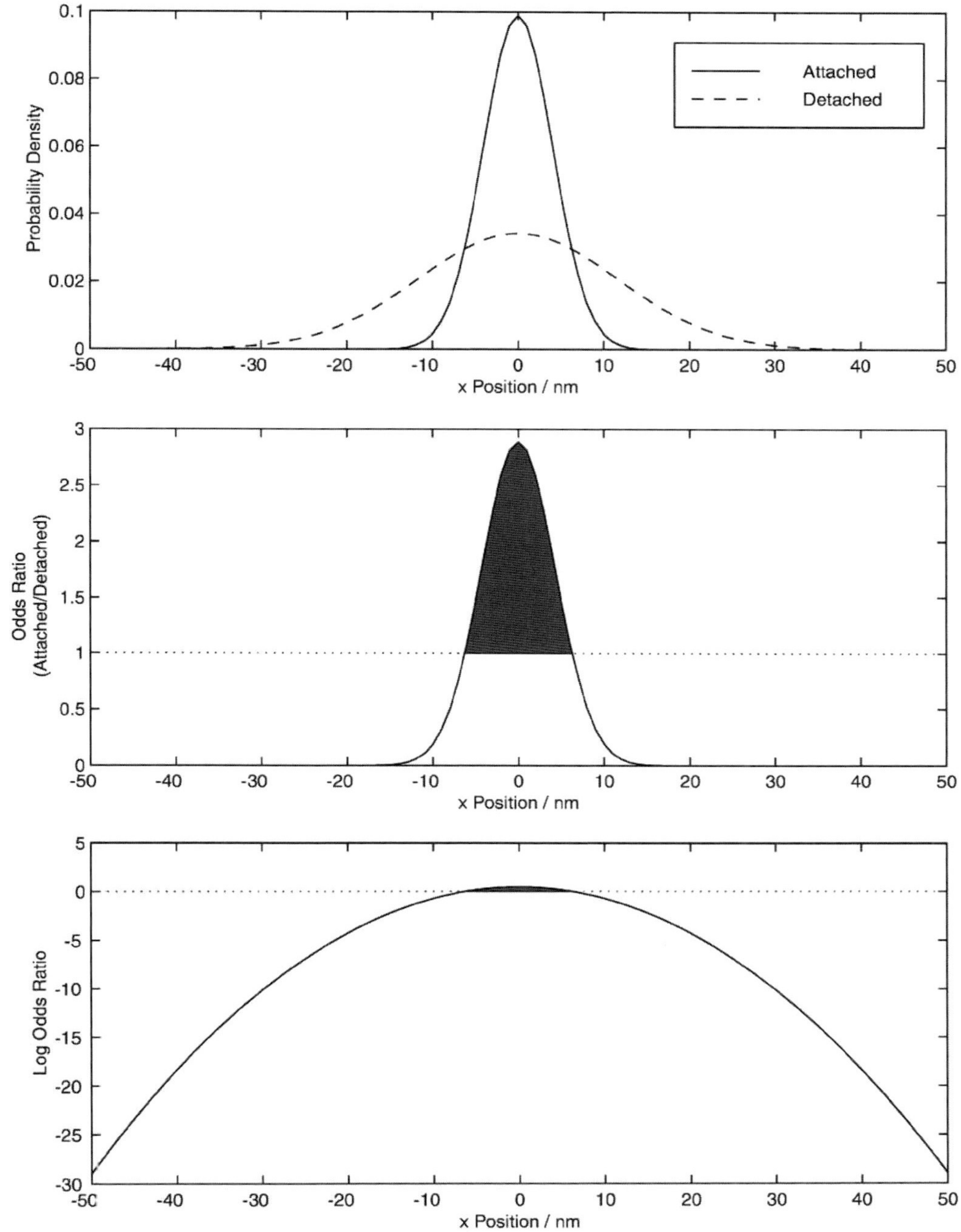

where f_{DET} and f_{ATT} are the probability densities for the detached and attached states, respectively.

The value of g will then be positive where the probability is highest that a value belongs to the attached distribution, and negative where a value is probably from the detached distribution.

During a detached phase, there will be a mixture of a few positive and many negative values of g; however, during an attachment most of the values will be positive (Fig. 6). To detect events, a zero thresholded cumulative sum (or *cusum*) is calculated for g, such that values below zero are discarded

$$Z_0 = 0,$$
$$Z_n = \max\{0, Z_{n-1} + g(x_n)\}, \tag{7}$$

where x_n is the x position for data point n and Z_n is the corresponding cusum. During an event, this cusum increases because of the predominance of positive values of g. At the end of the event, the cusum decreases steeply towards zero. Peak values in the cusum statistic therefore correspond to the ends of events. Similarly, the start of events can be detected by calculating the cusum in the reverse direction. Since the height of the peaks obtained is related to the duration of the events, this means that shorter events cannot reliably be distinguished from noise; therefore an empirically determined threshold is applied below which peaks are ignored. This method overcomes some of the limitations of the simple thresholding method described above. In practice it is more robust to small variations in experimental conditions and also has fewer arbitrary parameters to be set by the user.

4.7. Comparison of analysis methods

We evaluated the Page test method by comparing results with those obtained with the "running variance" method and by testing both methods on large simulated data sets (see Section 3) and on real data obtained using skeletal muscle myosin II. The comparison with known start and endpoints in simulated data gave useful statistics on the timing precision of the methods. The experimental data allowed comparison of robustness of the methods when confronted with real data sets (with warts and all). The results of this analysis are summarised in Figs. 7 and 8, and Tables 2 and 3, and show that the methods converge on the same basic parameters but that the Page's test method has slightly better time resolution.

When analysing real data sets we have found it useful to plot out the measured event amplitude versus experiment time to enable us to check for heterogeneity in the data set. Heterogeneity might arise for several reasons: e.g. preferential binding by myosin to one preferred site on the F-actin pseudo-helical repeat;[1] or interaction with a "dead" myosin head (e.g. oxidation of myosin II reactive sulphydryls disables its motor function). Different regions of the third bead, when sampled, might have patches where multiple myosin molecules are able to access the actin filament, giving rise to larger displacements; or different myosin heads may give statistically different mean step amplitudes e.g. due to different myosin molecule orientation relative to actin (Tanaka et al., 1998).

[1] The rotational Brownian motion of an optically-trapped bead is given by $\theta = \sqrt{kT\tau/4\pi\eta r^3}$ where θ is the r.m.s. angle (in radians) rotated after a time τ (in s); k is the Boltzmann constant ($1.380662 \times 10^{-23}\,\mathrm{J\,K^{-1}}$) and T is the absolute temperature; η is the viscosity of the medium in $\mathrm{N\,s\,m^{-2}}$ (for water, $0.001\,\mathrm{N\,s\,m^{-2}}$); and r is the radius in m (Einstein, 1906). For a 1 μm diameter sphere suspended in water at 21°C, in 1 s the r.m.s. rotation would be 1.6 radians or 92°. One would expect the rotation of the bead—actin dumbbell to be somewhat slower, but in the duration of a typical experiment (assuming that the myosin is mostly detached from the actin) the axial position of the myosin binding site would become effectively randomised.

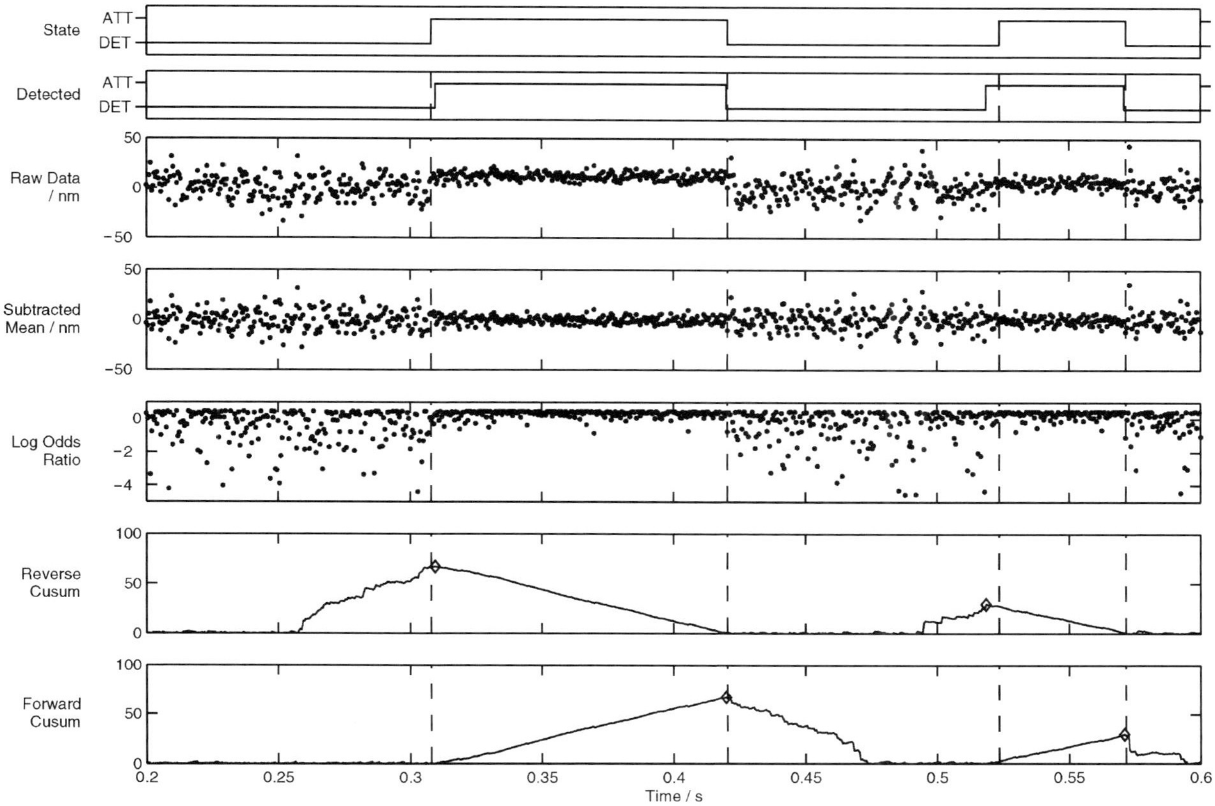

Fig. 6. Example of detection of simulated events by Page's test. To test the performance of the adapted Page's test method, simulated data were generated as described in Section 3. The top panel shows the state of the simulation; the vertical dashed lines in subsequent panels indicate the start and stop times of the simulated events. The second panel shows the results of the analysis. Note that the determination of start and endpoints is good but there is some timing error due to the stochastic nature of the signals. The next panel shows the raw data output by the simulation. Note that the two events show a distinct drop in variance, and that both are displaced from the baseline, but by differing amounts. In the following panel, a running mean has been subtracted from the data to eliminate this baseline shift. The data points now correspond to samples from two Gaussian distributions with zero mean but different variances. For each data point, we now calculate the log odds ratio, which expresses the likelihood that the point is a sample from the attached distribution (Fig. 5, Section 4.6). During detachment, many of the values of $g(x)$ are negative, although a few have small positive values. In contrast, during an attachment, many more points have a positive value and the negative values tend to be smaller. These properties of the log odds ratio are exploited in the calculation of the cusum statistic, which allows us to identify when events have occurred and to determine the start and end points. The cusum is simply a zero thresholded cumulative sum (Eq. (7)). For each point, the current value of $g(x)$ is added to the preceding value. The cusum is then the highest of the result or 0. This means that during detached intervals, when values of $g(x)$ are mostly negative, the cusum tends to remain close to zero. During an event, the values are mostly positive and so the cusum steadily increases. At the end of the event, the value of the cusum falls off sharply. The position of the cusum peak then gives an estimate of the time of the change in state accompanying the end of the event. By calculating the cusum in the reverse direction, the beginnings of events can also be identified by the same means. The bottom two panels show the reverse and forward cusums, respectively. The diamonds indicate the peak positions.

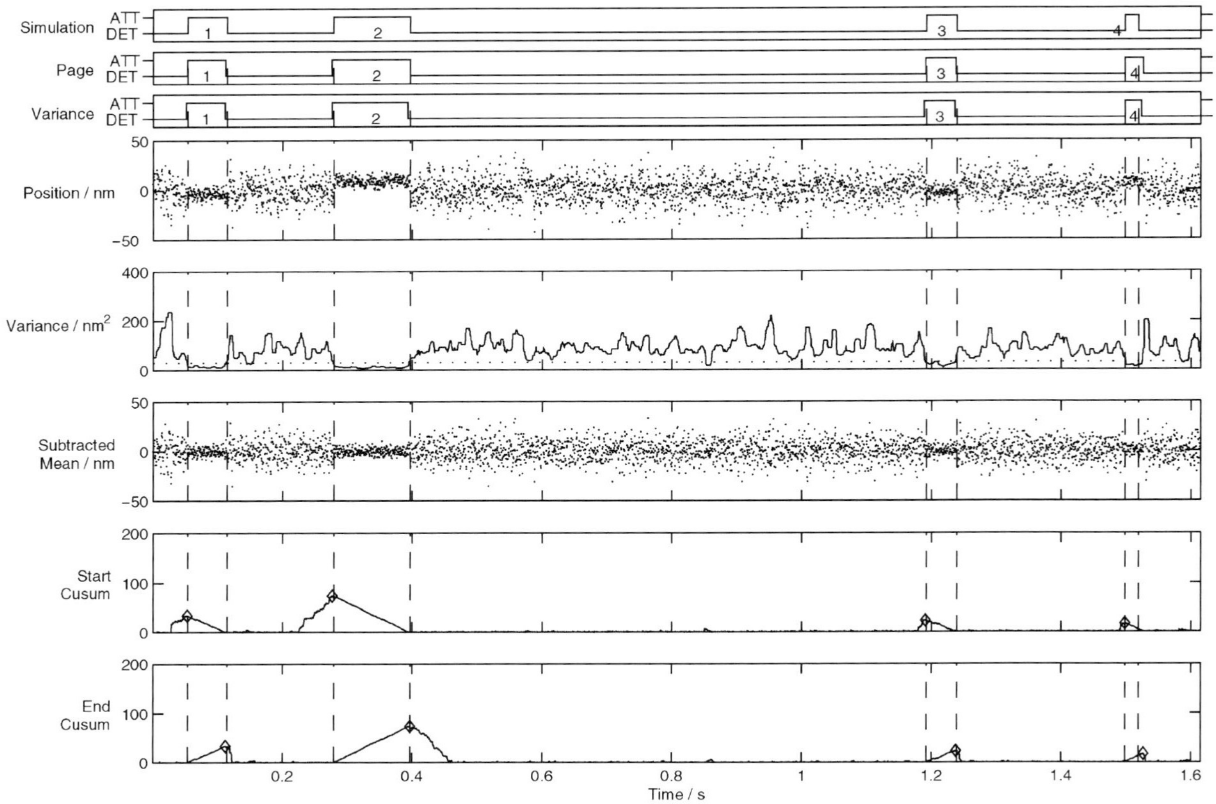

Fig. 7. Analysis of simulated optical trap data. These plots show simulated optical trap data and a comparison of the methods for analysis described in this paper. The top three panels show the state of the simulation (either attached or detached) and the state predicted by, respectively, the Page's test method and the running variance method. The vertical dashed lines indicate the beginning and end of the simulated events. In this example, all the events are accurately identified by both methods. The next panel shows the simulated data trace. Note the narrower width of the noise during events, and the variety of displacements. The median filtered variance of a sliding window is shown in the next panel. The dotted line shows the variance threshold used to identify events. The next panel shows the position data with the running mean subtracted. Note that the mean position is now zero throughout, but that the events can still be distinguished from the drop in thermal noise. The bottom two panels show the Page test cusums calculated from this data. Note that the peaks in the cusums correspond to the starts and ends of the events. The height of the peaks is related to the duration of the events.

5. Analysis of acto-myosin event data

The ultimate goal of our single-molecule experiments is to try to detect the separate phases that might make up the acto-myosin "power-stroke" and to measure the strain dependence of these separate phases or biochemical states. To do this, individual events must be analysed further and averaged in a way that preserves some aspect of the stochastically timed sequence of events that underlie each single interaction. In the following section we discuss how some of the details of the acto-myosin interaction may be extracted from single-molecule data records.

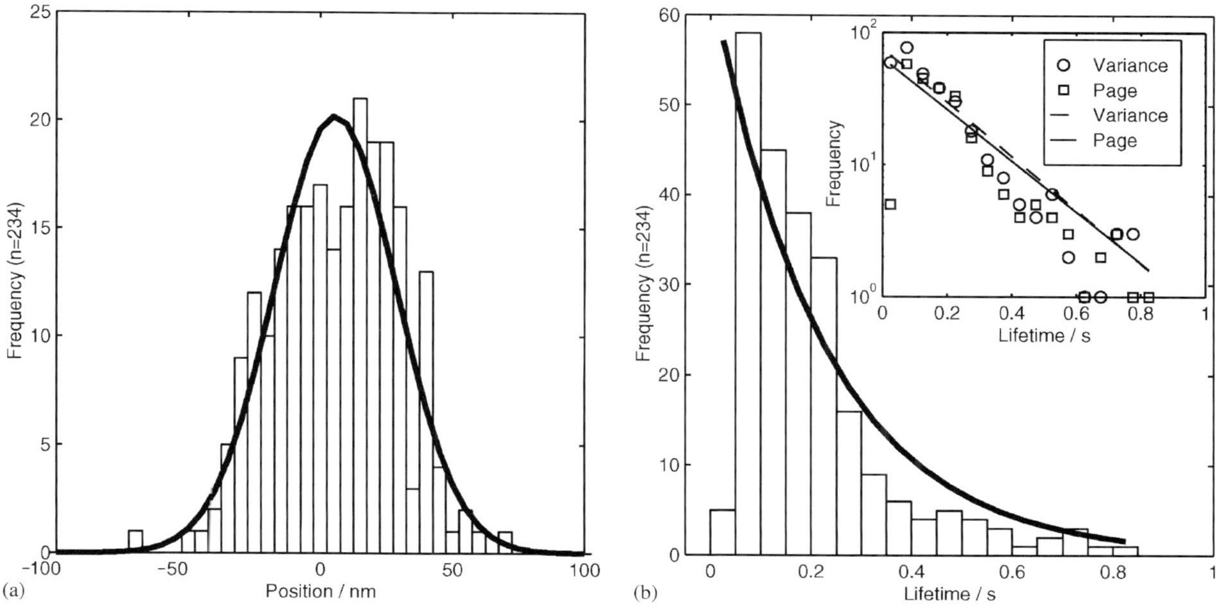

Fig. 8. Analysis of experimental data. The data set used in this analysis consisted of eleven, 12 s duration experimental records collected over a period of about 10 min (same experimental run as other figures). All records were collected using the same actin–bead pair assembly, in the same orientation, from a surface sparsely coated with rabbit skeletal HMM. The attached stiffness was derived in Fig. 2: $0.058\,pN\,nm^{-1}$, and the detached stiffness, $0.012\,pN\,nm^{-1}$. The threshold value used for the running variance method was $0.031\,pN\,nm^{-1}$. Both methods were used to analyse the data and the results are summarised in Table 3. Panel (a) shows the displacement histogram obtained by Page's test. Note that although the data set is relatively small the data are well approximated by the Gaussian curve calculated from the mean (5.65 nm) and standard deviation (23.12 nm) of the attached positions. Note that this standard deviation corresponds to a detached stiffness of $0.018\,pN\,nm^{-1}$, (Eq. (1)), close to the expected value. The standard error of the mean is 1.51 nm; to improve the precision of this measurement we would need either more data or a higher trap stiffness (Eq. (4)). Panel (b) shows a histogram of the event lifetimes obtained by Page's test. The curve calculated from the mean lifetime, $\bar{t}$ (155 ± 10 ms) is superimposed. The mean lifetime was calculated as in Eq. (8), with the standard error given by $\bar{t}/\sqrt{n}$, where n is the number of lifetime values used to calculate the mean. The inset shows the data from both methods (symbols ○, □, see legend) and the calculated curves (solid and broken lines, see legend) as semi-log plots. Note that the calculated lines for both methods are similar and that they fit the data well (the calculation is effectively weighted towards the more populated bins at short lifetimes).

5.1. *Lifetime of attachments*

The lifetime of a single-molecule "event" depends upon the probability per unit time of the event ending. This probability is governed by the apparent rate constant for an underlying biochemical transition. Therefore, lifetimes of individual events have a stochastic nature but the average lifetime obtained from many observations will be characteristic of some bulk-measured rate constant. It now becomes clear that we need to define what is meant by an "event": Here we define an event as an experimentally observed phenomenon, which therefore might be made up of several biochemical states; for example, in this paper we have termed the period while myosin is bound to actin as an attachment "event", but we know that during this period several biochemical

Table 2
Comparison of analysis methods on simulated data[a]

		Model	Running variance	Page's test
Results				
Events	Total	1600	1190 (74%)	1128 (71%)
	Merged		11	18
	Spurious		7	0
Displacement (nm)	Mean	4.74	4.37	4.37
	SD	11.79	11.41	11.31
Lifetime (ms)	Mean	50.23	49.94	53.57
		Running variance	**Page's test**	***F***
Timing errors (ms)				
Start	Median	1.5	0.5	0.32
	SD	6.63	3.73	
End	Median	3.5	0.5	0.21
	SD	6.68	3.03	
Duration	Median	1.5	− 0.5	0.27
	SD	9.39	4.92	
95% Confidence limits				0.89–1.12

[a] Simulated optical trap records were generated as described in the text, (Section 3) using the parameters in Table 1. The resulting records were then analysed by the running variance method (Section 4.5) and the Page's test method (Section 4.6). The performance of the two methods was then compared. The parameters of the actomyosin interactions are given in the upper panel. Values of displacement and mean lifetime produced by both methods are similar and close to the model parameters, as one would expect. The lower panel shows the timing errors (analysis time—model time) for both methods. The errors have two components—the systematic bias (given by the median) and the random component (given by the standard deviation). The systematic errors for both methods are of the order of the sampling interval (0.5 ms), although Page's test performs slightly better. In the random errors, Page's test performs significantly better, as is confirmed by the variance ratio, *F*, which is statistically significant as shown by the 95% confidence limits for the F-distribution. It is worth noting that the performance of each method depends on the exact parameters used; for example, the sensitivity can be increased with the penalty of an increased rate of false events detected. One advantage of the Page's test method is that there are fewer parameters that can be adjusted, which helps to improve reproducibility. Note that for the displacement measurements, the standard deviation is approximately 11 nm, which corresponds to a stiffness of 0.03 pN nm^{-1}.

steps must occur, including phosphate release, ADP release and perhaps other slow isomerisations. Hence, just as one might measure a composite or "apparent" rate constant in some solution biochemical experiments, here we measure a lumped or composite "event". The lifetime will be controlled by several biochemical state transitions. If the event lifetime is dominated by a single biochemical step, then the distribution of observed lifetimes will fit to a single exponential process. The single-molecule mechanical events measured in this paper were performed at limiting ATP concentration (3 μM ATP) so the average attached lifetime, $\bar{t}$, is characteristic of the first order binding of ATP to the acto-myosin rigor complex. The average should be computed such that missing events (of very short duration) do not interfere with this estimate. This is achieved by excluding data of shorter duration than a minimum detection threshold, t_{min}; determining the mean of the events of duration $> t_{min}$ and then subtracting t_{min}

Table 3
Comparison of analysis methods on experimental data[a]

	Running variance	Page's test
Events	315	234
Displacement (nm)	5.26	5.65
Standard deviation (nm)	21.61	23.12
Standard error (nm)	1.22	1.51
Lifetime (ms)	148	155
Standard error (ms)	9	10
Rate constant (s^{-1})	4.70	4.47

[a] This table compares the results obtained with the running variance and Page's test methods on a set of experimental data (see Section 4.7 and Fig. 8). Note the close similarity between the results.

from that mean to give the "true" mean (Eq. (8)). The apparent rate constant, k_{obs}, can then be calculated from this mean lifetime (Eq. (9)). The danger of this method is that it assumes a single exponential process. The original lifetime data should always be plotted as a histogram to ensure that the distribution shows a single exponential decay.

$$\bar{t} = \langle t > t_{min} \rangle - t_{min}, \tag{8}$$

$$k_{obs} = \ln(2)/\bar{t}. \tag{9}$$

5.2. *Ensemble averaging*

We know that the observed acto-myosin attachment event consists of several underlying biochemical states. Of greatest interest to us is the biochemical step responsible for the "power-stroke". In order for acto-myosin to generate external force this step must occur when myosin is tightly bound to actin. So, each attached event should show a small deflection (≈ 5 nm) as the power-stroke occurs. There are several reasons why such a deflection might be difficult to observe. For example: (1) the power-stroke might be produced as myosin "rocks" into its tightly bound state with actin; (2) the delay between the tight-binding step and power-stroke might be very brief (<2 ms); (3) the power stroke might be composed of many much smaller steps; (4) the power-stroke might dither back and forth on a rapid time scale and not appear as a single discrete step.

Once the start and end time points for each individual binding event have been determined accurately, several events may be synchronised to their start and end points and the position at each time point averaged between records. The resulting data record then gives the average position measured at different times throughout the binding event. We believe that such "ensemble averaging" has the potential to reveal small changes in position (e.g. the power-stroke) that would otherwise be obscured by residual thermal motion. However, we found that close inspection of our single-molecule mechanical data obtained using rabbit fast skeletal muscle myosin II failed to show movement occurring after tight-binding. Veigel et al. (1999) used a high frequency (1 kHz) perturbation to measure stiffness with higher time resolution. Oscillating the position of one of the optical traps, in the detached state allows the signal to propagate to the other bead, but myosin attachment causes the intensity of the signal at the other bead to drop rapidly due to increased system stiffness (Section 2.6). They found that part of the power-stroke produced by slow, unconventional myosin Is during the attachment event was clearly observed. In contrast, skeletal muscle myosin II still did not show a difference in position between the start and end of each event. The authors believed that the skeletal muscle myosin II produces its power stroke within the time resolution of the measurement. In fact, the initial ≈ 5 nm motion produced by all myosin isotypes tested occurred within the time resolution of current methods and this leaves possibilities 1 and 2 above open to debate.

In contrast to these measurements, Kitamura et al. (1999) obtained contradictory results using a modified atomic force microscope transducer (1999). They found that skeletal muscle myosin produces several processive steps (up to 5) of 5.3 nm amplitude each. The observed runs of steps required hydrolysis of just a single ATP molecule which would imply that under some conditions myosin might step along adjacent actin monomers, in contradiction to existing data (see for example, Sase et al., 1997). The underlying mechanism to this must be quite distinct from the "power-stroke" idea. The difference between these results may be due to the different experimental techniques used. Further research is necessary to resolve this important controversy.

6. Conclusion

Optical tweezers based mechanotransducers have the correct sensitivity and mechanical properties to measure single biomolecular interactions. Key to these measurements is the determination of the start and end of the interaction "event". Good time resolution is important both because it allows the kinetics to be measured accurately and because observed events may then be further analysed to look for other underlying processes. Identifying these separate phases occurring within an event should greatly increase our understanding of the molecular mechanism being studied. In this paper we have presented the use of a modified form of Page's test that we think will improve the timing precision and robustness of event detection. This technique should be applicable not only to myosins but also to other non-processive motors and transient biomolecular interactions. Single-molecule experiments have the potential to unravel the coupling mechanism between biochemistry and mechanics in an unequivocal way. However, many of the important details of the acto-myosin mechanical mechanism are still up for grabs!

Acknowledgements

We thank BBSRC, Wellcome Trust and Royal Society for financial support.

References

Ashkin, A., Dziedzic, J.M., Bjorkholm, J.E., Chu, S., 1986. Observation of a single-beam gradient force optical trap for dielectric particles. Opt. Lett. 11, 288–290.

Block, S.M., Goldstein, L.S.B., Schnapp, B.J., 1990. Bead movement by single kinesin molecules studied with optical tweezers. Nature 348, 348–352.

Cooke, R., 1998. Actomyosin interaction in striated muscle. Physiol. Rev. 77, 671–697.

Davenport, R.J., Wuite, G.J.L., Landick, R., Bustamante, C., 2000. Single-molecule study of transcriptional pausing and arrest by E-coli RNA polymerase. Science 287, 2497–2500.

Dickson, R.M., Cubitt, A.B., Tsien, R.Y., Moerner, W.E., 1997. On/off blinking and switching behaviour of single molecules of green fluorescent protein. Nature 388, 355–358.

Dupuis, D.E., Guilford, W.H., Wu, J., Warshaw, D.M., 1997. Actin filament mechanics in the laser trap. J. Muscle Res. Cell Motility 18, 17–30.

Einstein, A., 1906. On the theory of the Brownian movement. Ann. Phys. 4, 371–381.

Finer, J.T., Simmons, R.M., Spudich, J.A., 1994. Single myosin molecule mechanics: piconewton forces and nanometre steps. Nature 368, 113–119.

Graps, A., 1995. An introduction to wavelets. IEEE Comput. Sci. Eng. 2, 50–61.

Guilford, W.H., Dupuis, D.E., Kennedy, G., Wu, J.R., Patlak, J.B., Warshaw, D.M., 1997. Smooth muscle and skeletal muscle myosins produce similar unitary forces and displacements in the laser trap. Biophys. J. 72, 1006–1021.

Hall, K., Cole, D.G., Yeh, Y., Scholey, J.M., Baskin, R.J., 1993. Force–velocity relationships in kinesin-driven motility. Nature 364, 457–459.

Han, C.M., Willett, P.K., Abraham, D.A., 1999. Some methods to evaluate the performance of Page's test as used to detect transient signals. IEEE Trans. Signal Process. 47, 2112–2127.

Hill, T.L., 1974. Theoretical formalism for the sliding filament model of contraction of striated muscle: part I. Prog. Biophys. Mol. Biol. 28, 267–340.

Hunt, A.J., Gittes, F., Howard, J., 1994. The force exerted by a single kinesin molecule against a viscous load. Biophys. J. 67, 766–781.

Huxley, A.F., 1957. Muscle structure and theories of contraction. Prog. Biophys. Mol. Biol. 7, 255–318.

Huxley, H.E., 1969. The mechanism of muscular contraction. Science 164, 1356–1366.

Huxley, A.F., Simmons, R.M., 1971. Proposed mechanism of force generation in striated muscle. Nature 233, 533–538.

Ishijima, A., Doi, T., Sakurada, K., Yanagida, T., 1991. Sub-piconewton force fluctuations of actomyosin in vitro. Nature 352, 301–306.

Ishijima, A., Kojima, H., Funatsu, T., Tokunaga, M., Higuchi, H., Tanaka, H., Yanagida, T., 1998. Simultaneous observation of individual ATPase and mechanical events by a single myosin molecule during interaction with actin. Cell 92, 161–171.

Kitamura, K., Tokunaga, M., Iwane, A.H., Yanagida, T., 1999. A single myosin head moves along an actin filament with regular steps of 5.3 nanometres. Nature 397, 129–134.

Kojima, H., Muto, E., Higuchi, H., Yanagida, T., 1997. Mechanics of single kinesin molecules measured by optical trapping nanometry. Biophys. J. 73, 2012–2022.

Kron, S.J., Spudich, J.A., 1986. Fluorescent actin filaments move on myosin fixed to a glass surface. Proc. Natl. Acad. Sci. USA 83, 6272–6276.

Lymn, R.W., Taylor, E.W., 1971. Mechanism of adenosine triphosphate hydrolysis by actomyosin. Biochemistry 21, 1925–1928.

Mehta, A.D., Finer, J.T., Spudich, J.A., 1997. Detection of single-molecule interactions using correlated thermal diffusion. Proc. Natl. Acad. Sci. USA 94, 7927–7931.

Mehta, A.D., Rief, M., Spudich, J.A., Smith, D.A., Simmons, R.M., 1999a. Single-molecule biomechanics with optical methods. Science 283, 1689–1695.

Mehta, A.D., Rock, R.S., Rief, M., Spudich, J.A., Mooseker, M.S., Cheney, R.E., 1999b. Myosin-V is a processive actin-based motor. Nature 400, 590–593.

Molloy, J.E., 1998. Optical chopsticks: digital synthesis of multiple optical traps. Methods Cell Biol. 55, 205–216.

Molloy, J.E., Burns, J.E., Kendrick-Jones, J., Tregear, R.T., White, D.C.S., 1995. Movement and force produced by a single myosin head. Nature 378, 209–212.

Noji, H., Yasuda, R., Yoshida, M., Kinosita, K., 1997. Direct observation of the rotation of F-1-ATPase. Nature 386, 299–302.

Oiwa, K., Chaen, S., Kamitsubo, E., Shimmen, T., Sugi, H., 1990. Steady-state force velocity relation in the ATP-dependent sliding movement of myosin-coated beads on actin cables in vitro studied with a centrifuge microscope. Proc. Natl. Acad. Sci. USA 87, 7893–7897.

Page, E.S., 1954. Continuous inspection schemes. Biometrika 41, 100–115.

Sakakibara, H., Kojima, H., Sakai, Y., Katayama, E., Oiwa, K., 1999. Inner-arm dynein c of Chlamydomonas flagella is a single-headed processive motor. Nature 400, 586–590.

Sase, I., Miyata, H., Ishiwata, S., Kinosita, K., 1997. Axial rotation of sliding actin filaments revealed by single-fluorophore imaging. Proc. Natl. Acad. Sci. USA 94, 5646–5650.

Schafer, D.A., Gelles, J., Sheetz, M.P., Landick, R., 1991. Transcription by single molecules of RNA polymerase observed by light microscopy. Nature 352, 444–448.

Scholey, J.M., 1993. Motility assays for motor proteins. In: Wilson, L., Matsudaira, P. (Eds.), Methods in Cell Biology, Vol. 39. Academic Press, San Diego, pp. 304.

Shingyoji, C., Higuchi, H., Yoshimura, M., Katayama, E., Yanagida, T., 1998. Dynein arms are oscillating force generators. Nature 393, 711–714.

Svoboda, K., Block, S.M., 1994. Biological applications of optical forces. Annu. Rev. Biophys. Biomol. Struct. 23, 247–285.

Tanaka, H., Ishijima, A., Honda, M., Saito, K., Yanagida, T., 1998. Orientation dependence of displacements by a single one-headed myosin relative to the actin filament. Biophys. J. 75, 1886–1894.

Tyska, M.J., Dupuis, D.E., Guilford, W.H., Patlak, J.B., Waller, G.S., Trybus, K.M., Warshaw, D.M., Lowey, S., 1999. Two heads of myosin are better than one for generating force and motion. Proc. Natl. Acad. Sci. USA 96, 4402–4407.

Uppenbrink, J., Clery, D., 1999. Single molecules—introduction. Science 283, 1667.

Vale, R.D., Funatsu, T., Pierce, D.W., Romberg, L., Harada, Y., Yanagida, T., 1996. Direct observation of single kinesin molecules moving along microtubules. Nature 380, 451–453.

Vale, R.D., Schnapp, B.J., Reese, T.S., Sheetz, M.P., 1985. Organelle, bead, and microtubule translocations promoted by soluble factors from the squid giant-axon. Cell 40, 559–569.

Veigel C. et al., Processivity of myosin V—interactions between single and double headed myosin V molecules and F-actin, in preparation.

Veigel, C., Bartoo, M.L., White, D.C.S., Sparrow, J.C., Molloy, J.E., 1998. The stiffness of rabbit skeletal actomyosin cross-bridges determined with an optical tweezers transducer. Biophys. J. 75, 1424–1438.

Veigel, C., Coluccio, L.M., Jontes, J.D., Sparrow, J.C., Milligan, R.A., Molloy, J.E., 1999. The motor protein myosin-I produces its working stroke in two steps. Nature 398, 530–533.

Visscher, K., Block, S.M., 1998. Versatile optical traps with feedback control. Methods Enzymol. 298, 460–489.

Visscher, K., Gross, S.P., Block, S.M., 1996. Construction of multiple-beam optical traps with nanometer-resolution position sensing. IEEE J. Selected Top. Quantum Electron. 2, 1066–1076.

Visscher, K., Schnitzer, M.J., Block, S.M., 1999. Single kinesin molecules studied with a molecular force clamp. Nature 400, 184–189.

Wang, M.D., Schnitzer, M.J., Yin, H., Landick, R., Gelles, J., Block, S.M., 1998. Force and velocity measured for single molecules of RNA polymerase. Science 282, 902–907.

White, D.C.S., Thorson, J.W., 1973. The kinetics of muscle contraction. Prog. Biophys. Mol. Biol. 27, 173–255.

Biology at the Single Molecule Level
S.H. Leuba an J. Zlatanova (Eds)

Review

Structural and functional imaging with carbon nanotube AFM probes

J.H. Hafner, C.-L. Cheung, A.T. Woolley, C.M. Lieber*

Department of Chemistry and Chemical Biology, Harvard University, Cambridge, MA 02138, USA

Abstract

Atomic force microscopy (AFM) has great potential as a tool for structural biology, a field in which there is increasing demand to characterize larger and more complex biomolecular systems. However, the poorly characterized silicon and silicon nitride probe tips currently employed in AFM limit its biological applications. Carbon nanotubes represent ideal AFM tip materials due to their small diameter, high aspect ratio, large Young's modulus, mechanical robustness, well-defined structure, and unique chemical properties. Nanotube probes were first fabricated by manual assembly, but more recent methods based on chemical vapor deposition provide higher resolution probes and are geared towards mass production, including recent developments that enable quantitative preparation of individual single-walled carbon nanotube tips [J. Phys. Chem. B 105 (2001) 743]. The high-resolution imaging capabilities of these nanotube AFM probes have been demonstrated on gold nanoparticles and well-characterized biomolecules such as IgG and GroES. Using the nanotube probes, new biological structures have been investigated in the areas of amyloid-beta protein aggregation and chromatin remodeling, and new biotechnologies have been developed such as AFM-based haplotyping. In addition to measuring topography, chemically functionalized AFM probes can measure the spatial arrangement of chemical functional groups in a sample. However, standard silicon and silicon nitride tips, once functionalized, do not yield sufficient resolution to allow combined structural and functional imaging of biomolecules. The unique end-group chemistry of carbon nanotubes, which can be arbitrarily modified by established chemical methods, has been exploited for chemical force microscopy, allowing single-molecule measurements with well-defined functionalized tips.

Keywords: Atomic force microscopy; Carbon nanotube; Chemical force microscopy; Single molecule spectroscopy; Structural biology

*Corresponding author. Tel.: +1-617-495-9833; fax: +1-617-496-5442.
E-mail address: cml@cmliris.harvard.edu (C.M. Lieber).

Contents

1. Introduction

Elucidating the three dimensional structures of proteins, nucleic acids and assemblies of these macromolecules is critical to developing a molecular level understanding of biological function. X-ray diffraction, electron diffraction, and nuclear magnetic resonance (NMR) are well-established and powerful tools for the analysis of macromolecular structure with atomic precision, which with advances in technology, have continued to push the limits of the size and complexity of systems that can be structurally illuminated (Beauchamp and Isaacs, 1999; Chiu et al., 1999; Glaeser, 1999; Glusker, 1993; Henderson, 1995; Siegal et al., 1999; Stowell et al., 1998; Wuthrich, 1995). As one looks to the future, the needs for and impact of macromolecular structure analysis are expected to expand significantly in several areas, including: (1) increased throughput of structural analysis of gene products discovered by large scale genome sequencing (Kim, 1998); (2) routine characterization of large multimeric structures, such as protein assemblies and protein–nucleic acid complexes involved in signaling and regulation; and (3) understanding dynamic features of protein–protein and protein–nucleic acid assemblies. The standard structural tools can impact each of these areas but also have limitations. For example, difficulties in crystallization of large multimeric signaling and regulatory complexes may limit the general use of X-ray and electron diffraction methods for these critical problems, while solution NMR studies of macromolecular dynamics (and structure) have size restrictions that limit the systems that can be studied (Wuthrich, 2000).

It thus seems clear that future progress in elucidating the complex machinery of the cell will depend in part on additional and perhaps revolutionary developments in structural biology tools. In this regard, two complementary techniques are receiving increased attention: (1) cryogenic transmission electron microscopy (cryoEM) and (2) atomic force microscopy (AFM). CryoEM allows the direct visualization of biological structures, and can produce high-resolution images of isolated proteins with resolution of 10–25 Å (Grigorieff, 1998; Kuhlbrandt and Williams, 1999). Because contrast is low in biological EM, the high resolution achieved by cryoEM requires computer averaging and processing of thousands of images. The ability to determine structure from single macromolecule images represents a distinct advantage for the large multimeric systems discussed above. AFM allows direct visualization of biological structures like cryoEM, and moreover, can probe structural dynamics in vitro since structural imaging can be carried out in solution (for reviews, see Shao et al., 1996; Engel et al., 1997; Fritz et al., 1995; Bustamante et al., 1997). This latter feature of AFM represents a distinct advantage compared to cryoEM.

In previous work, AFM has been used to probe the structures of a number of protein and nucleic acid systems. Key results obtained from past studies include (1) resolution of the packing and often subunit arrangement of membrane proteins (Muller et al., 1997, 1999b; Scheuring et al., 1999; Shao et al., 1996), (2) determination of the overall structures of protein–DNA complexes (Bustamante and Rivetti, 1996), and (3) elucidation of dynamic processes (Kasas et al., 1997; Guthold et al., 1999). In addition, in chemical force microscopy (CFM), it has been shown that the AFM tip can be modified to present well-defined chemical or biological functionality, which provides specific chemical contrast as well as topography (Frisbie et al., 1994; Fujihira and Ohzono, 1999; Green et al., 1995; Noy et al., 1997b; Vezenov et al., 1997). These capabilities of AFM suggest a great potential to impact many areas of structural biology, although to date the impact on major structural and dynamic problems has been much less than that achieved by electron microscopy as well as X-ray diffraction and NMR.

When we ask how the significant potential of AFM might be better exploited in structural biology, one needs to look no further than the structure of the tip used for imaging. Since an AFM image is a convolution of the structure of the macromolecule or macromolecular assembly under investigation and the tip structure, a well-characterized tip is essential for accurately interpreting an image, and moreover, the size of the tip will define the resolution of the image. The situation is somewhat analogous to the need for well-collimated and monoenergetic electron beams in cryoEM and well-defined X-ray sources in X-ray diffraction. In the following review, we will focus on recent and significant progress in this area of AFM technology and new and potential science enabled by the advances in AFM probes. The structure of the review will be as follows. First, we will discuss briefly the basics of AFM and imaging biological structures with an emphasis on highlighting key issues for probe tips. Second, we will describe the fabrication and characterization of a new generation of tips based on carbon nanotubes. Third, we will examine the reproducibility and resolution of these probes in imaging relatively well-defined inorganic and biological structures. Fourth, we will describe studies that are beginning to exploit the unique features of these new probes for addressing new structural biology and other problems. Fifth, we introduce the concept of CFM and describe the unique potential of carbon nanotube tips for functional imaging at the nanometer scale. Lastly, we briefly summarize and speculate on prospects of carbon nanotubes probes in the future.

2. AFM imaging and probe tips

2.1. AFM basics

To understand the strengths and weaknesses of AFM, we first consider the principles of operation. A typical AFM (Fig. 1) consists of an integrated cantilever-tip assembly, a detector to

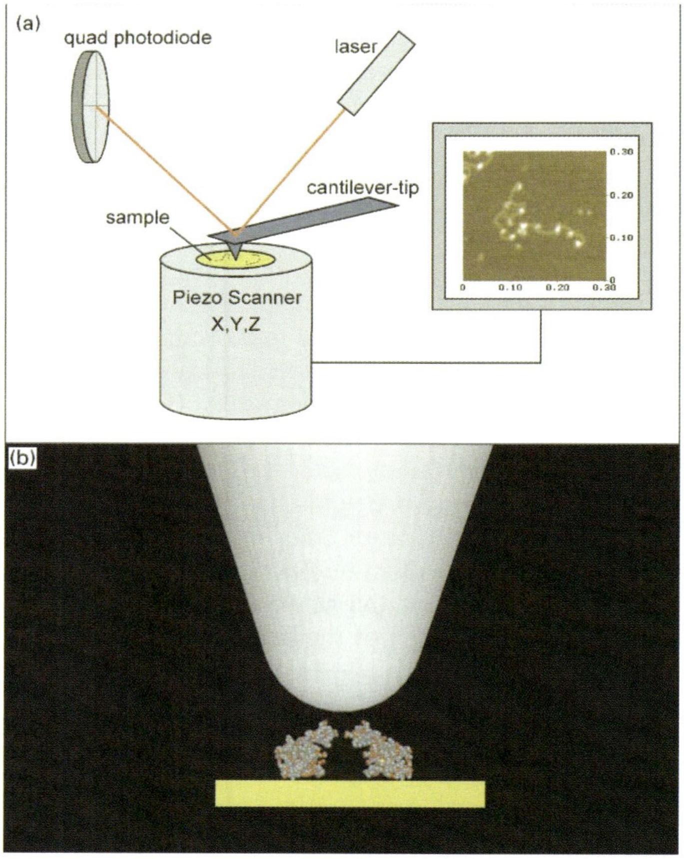

Fig. 1. A schematic of the atomic force microscope: (a) The cantilever-tip assembly, piezo tube scanner, laser deflection system, and computer are depicted. (b) Comparison of a 5 nm radius of curvature cone-shaped tip to a cross section of an individual GroES molecule, which highlights the limitations of such tips in high-resolution imaging.

measure cantilever displacement and feedback electronics to maintain a constant imaging parameter, such as force or tip–sample separation. The sample, which is supported on a solid substrate surface, can be scanned in three-dimensions (X, Y, Z) with angstrom precision using a three-axis piezoelectric translation stage. The tip position is monitored with an optical beam deflection system. Tip deflections on the order of angstroms, which correspond to forces of approximately 10 pN, can be reliably measured using standard cantilevers. Recently, higher resonance frequency cantilevers have been developed (Viani et al., 1999a, b). These cantilevers extend both the potential imaging speed and force resolution by approximately one order of magnitude, and thus should be attractive for biological studies.

There are two basic modes with which structural images can be acquired. In contact mode, the tip remains in contact with the sample/substrate, while scanning in a (X, Y) raster pattern. Typically, the Z position is adjusted during scanning to maintain constant cantilever deflection (i.e., constant force) at a set point value. The resulting Z data represent a topographic image of the surface. Contact mode imaging is best for relatively flat and hard samples since the scanning tip applies large lateral forces. Attempts to image isolated biological structures usually result in poor resolution and sample movement or damage. If the forces are very carefully controlled, however, contact mode can provide high-resolution images of 2-D protein arrays (Muller et al., 1998, 1999a). Alternatively, images can be acquired using intermittent contact or tapping mode. In this mode, the cantilever-tip assembly is oscillated near resonance and when brought to the surface intermittently contacts—taps—the sample. To image in tapping mode, the Z position is adjusted to maintain the oscillation amplitude at a constant set point during the (X–Y) scan. The significant advantage of tapping mode is that there are essentially no lateral forces applied to a sample during imaging, and thus it is the technique of preference for probing isolated macromolecules, macromolecular assemblies and the dynamics of these systems.

2.2. *Imaging and the probe tip*

The structural resolution of AFM will depend strongly on the nature of the probe tip, and thus it is instructive to reference state-of-the-art microfabricated tips to known protein structures. For example, Fig. 1b compares the relative sizes (in cross section) of a 5 nm radius of curvature silicon tip and the ca. 8 nm diameter heptameric protein GroES. This comparison clearly suggests that conventional tips will not provide significant structural detail.

Experimental studies of isolated proteins and nucleic acids, and two-dimensional protein arrays indicate that the situation is more complex. AFM investigations of isolated DNA have shown several features, including kinks (Han et al., 1997) and supercoiled structures (Lyubchenko and Schlyakhtenko, 1997). Resolution on the order of 1–2 nm has been reported in these studies, although the 3.4 nm helical pitch of DNA was not resolved. Earlier studies of isolated DNA (Hansma et al., 1995) and condensed DNA arrays (Mou et al., 1995) did show the DNA helical pitch, although this structure has not been observed reproducibly in subsequent studies. Moreover, it should be noted that the best reported lateral resolution was achieved using tips having average radii of curvature on the order of 50 nm. AFM imaging of other isolated biological structures have produced various levels of resolution. Membrane bound *E. coli* F1F0 ATP synthase revealed a ring-like structure with an off center mass at resolution of ca. 3 nm (Singh et al., 1996). Several studies of immunoglobulin G (IgG) failed to resolve the characteristic 15 nm

Y-shape of this macromolecule (Bergkvist et al., 1998), while others imaged the Y-shape as a heart-like structure (Fritz et al., 1997). However, the Y-shape was apparent at cryogenic temperatures (Zhang et al., 1996). These widely varying results illustrate the sensitivity of high-resolution biological AFM to imaging conditions and tip quality. Comparison of these different studies shows that the highest resolutions were found in low-height structures (i.e., DNA and the membrane bound protein), while isolated proteins (e.g., IgG) showed significantly lower resolution with conventional tips. Hence, these results suggest that roughness or asperities at the tip end might be responsible for the high-resolution images of DNA and membrane structures, although such asperities have not been independently characterized.

In addition, very high resolution has been observed in AFM studies of densely packed two-dimensional arrays of proteins (Muller et al., 1998, 1999a). For example, 5–6 Å lateral resolution has been reported for images of bacteriorhodopsin (Muller et al., 1999b), even though the silicon nitride tips used for imaging have radii of 20–60 nm. To explain this apparent contradiction, it has been hypothesized that the image contrast is due to short-range repulsive forces on a small nano-protrusion at the tip end (Muller et al., 1999b). This explanation is similar to that used to explain high-resolution images obtained on isolated low-height molecules, such as DNA. Another possibility is that the subnanometer resolution corresponds to Fourier components from the periodic protein lattice. This latter idea is supported by the fact that the highest resolution (Muller et al., 1999b) has been achieved for proteins in a crystalline 2-D array, while approximately two-fold lower resolution has been observed in less well-ordered arrays (Mou et al., 1996); that is, the reduced positional and orientational order would produce lower resolution in the Fourier model. Lastly, the observation of missing-molecule defects may be consistent with both models, since the highly nonlinear tip–sample interaction can still produce an apparent defect in the Fourier model.

2.3. Key role of the tip

The studies reviewed above show that microfabricated tips can provide resolution much higher than expected based on the smooth end radii of curvature of tips. While observed high resolution can be reasonably attributed to asperities at the tip ends, this poses several important issues for the general use of AFM in high-resolution structural biology. First, the presence of these asperities has not been characterized independently by techniques such as transmission electron microscopy. Second, the observation of high resolution is highly dependent on obtaining a "good" tip and cannot be predicted a priori. Moreover, current tips have not been able to image large, isolated protein structures, which form the basis of much of the cell signaling and regulatory machinery, with high resolution. We believe that these uncertainties in the detailed nature of the probe tip must be addressed in order to move AFM to the level of other structural biology techniques for addressing new scientific problems.

Let us take a step back and examine microfabricated tips and the requirements for an ideal tip. Integrated AFM cantilever-tip assemblies are fabricated from silicon (Si) or silicon nitride (Si_3N_4) (Albrecht et al., 1989). The tips on these assemblies are pyramidal in shape, have cone angles of 20–30° and radii of curvature of 5–10 nm (Si) or 20–60 nm (Si_3N_4). Several techniques have been developed to improve these geometrical factors. Si_3N_4 tips can be oxide sharpened to improve their aspect ratio and reduce the tip radii to as low as 5 nm. Silicon tips can be machined with focused ion beam (FIB) instruments to improve the aspect ratio, and sharper Si tips are now available with a high yield of radii less than 5 nm. Despite these technological advances, there remain important limitations. For example, the variation in tip-to-tip properties can be quite large, and will always be difficult to control at the scale relevant to high-resolution structural imaging. In addition and critical to reproducibility, tips wear during scanning (Hansma et al., 1995), making it quite difficult to account quantitatively for tip contributions to image broadening. The problem of tip wear will only increase for sharper tips due to increased pressure at the tip–sample interface.

Consider the ideal AFM tip. It should have a high aspect ratio with 0° cone angle, have a radius as small as possible with well-defined and reproducible molecular structure, and be mechanically and chemically robust such that its structure is not altered while imaging in air or fluid environments. Carbon nanotubes are the only known materials that can satisfy all of these critical criteria, and thus have the potential to create ideal probes for AFM imaging. Below we review the key properties of carbon nanotubes and methods to create tips with this unique class of material.

3. Carbon nanotube probes: fabrication and properties

3.1. Carbon nanotube: structural and mechanical properties

Carbon nanotubes consist of a honeycomb sp^2 hybridized carbon network (termed a graphene sheet) that is rolled up into a seamless cylinder (Fig. 2a), which can be microns in length. There are two basic structural classes of carbon nanotubes: single-walled nanotubes (SWNTs) and multi-walled nanotubes (MWNTs). SWNTs consist of a single seamless cylinder (Fig. 2a) with radii ranging from 0.35 to 2.5 nm, while MWNTs consist of multiple concentric graphene cylinders with radii ranging from 3 to 50 nm. Transmission electron microscopy (TEM) images of an individual MWNT and SWNT are displayed in Figs. 2b and c, respectively. These TEM images show clearly the tubular structure of the nanotubes, and in the case of the SWNT (Fig. 2c) highlight the extremely small diameter and high aspect ratio, which makes this material uniquely suited as a high-resolution probe tip.

Nanotubes have exceptional mechanical properties. The characteristics relevant to the use of nanotubes as AFM tips are the stiffness or Young's modulus and the ability to buckle elastically under large loads. Recent calculations of SWNT and MWNT Young's moduli suggest that both classes of nanotube have moduli on the order of 1 TPa (Lu, 1997). Experimental measurements of MWNTs Young's moduli made by recording thermal vibration amplitudes in a TEM (Treacy et al., 1996) and by direct bending with AFM (Wong et al., 1997) revealed Young's moduli of 1.8 and 1.3 TPa, respectively. More recent studies of individual 1.0–1.5 nm diameter SWNTs found a similar Young's modulus, 1.25 TPa (Krishnan et al., 1999). Hence, the theoretical and experimental results are in good agreement, and show that nanotubes are stiffer than any other known material. The extremely high Young's modulus of nanotubes is critical to the creation of high aspect ratio, sub-nanometer radius tips with high resolution—if the modulus were significantly smaller, then the amplitude of thermal vibrations would degrade the resolution of tips.

Carbon nanotubes buckle elastically under large loads unlike all conventional materials, which either fracture or plastically deform. MWNTs and SWNTs have been observed in the

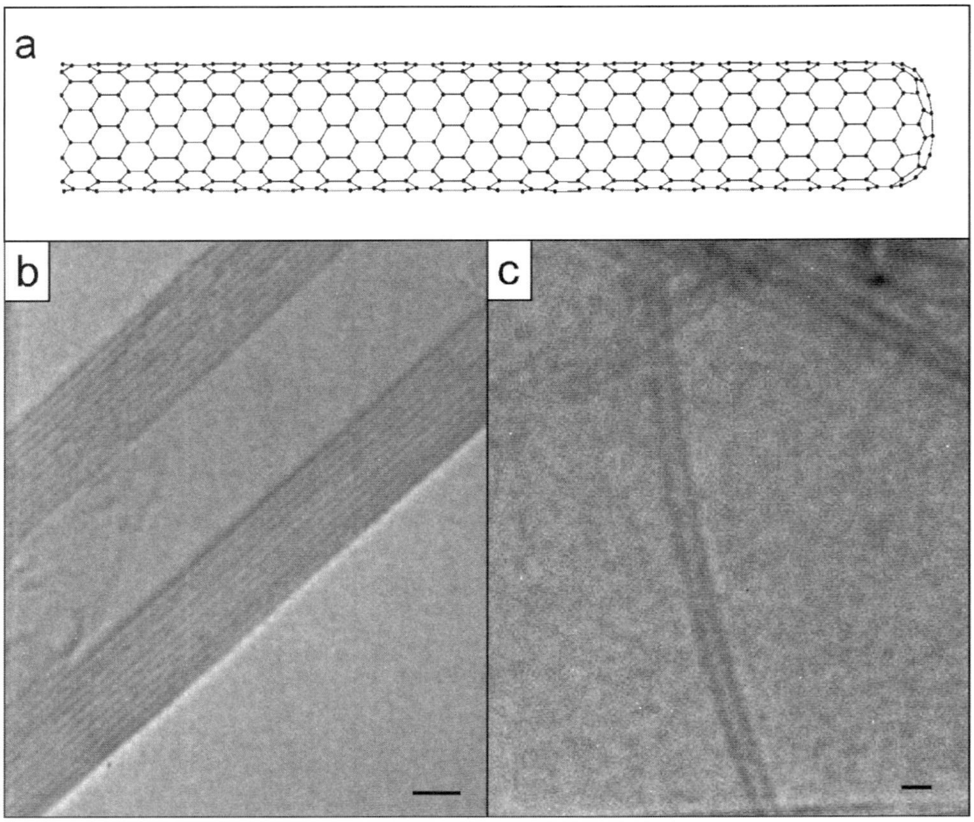

Fig. 2. The structure of carbon nanotubes: (a) Fragment of a stick model of a single-walled carbon nanotube. Each vertex in the honeycomb mesh corresponds to a sp^2-bonded carbon atom. Transmission electron micrographs show the cross-sectional structures of (b) a multi-walled nanotube and (c) a single-walled nanotube.

buckled state by TEM (Iijima et al., 1996), and appear to buckle in a manner strikingly similar to macroscopic tubes made of elastic materials. The first direct experimental evidence that carbon nanotube buckling is elastic came from the use of nanotubes as AFM tips (Dai et al., 1996). The buckling force of the nanotube tip could be measured from the deflection of the AFM cantilever as the nanotube tip–surface separation was reduced. The nanotube would then return to its original configuration when the tip was removed from the surface. A more direct measurement of elastic nanotube buckling was achieved by AFM measurements of force versus displacement of nanotubes pinned at one end on a surface (Wong et al., 1997). Both types of experiments demonstrated that nanotubes can be bent close to 90° many times without observable damage, and thus should be highly robust probes for AFM imaging.

3.2. Manual assembly of nanotube probe tips

The first carbon nanotube AFM tips were produced by manually attaching MWNTs to the pyramids of conventional tips (Dai et al., 1996). In this process, micromanipulators are used to control independently the positions of a commercial cantilever-tip assembly and nanotubes, while viewing in an optical microscope. The nanotubes and tip are readily seen under dark-field illumination at 500–1000 × magnification. A thin layer of adhesive is applied to the pyramidal silicon tip by touching the surface of adhesive tape that supports the nanotube material. The pyramidal tip is then used to pull an MWNT or bundle of MWNTs from the tape surface. Micromanipulators allow the fabrication of nanotube tips that are well aligned for AFM imaging; that is, they are parallel to the tip axis and therefore perpendicular to the sample surface. However, the attached nanotube tips are typically too long to permit high-resolution imaging. The vibration amplitude at the end of the tip, X_{tip}, can be readily estimated from $X_{\text{tip}} = [(4k_{\text{B}}TL^3)/(3\pi Er^4)]^{1/2}$, where k_{B} is the Boltzmann's constant, T is the temperature in Kelvin, L is the length, E is the Young's modulus and r is the radius of the nanotube tip. This estimate of the vibration amplitude, which assumes $1/2k_{\text{B}}T$ thermal energy in the nanotube tip, works well for both individual and bundle nanotube tips since the Young's modulus of a bundle of nanotubes is essentially the same as an individual nanotube (Lu, 1997). The length of a nanotube tip can be decreased by electrical etching on a conductive surface (Dai et al., 1996; Wong et al., 1998a) to reduce the amplitude of vibration to a level where it does not affect resolution.

Mechanically assembled MWNT tips have demonstrated several important features. First, the high aspect ratio of the tips enables more accurate images of structures with steep sidewalls such as silicon trenches (Dai et al., 1996). Second, the tip–sample adhesion is greatly reduced due to the small size and cylindrical geometry of the nanotube (Dai et al., 1996; Wong et al., 1998a). Reduced adhesion allows imaging with lower cantilever energies and less potential sample damage. Third, they have demonstrated clearly the elastic buckling property of nanotubes. For example, as an oscillating nanotube tip is brought into contact with a sample, the cantilever oscillation amplitude is damped strongly but then partially recovers when the nanotube buckles as shown in Fig. 3. The deflection at the buckling point corresponds to the Euler buckling force, $F_{\text{B}} = \pi^2 EI/L^2$, where E is the Young's modulus, $I = \pi r^4/4$ is the moment of inertia, L is the nanotube tip length, and r is the nanotube radius. This elastic buckling behavior has tremendous consequences for nanotube tip robustness, and arises from the sp^2 bonded cylindrical structure of carbon nanotubes. In deflecting carbon nanotubes and carbide nanorods with an AFM tip, it was found that nanotubes elastically buckle while nanorods fracture (Wong et al., 1997).

MWNT tips were found to provide only a modest improvement in resolution compared with standard silicon tips when imaging isolated amyloid fibrils (Wong et al., 1998a). The clear route to higher resolution is to use SWNTs, since they typically have 0.5–2 nm radii. SWNT material produced by laser vaporization of a graphite target doped with metal catalyst consists of high-purity SWNTs (Thess et al., 1996). Unfortunately, the SWNTs produced in this way are aligned in bundles approximately 10 nm wide containing up to hundreds of nanotubes each. These bundles can be readily attached to silicon AFM tips (Wong et al., 1998a, c, d), and their lengths adjusted by the electrical etching procedure described above for optimal imaging. Significantly, this etching process was found to occasionally produce very high-resolution tips that likely resulted from the

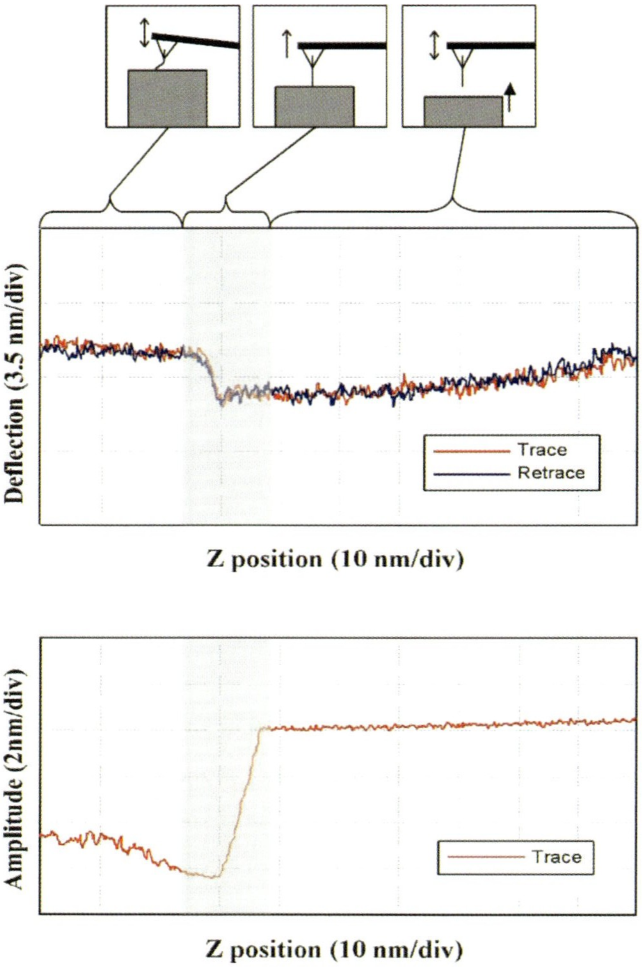

Fig. 3. Nanotube tip buckling: The oscillation amplitude and deflection of a nanotube tip are plotted as a function of tip–sample separation. To the right of the shaded region, the tip and sample are out of contact, so the amplitude is constant and the deflection is zero. In the shaded region, the tip taps the sample resulting in reduced amplitude and tip deflection. To the left of the shaded region, the nanotube has buckled, which results in a leveling off of the cantilever deflection and an increase of the oscillation amplitude.

exposure of only a small number of SWNTs at the apex (Wong et al., 1998a, d). It was not possible in these studies to prepare reproducibly individual SWNT tips for imaging.

The manual assembly method of producing nanotube tips is conceptually straightforward but also has several important limitations. First, this assembly procedure inherently leads to the selection of thick bundles of nanotubes since these are easiest to observe in the optical microscope. Recently, mechanical assembly of nanotube tips was also performed inside a scanning electron microscope (SEM) (Nishijima et al., 1999). The use of an SEM still limits assembly to nanotube bundles or individual tubes with diameters greater than 5–10 nm, and moreover, increases greatly the time required to make one tip. Second, well-defined and reproducible tip etching procedures designed to expose individual SWNTs at the tip apex do not exist. Third, a relatively long time is required to attach each nanotube to an existing cantilever. This not only inhibits carrying out the research needed to develop these tips but also precludes mass production required for general usage. Nevertheless, nanotube tips are now being produced by this method and sold commercially (www.piezomax.com), although each tip is quite expensive.

3.3. Direct growth of nanotube probes by chemical vapor deposition

All of the problems associated with manual assembly potentially can be solved by directly growing nanotubes onto AFM tips by metal-catalyzed chemical vapor deposition (CVD). The key features of the nanotube CVD process are illustrated in Fig. 4. In the CVD synthesis of carbon nanotubes, metal catalyst particles are heated in a gas mixture containing hydrocarbon or CO. The gas molecules dissociate on the metal surface and carbon is adsorbed into the catalyst particle. When this carbon precipitates, it nucleates a nanotube of similar diameter to the catalyst particle. Hence, CVD allows control over nanotube size and structure including the production of SWNTs (Hafner et al., 1998), with radii as low as 3.5 Å (Nikolaev et al., 1999).

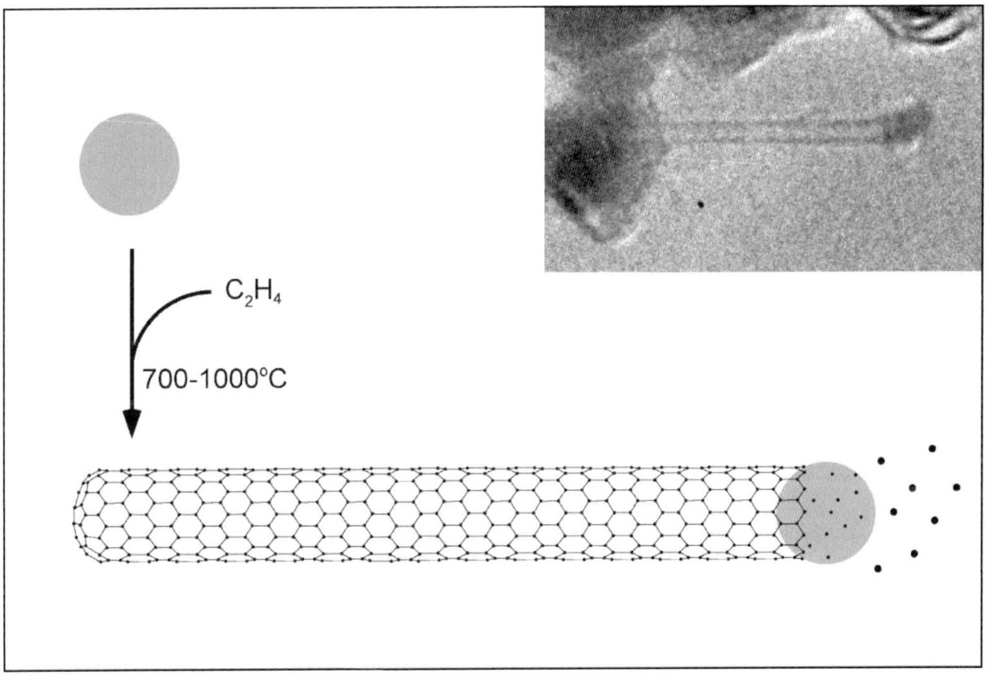

Fig. 4. Nanotube production by CVD: Nanometer-scale catalyst particles are heated in the presence of a hydrocarbon gas or carbon monoxide. The reactant molecules dissociate on the surface of the catalyst particle, carbon is adsorbed into the particle, and precipitates from the particle as a nanotube. In the upper right is a TEM image of a ca. 2 nm diameter single-walled nanotube with the catalyst nanoparticle attached as shown schematically in the diagram.

3.3.1. Pore growth

The central issues in the growth of nanotube AFM tips by CVD are (1) how to align the nanotubes at the tip such that they are well positioned for imaging and (2) how to ensure there is only one nanotube or nanotube bundle at the tip apex. In previous studies of bulk nanotube production, it was found that nanotubes grew perpendicular to a porous surface containing embedded catalyst (Li et al., 1996). We exploited this approach in the first demonstration of CVD nanotube tips (Hafner et al., 1999a) as illustrated in Fig. 5. A flattened area of approximately 1–5 μm^2 was created on Si tips by high-load scanning on a hard, synthetic diamond surface. The tip was then anodized in HF to create 100 nm diameter pores in this flat surface. Next, iron was electrochemically deposited into the pores to form catalyst particles. Tips prepared in this way were heated in low concentrations of ethylene at 800°C, which is known to favor the growth of thin nanotubes (Hafner et al., 1998). When imaged by SEM, nanotubes prepared in this way were found to grow perpendicular to the surface from the pores as desired (Fig. 5). Moreover, TEM observations revealed thin individual multiwalled nanotubes protruding from the ends of the silicon tips, with typical radii ranging from 3 to 5 nm (Fig. 5). It is also worth noting, if tubes

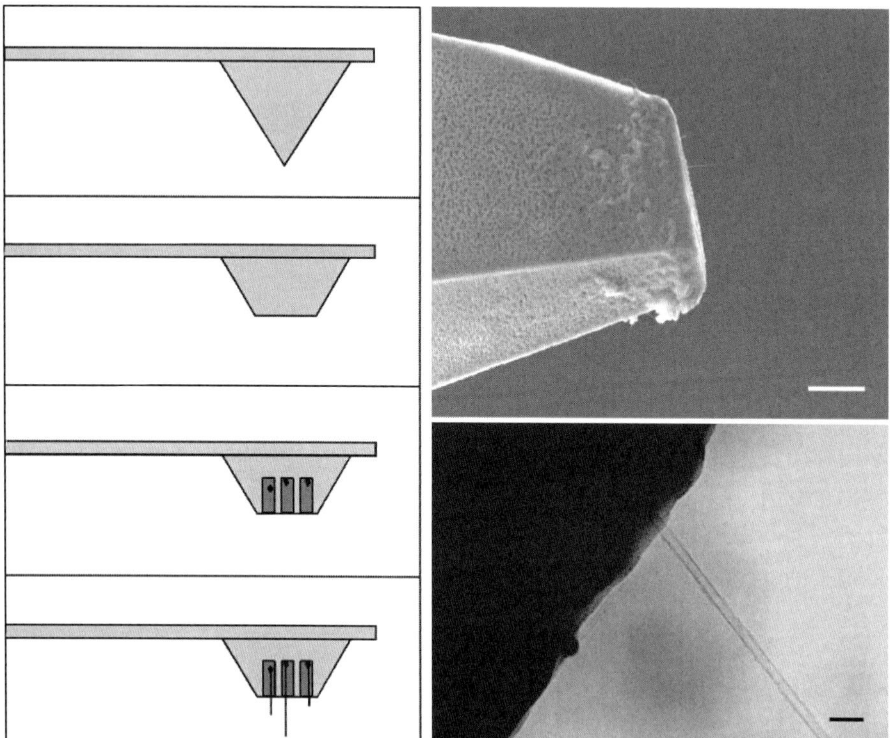

Fig. 5. The pore-growth technique for CVD nanotube tip production: The panels on the left illustrate the steps involved in tip production: flattening, pore formation, catalyst deposition, and CVD nanotube growth. SEM (top right, 1 μm scale bar) and TEM (bottom right, 20 nm scale bar) of individual carbon nanotube tips produced by the pore-growth method (Hafner et al., 1999a).

did not grow in an acceptable orientation or when they ultimately failed, that the carbon could be removed by oxidation and then CVD repeated to grow new nanotube tips (Hafner et al., 1999a).

These "pore-growth" CVD nanotube tips exhibit the favorable mechanical and adhesion properties found earlier with manually assembled nanotube tips, that is, the CVD tips buckled elastically, were very robust in imaging, and had low tip–sample adhesion. In addition, the ability to produce reproducibly thin, individual nanotube tips has enabled improved resolution (Hafner et al., 1999a) on isolated proteins. Continued development of the pore-growth nanotube tips has focused on the development of more controlled catalysts, such as well-defined iron oxide nanocrystals, to enable greater control over nanotube growth. In particular, this effort has enabled the controlled growth of thin SWNT bundles of 1–3 nm in diameter from pores made at the silicon tip ends (Cheung et al., 2000b). The pore-growth method has demonstrated the great potential of CVD to grow directly controlled diameter nanotube tips, although it still has some limitations. In particular, the preparation of a porous layer can be time consuming and may not place individual SWNTs at the optimal location on the flattened apex. One solution to this issue involves using focused ion beam milling, which can be carried out with 5–7 nm resolution, to create a single pore at the tip apex, which would then necessitate growth from only this one optimal position. This approach could also be automated on a wafer wide scale using commercial ion milling instruments.

3.3.2. Surface growth

An elegant yet simple alternative to produce CVD SWNT tips involves direct growth of SWNTs tips from the pyramid of a silicon cantilever-tip assembly. In this "surface growth" approach, we utilize the trade-off between the energy gain of the nanotube–surface interaction and energy cost to bend nanotubes to grow reproducibly SWNTs from the silicon pyramid apex in the ideal orientation for high-resolution imaging. This perhaps unintuitive result can be understood by considering the fate of a nanotube during growth. When a growing nanotube reaches an edge of the pyramid, it can either bend to align with the edge or protrude from the surface. The pathway followed by the nanotube is determined by a trade-off in the energetic terms introduced above: if the energy required to bend the tube and follow the edge is less than the attractive nanotube–surface energy, then the nanotube will follow the pyramid edge to the apex; that is, nanotubes are steered towards the tip apex by the pyramid edges. At the apex, the nanotube must protrude along the tip axis since the energetic cost of bending is too high. A schematic of this approach is shown in Fig. 6.

This steering of nanotubes to the pyramid apex to form ideal probe tips has been demonstrated experimentally (Hafner et al., 1999b). For example, SEM investigations of nanotube tips produced by the surface growth method show that a very high yield of tips contains nanotubes only at the apex, with very few protruding elsewhere from the pyramid (Fig. 6). These tips, which have been produced under conditions expected to yield SWNTs (Hafner et al., 1998), are sufficiently small that SEM cannot provide an accurate measure of their diameters. TEM analysis was able to demonstrate that the tips consist of individual SWNTs and small SWNT bundles. In the case of the small SWNT bundles, the TEM images show that the bundles are formed by nanotubes coming together from different edges of the pyramid to join at the apex, thus confirming the surface growth model described above (Fig. 6). We also believe that the surface-

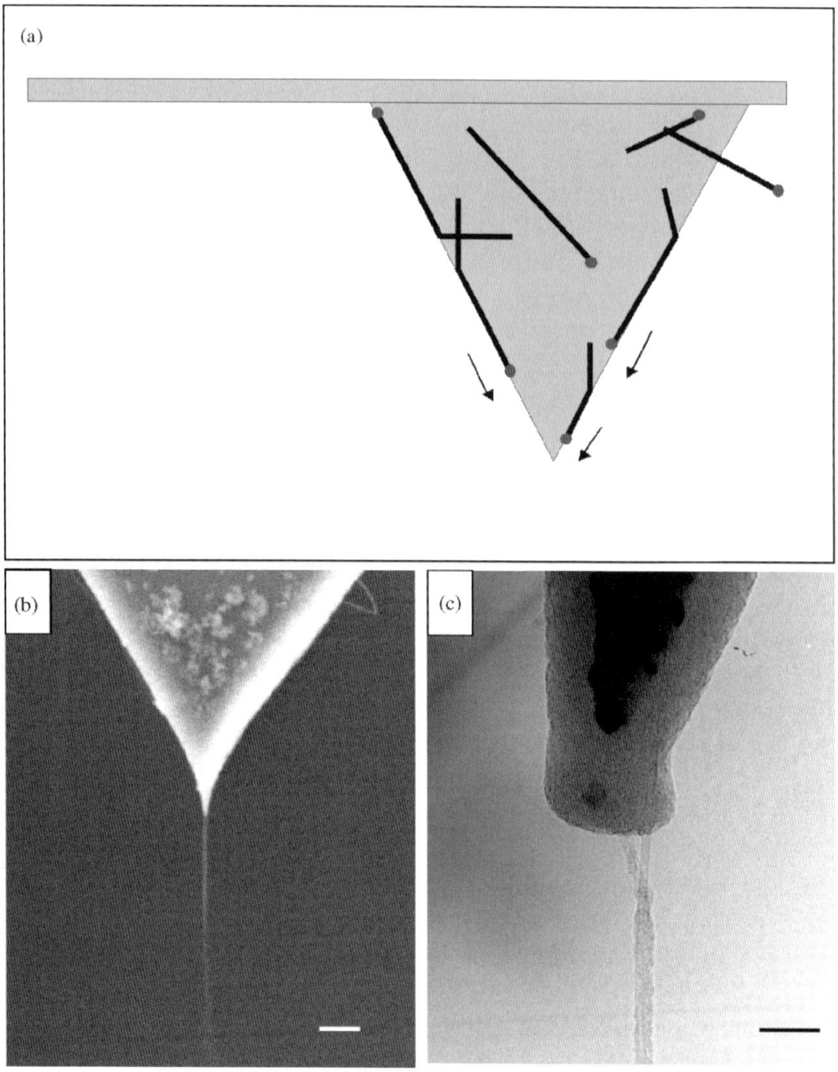

Fig. 6. Surface growth method for CVD nanotube tip preparation: (a) Schematic illustration of surface growth process, where nanotubes grow on the pyramidal surface, guided along the edges towards the tip apex. (b) SEM and (c) TEM images of an SWNT surface growth tip consisting of two SWNTs (Hafner et al., 1999b). The scale bars in (b) and (c) are 200 and 20 nm, respectively.

growth approach is especially important since it provides a conceptual and practical framework for preparing individual SWNT tips by lowering the catalyst density on the surface such that only 1 nanotube reaches the apex. Hence, by controlling reproducibly the catalyst density used for growth it is possible to produce well-defined individual SWNT tips, as shown in Fig. 7 (Hafner et al., 1999b, 2000a; Cheung et al., 2000c). Lastly, we note that these surface growth SWNT tips

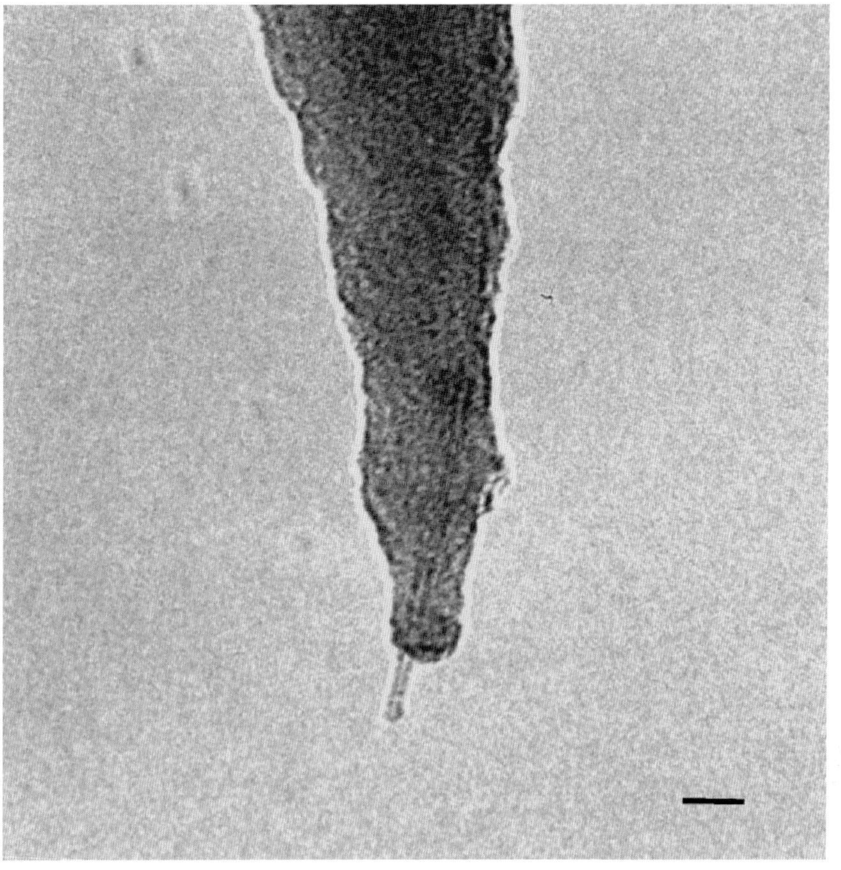

Fig. 7. TEM image of an individual SWNT tip produced by controlling carefully the catalyst density (Cheung et al., 2000c). The scale bar equals 10 nm.

display the elastic buckling behavior characteristic of other nanotube tips, and can be recycled by oxidation as with the pore-growth tips.

The synthesis of carbon nanotube AFM tips by CVD clearly resolves the major limitations of nanotube tips that arise from the manual assembly method. Rather than requiring tedious micromanipulation for each tip, it is possible to envision production of an entire wafer of SWNT tips. In addition to ease of production, CVD yields thin, individual SWNT tips that cannot be made by other techniques and perhaps represent the ultimate AFM probe for structural biology.

4. Imaging with and resolution of nanotube probe tips

4.1. Evaluation of nanotube tip resolution with gold nanoparticles

To accurately assess the structure and resolution of nanotube tips, an imaging standard is needed. Monodisperse gold nanoparticles, which are essentially incompressible in AFM, are ideal

Table 1
Comparison of tip radii of different types of carbon nanotube probes

Nanotube type	Tip radius (nm)[a]
Mech. assembled MWNT	$\geqslant 6$
Mech. assembled SWNT	$\geqslant 3.5$
Pore-growth MWNT	3.5–6
Pore-growth SWNT	2–4
Surface-growth SWNT	2–4
Low density surface growth[b]	$\leqslant 2$

[a] Tip radii were determined from images of 5.7 nm gold nanoparticles using the two-sphere model (Bustamante et al., 1993).

[b] The lower limit of tips produced by this method is expected to be on the order of 0.35–0.5 nm.

for this purpose (Vesenka et al., 1993). We have used 5.7 nm diameter gold nanoparticles to characterize different types of carbon nanotube tips. This size of gold nanoparticle represents an appropriate one for testing tips that will be used for imaging large isolated proteins, which have been the focus of our work. The effective tip radius is calculated from the particle image using the two-sphere model (Bustamante et al., 1993). Manually assembled MWNT tips typically have shown radii as small as 6 nm, which is expected for the size of arc-produced MWNT material (Wong et al., 1998a). Manually assembled SWNT tips are composed of thick SWNT bundles. Since the SWNT bundles are made from 1.4 nm nanotubes, the etching procedure would at times produce high resolution on gold nanoparticles, down to 3 nm radius (Wong et al., 1998a, d). Pore-growth CVD MWNT have radii ranging from 3 to 6 nm when measured from gold nanoparticle images (Hafner et al., 1999a). These tips provide better resolution than manually assembled MWNT tips and typical assembled SWNT tips, because the CVD MWNT tips are relatively thin and always consist of individual nanotubes. Both the pore-growth (Cheung et al., 2000b) and surface growth (Hafner et al., 1999b, 2000a; Cheung et al., 2000c) SWNT tips have shown radii of less than 4 nm, reflecting the very thin SWNT bundle or individual SWNT tip structures possible with these methods. These results are summarized in Table 1.

Another very significant point with the nanotube tips is that the range of tip radii measured by AFM agreed with the range measured by TEM. This demonstrates that, unlike in the case of microfabricated Si and Si_3N_4 tips, we are imaging with very well-defined probes, and that the probe contribution to images can be readily removed by deconvolution techniques. Lastly, we reiterate that the levels of resolution achieved with the nanotubes were obtained on gold nanoparticles nearly 6 nm in height, rather than on 1–2 nm features which at times indicate high resolution with tenuous, asperities on microfabricated Si and Si_3N_4 tips.

4.2. *Studies of well-characterized biological structures*

To determine the potential of nanotube tips in structural biology, DNA and several well-characterized proteins have been imaged. DNA was imaged by manually assembled MWNT tips in air (Nishijima et al., 1999) and in fluid (Li et al., 1999). The fluid imaging experiments produced a measured height of 2 nm, the expected value based on the intrinsic DNA diameter, and the resolution in these studies was on the order of 3.5–5 nm. These values for the resolution are consistent with that expected for multiwalled nanotube material, but also similar to the best values observed with microfabricated tips (Han et al., 1997; Lyubchenko and Schlyakhtenko, 1997). However, as discussed above, we do not believe that imaging DNA provides a robust test of resolution and capabilities of tips that will be used for investigating relatively large proteins.

Studies of isolated proteins provide a much more stringent test of the capabilities of probe tips, and we believe demonstrate clearly the advantages of nanotube tips. Pore-growth MWNT CVD tips were used to image isolated IgG and Immunoglobulin-M (IgM) antibody proteins (Hafner et al., 1999a; Cheung et al., 2000b). IgG is a ca. 180 kDa molecule with a characteristic Y-shaped structure approximately 15 nm across as determined by single crystal X-ray diffraction measurement (Silverton et al., 1977). Previous AFM imaging studies either saw no structure (Bergkvist et al., 1998), a heart-shaped structure (Fritz et al., 1997), or the Y-shape only at cryogenic temperatures (Zhang et al., 1996). In our studies, thc Y-shaped structure was easily and reproducibly resolved by CVD MWNT tips at room temperature as shown in Fig. 8 (Cheung et al., 2000b). These IgG images are also compared (Fig. 8) with the crystal structure to emphasize the very small amount of tip-induced broadening in the observed images obtained with nanotube tips.

Pore-growth CVD MWNT tips have also used to examine the structure of IgM (Hafner et al., 1999a). This antibody is a ca. 1 MDa protein consisting essentially of 5 IgGs arranged in a pentameric structure. To date, it has not been possible to crystallize IgM and determine its atomic structure by X-ray diffraction methods. Previous structural analysis has relied primarily on electron microscopy analysis, which has elucidated the basic features of the pentameric structure (Perkins et al., 1991). Our room temperature studies with the pore-growth CVD MWNT nanotube tips showed clearly the pentameric structure, including five external pairs of antigen binding fragments (Fab domains) and five internal Fc fragments (Fig. 9). In addition, these images occasionally exhibited the loop structure connecting two of the five Fc domains that likely corresponds to the joining (J) loop (Hafner et al., 1999a). Because the exact structure of this region is still unclear (Perkins et al., 1991), our nanotube AFM investigations have shown clearly the potential of nanotube probes to reveal new structural features on large proteins that cannot be crystallized for diffraction studies.

To test the resolution of the smaller diameter CVD SWNT tips, we have focused a smaller protein GroES. GroES, which is a component of the GroEL/GroES chaperonin system involved in protein folding, is a hollow dome-shaped heptamer that is approximately 8 nm in outer diameter (Sigler et al., 1998). The seven 10 kDa subunits each consist of a core β-barrel with a β-hairpin loop at the top and bottom. The top β-hairpins point inward to form the top of the dome, while the bottom hairpins are disordered when not in contact with GroEL (Hunt et al., 1996). Previously, AFM images of GroES have only been obtained on densely packed arrays in contact mode (Mou et al., 1996). In this work, subunit resolution was observed when the samples were fixed using gluteraldehyde but lower resolution was obtained in the absence of fixation. In our studies (Cheung et al., 2000b), CVD SWNT tapping mode images of unfixed individual, well-separated GroES molecules on mica reveal two conformations (Fig. 10). One conformation shows a ring-like structure with an 11 nm outer diameter and internal structure to the ring. The other conformation simply looks like a dome of the same diameter. We have interpreted these two

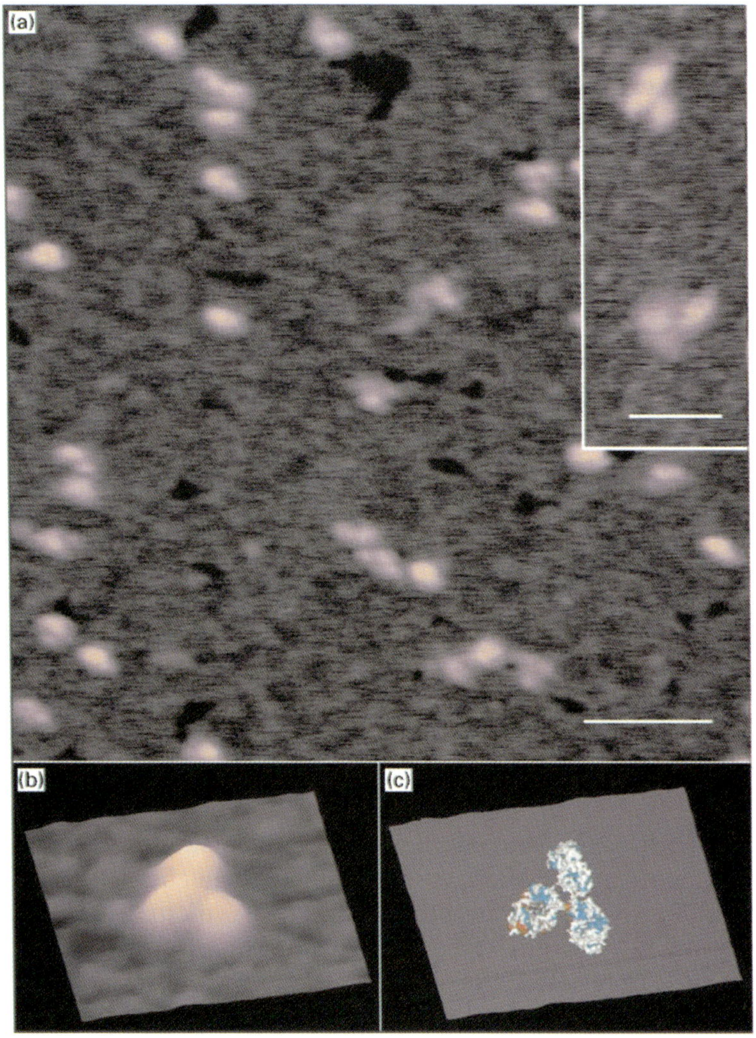

Fig. 8. Nanotube tip AFM images of Immunoglobulin G: (a) A field of several IgG molecules highlights the flexibility of the molecule, which leads to many conformations on the surface (scale bar equals 50 nm). The inset (scale bar equals 20 nm) shows two molecules exhibiting the well-defined Y-shaped structure expected for this protein. (b) Three-dimensional plot of an isolated IgG molecule, which exhibits the characteristic Y-shaped structure, compared to (c) the crystal structure to emphasize the minimal tip-induced broadening (Cheung et al., 2000b).

conformations as the two sides of the GroES heptamers. The domed image is the side with the ordered β-hairpins creating the dome roof, and the ringed image is the disordered β-hairpins open to the hollow cavity. The disordered hairpins have conformational flexibility, but it is possible to observe the expected seven-fold symmetry (Hunt et al., 1996) as shown in Fig. 10b. For comparison, Fig. 10c shows the bottom view of the GroES crystal structure with the top of the dome removed to make the pore visible. These results demonstrate the ability of the present CVD nanotube tips to achieve sub-molecular resolution on isolated protein assemblies without fixing, and thus clearly suggest the possibility of high-resolution in vitro work in the future.

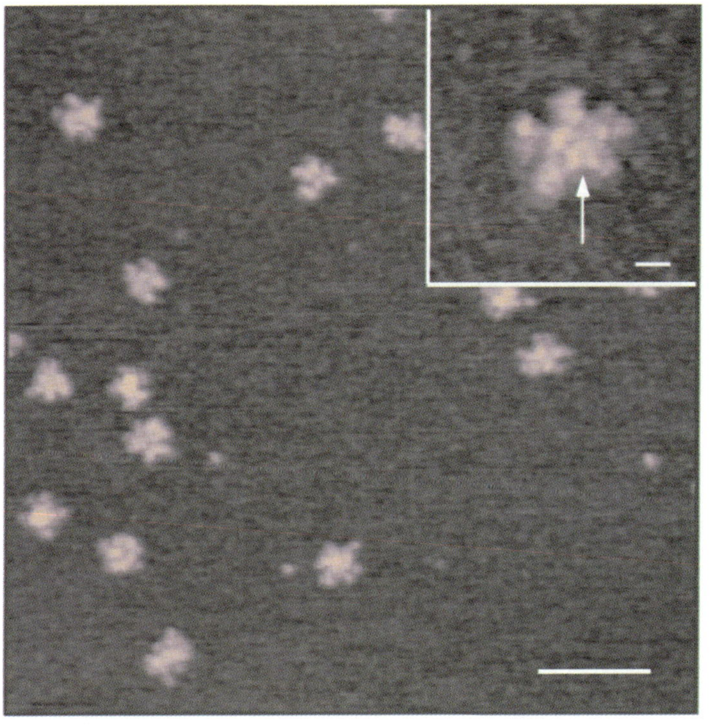

Fig. 9. Nanotube tip AFM images of IgM: The pentameric structure of IgM is not always apparent due to the high conformational flexibility of this molecule (scale bar is 100 nm). The inset shows an example with clear pentameric symmetry, including structural information on the central region (Hafner et al., 1999a). The inset scale bar equals 10 nm. The arrow highlights the position of the J-loop.

5. Structural biology and other applications of nanotube tips

We believe that the CVD nanotube tips described above represent powerful probes for structural biology studies of systems that have not been characterized by other methods, since these tips provide reproducible and predictable resolution that is comparable to that obtained with cryoEM. While the very recent advances in the growth of the SWNT probes have not yet enabled extensive studies of complex protein and protein/nucleic acid systems, the early results described below, which address (1) intermediates along the aggregation pathway of amyloid proteins, (2) ATP-dependent nucleosome remodeling, and (3) DNA sequence determination, indicate clearly the potential of these structural probes.

5.1. Aggregation and fibril formation of amyloid proteins

Cortical amyloid plaques, comprising a fibrillar form of the amyloid-β proteins Aβ40 and Aβ42, are a defining pathological feature of Alzheimer's disease (AD) brain (Lansbury, 1996; Selkoe, 1999), and moreover, the number of plaques at autopsy roughly correlates with the severity of AD symptoms (Cummings et al., 1996). These and other pathological observations

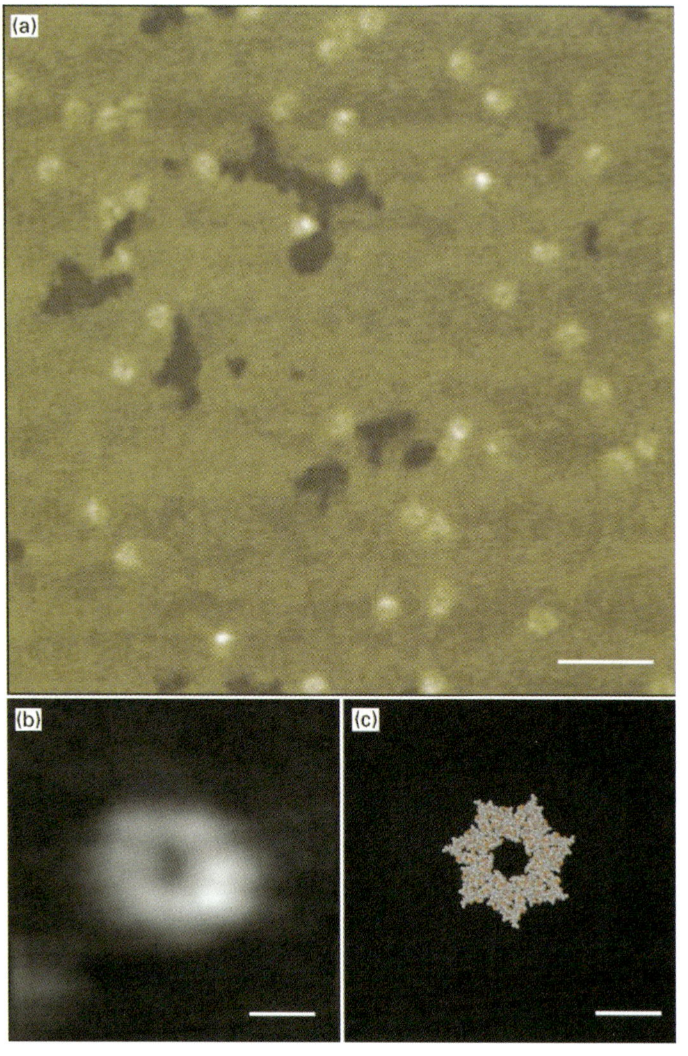

Fig. 10. Nanotube tip AFM images of isolated GroES molecules: (a) Field of many GroES molecules reveal dome and ring conformations, representing the two sides of GroES heptamer (scale bar equals 50 nm). (b) A high-resolution image of the ring conformation shows clearly the heptameric symmetry, and little tip-induced broadening relative to the crystal structure (c) (Cheung et al., 2000b). The scale bars in (b) and (c) equal 5 nm.

suggest that amyloid fibril formation is an important event in the AD process (Cummings et al., 1996; Lemere et al., 1996), although it is not known whether the fibrils or a precursor intermediate is responsible for neuronal cell death. Understanding this latter point is especially important since therapeutic agents designed to inhibit fibril formation at too late a stage could result in the build up of an intermediate that accelerates the onset of AD. In this regard, it is clearly important to define the pathway of fibrillization for monomeric Aβ proteins, including the structural characterization of intermediates.

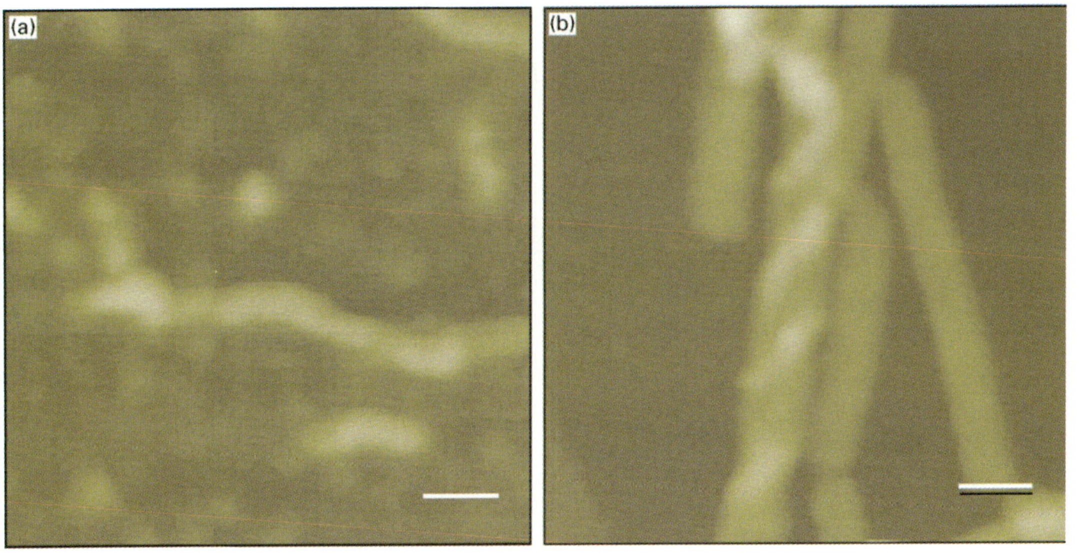

Fig. 11. High-resolution nanotube tip images of Aβ aggregation: Images of (a) protofibrils and (b) helical fibrils and rigid filaments. The scale bars in (a) and (b) are 25 nm.

AFM represents a powerful tool for studying the fibrillization pathway since the wide range and coexistence of relatively large structural species are difficult to probe using diffraction and NMR methods (Harper et al., 1997). Our previous studies of Aβ fibril formation revealed the existence of a discrete, dynamic intermediate, which we termed the amyloid protofibril (Harper et al., 1997, 1999). Oligomeric species which are probably the same as the protofibrils, have also been observed by EM (Walsh et al., 1997). Images of Aβ protofibrils and fibrils obtained with nanotube tips are shown in Fig. 11 (Hafner et al., 2000b). These data show the same overall features as reported in our earlier studies using microfabricated silicon tips, although the structural resolution is higher. For example, the helical structure observed in fibrils is much clearer and more detailed in images recorded with SWNT tips.

In addition, Aβ oligomers, which are smaller than typical protofibrils, were observed in previous AFM studies (Harper et al., 1997, 1999; Lambert et al., 1998). Height measurements made on these oligomers were interpreted as diameters of a globular morphology, due to the resolution limitations of conventional tips. To obtain greater structural information about these and other potentially pathogenic Aβ species that precede and/or coexist with protofibrils, we have studied the early stages of fibrillization using high-resolution CVD SWNT tips. Images of the Aβ40 protein obtained with SWNT probes after 1–3 days of incubation show a new structural species with a 5–6 nm height, a ca. 19 nm diameter and four-fold internal structure as shown in Fig. 12 (Hafner et al., 2000b). The height of this species is similar to that found in previous low-resolution studies, where it was assumed to correspond to a globular structure. Our higher resolution images with nanotube tips exhibit clear four-fold symmetry and a central depression ca. 1 nm deep. This depression could correspond to a pore through the center of the molecule that is too small to be accessed by the tip. Based on these structural features we have termed this new morphology the annular Aβ structure.

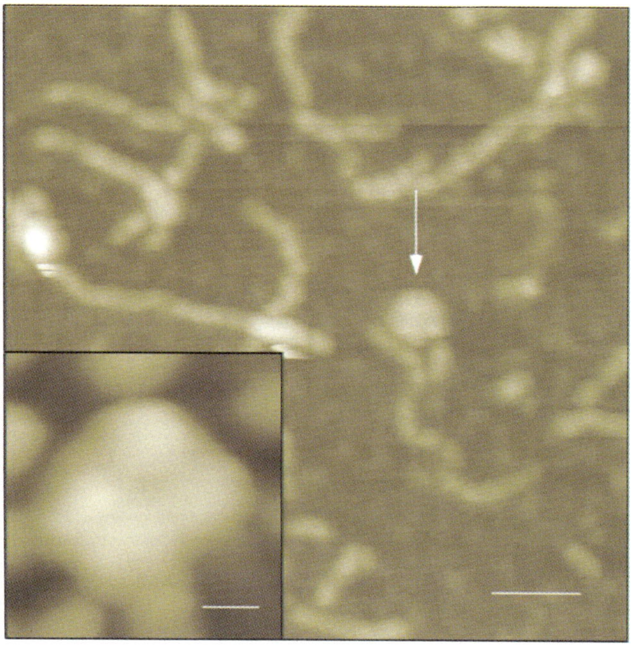

Fig. 12. Nanotube tip AFM images of the Aβ annular structure: The annular structure, highlighted by the white arrow, is seen among protofibrils deposited on mica (scale bar is 50 nm). The inset shows a high-resolution image of an Aβ annular structure (scale bar equals 5 nm).

We have outlined the possible relationship of the annular Aβ structure to other species characterized along the fibrillization pathway in Fig. 13 (Hafner et al., 2000b). The four-fold morphology of the annular structure is qualitatively similar in appearance to the subunits of elongated protofibrils, and thus it is reasonable to speculate that the annular structure is a head-to-tail loop of a short protofibril. However, this head-to-tail mechanism is hard to reconcile with the fact that (1) the annular Aβ structure has a constant 18–21 nm diameter, while protofibrils greater than 200 nm in length are commonly observed, and (2) the heights of the annular structure, 5.8 nm, and protofibril, 4–5 nm, are recognizably different. Hence, we have suggested that the molecular mechanisms for formation of the annular structure and protofibrils are distinct.

The annular Aβ structure may be relevant to AD pathology for several reasons. The possibility that nonfibrillar Aβ oligomers may be pathogenic has been suggested by several experiments. Nonfibrillar Aβ-derived species have been shown to have in vitro neurotoxic activity (Lambert et al., 1998) and may be responsible for neurodegeneration in AD (Hsia et al., 1999; Moechars et al., 1999). In addition, there is evidence that Aβ forms ion channels in lipid bilayers (Arispe et al., 1993; Lin et al., 1999) and this channel activity has been implicated in AD pathogenesis (Arispe et al., 1993). It is interesting to suggest that the annular Aβ structure is this proposed channel forming structure. As precedent, a recent cryoEM study of the membrane-active α-latrotoxin neurotoxin reveals a similar structure: a 15–18 nm diameter tetramer that contains a 1 nm diameter ion channel (Orlova et al., 2000). Our current efforts are focused on

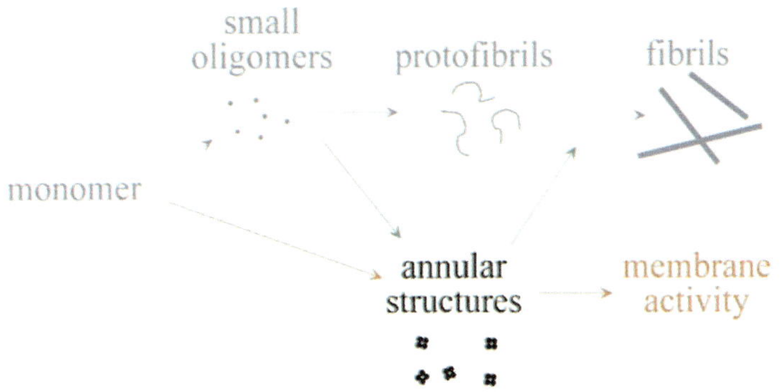

Fig. 13. Aβ aggregation pathways: The blue pathway represents that observed with standard AFM tips in which featureless oligomers lead to protofibrils, which ultimately form fibrils. The Aβ annular structure, which is shown on the red pathway, may exhibit distinct pathological activity or could return to the fibril pathway (green arrow).

isolating the annular Aβ structure so that its structure can be further characterized and its potential biological activity studied.

5.2. Nucleosome remodeling

Eukaryotic cells package their DNA by complexing it with a defined set of histone proteins. Approximately 146 base pairs of DNA wrap around an octamer of histones, composed of an H3–H4 histone tetramer and two H2A–H2B histone dimers, thereby forming a DNA–protein nucleosome structure (Fig. 14a). Additional structuring of the nucleosome arrays by linker histones produces the higher order chromatin structure that serves to further compact the DNA (Kornberg and Lorch, 1999; Zlatanova et al., 1998). Fundamental cellular processes such as DNA replication and gene transcription require the binding of transcription factor proteins to specific DNA sequences. In the past, the emphasis has been on understanding the interactions between specific protein transcription factors and naked DNA. However, it has become evident that histone and non-histone chromatin proteins can block access to DNA and thereby hinder replication. Hence, the cell must possess mechanisms to modify the accessibility of nucleosomal DNA to DNA binding factors. Two such critical mechanisms have been elucidated thus far. First, protein complexes that remodel chromatin structure and make the DNA more accessible can regulate activity (Kadonaga, 1998). Second, alterations in the chemical state of histones, which affect DNA–histone binding, can regulate activity (Hassig and Schreiber, 1997). As an example of the first mechanism, the 2 MDa protein complex SWI/SNF has been shown to possess ATP-dependent nucleosome remodeling activity (Pazin and Kadonaga, 1997; Schnitzler et al., 1998), while the recently discovered NuRD complex exhibits both ATP-dependent remodeling and deacetylation activity (Hassig and Schreiber, 1997).

Thus far, several homologous remodeling factors have been characterized. These are nucleotide triphosphate binding proteins that have conserved helicase/ATPase domains. Hydrolysis of ATP appears to be necessary to produce the energy needed to disrupt nucleosome structure. These remodeling factors have been shown to function in a variety of cellular processes, including the regulation of gene expression. For example, in yeast, the SWI/SNF chromatin remodeling

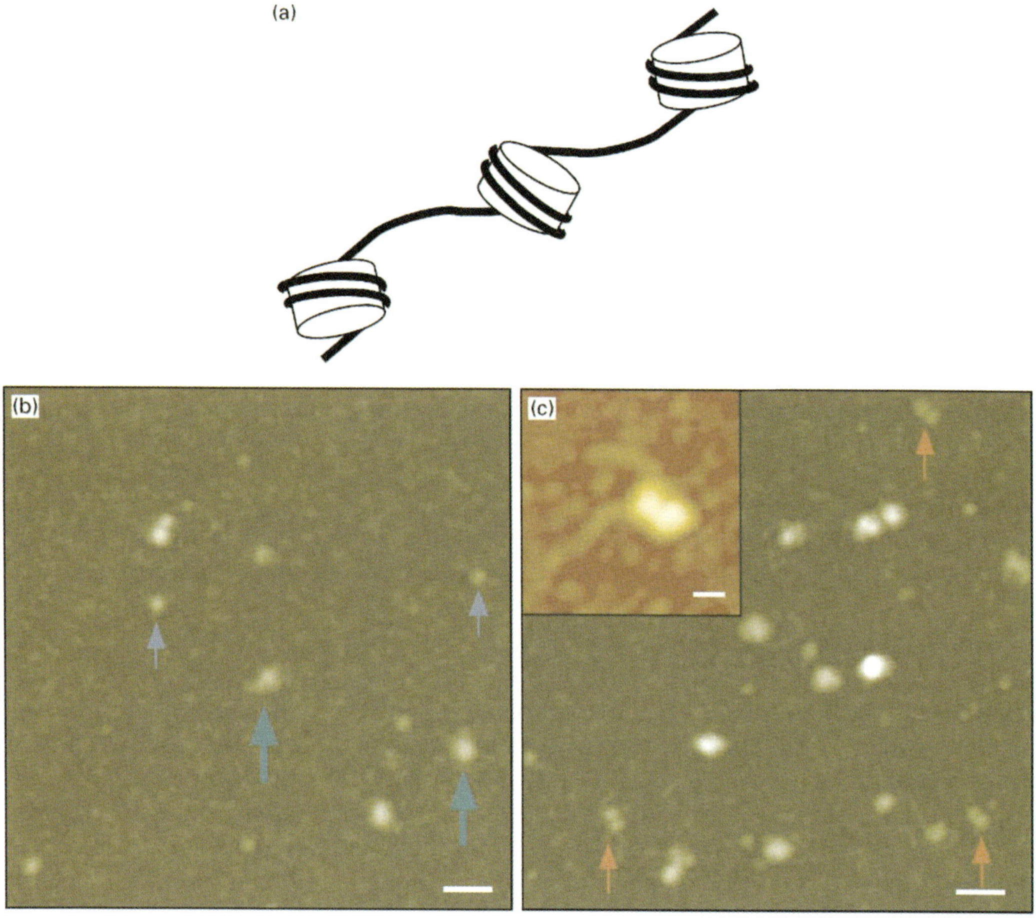

Fig. 14. Nucleosome remodeling: (a) Schematic representation of chromatin showing histone octamers (open cylinders) and DNA (solid black line). The nucleosome corresponds to the histone octamer wrapped with DNA ca. 146 base pairs long. CVD MWNT tip images of (b) mononucleosome and hSWI/SNF without ATP, and (c) mononucleosome and hSWI/SNF in the presence of ATP. The blue and green arrows in (b) highlight mononucleosome and SWI/SNF complex; the red arrows in (c) highlight the dimer product. The scale bars in (b) and (c) are 50 nm. The inset in (c) is a high-resolution image of the dimer product; scale bar = 10 nm.

complex is required to achieve transcription of a group of inducible genes. The RSC complex is essential for yeast growth. In *Drosophila*, the activity of the NURF complex is required for the binding of certain transcription factors to DNA. Each of these complexes contains SWI/SNF homologues which can disrupt nucleosomal structure in an ATP-dependent manner, as determined by increased accessibility to DNA cleaving enzymes and improved binding of transcription factors. However, the exact mechanism by which nucleosome structure is altered remains unclear (Pazin and Kadonaga, 1997).

Two recent reports have provided biochemical insight into the mechanism of chromatin remodeling (Lorch et al., 1998; Schnitzler et al., 1998). It has been demonstrated that the SWI/SNF and RSC chromatin remodeling complexes can produce a stable altered nucleosome structure. This altered form is approximately twice the size of a single mononucleosome, contains only the core histones and DNA, and has an increased susceptibility to cleavage by DNase I and restriction enzymes. It has been suggested based on biochemical evidence that the remodeled state may represent a "dimerized" form of mononucleosomes, although clearly there exists a need for direct structural information. In addition, there is very little understanding of the relationship of the remodeled model system "dimer" to remodeling in polynucleosomes and chromatin. For example, do adjacent or distant nucleosome units become dimerized or are there new structural changes in polynucleosomes. High-resolution SWNT AFM imaging represents a promising approach for obtaining much needed structural information on these large systems.

Our initial efforts have focused on using AFM to examine structurally the proposal that the human SWI/SNF (hSWI/SNF) remodeling of mononucleosomes creates a dimer with altered DNA–histone contacts (Schnitzler et al., 1998). To address carefully this problem we have used CVD nanotube probes to characterize mononucleosomes and dinucleosomes isolated from human HeLa cells and the products obtained from the reaction of hSWI/SNF and mononucleosomes with and without ATP (Hafner et al., 2000c). Images characteristic of the products obtained from the reaction in the absence and presence of ATP are shown in Fig. 14. These data suggest that no structural changes occur in the absence of ATP, while dimer products are observed in the presence of ATP. While these observations are expected on the basis of biochemical studies, we believe they also provide a taste of what can be expected in the future as one exploits higher resolution probes and investigates remodeling of polynucleosomes.

In addition, these studies of mononucleosome remodeling have shown substructure in the hSWI/SNF complex. An image, which was obtained with a CVD MWNT tip, is shown in Fig. 15. The images show that the SWI/SNF complex is a multilobed (subunit) structure, and appears to have a 2-fold axis (Hafner et al., 2000c). Our ability to resolve substructure in SWI/SNF, which was achieved even without the current, higher resolution CVD SWNT tips, opens up a number of possibilities for future studies. These opportunities include (1) probing for distinct conformational states in this ATP-dependent machine and (2) using functional imaging to probe for DNA and/or histone binding sites.

5.3. DNA sequence analysis and haplotyping

The reproducibly high resolution afforded by SWNT probe tips can also open up unexpected opportunities, and in this regard we have investigated their possible uses in DNA sequence analysis and related genomics problems (Woolley et al., 2000). This effort is motivated by the recognition that the haplotype of a subject—the specific alleles associated with each chromosome homologue—is critical to maximizing the impact of single nucleotide polymorphism (SNP) mapping (Brookes, 1999). However, the current methods for determining haplotypes have significant limitations that have prevented their use in large-scale genetic screening. For example, parental genotyping can be used to infer haplotypes in a family study (Hodge et al., 1999; Sobel and Lange, 1996), although in many cases it is impractical or impossible to obtain parental DNA.

Our method utilizes high-resolution SWNT probes to read directly multiple polymorphic sites in DNA fragments containing from ca. 100 to at least 10,000 bases. The approach involves specific hybridization of labeled oligonucleotide probes to target sequences in DNA fragments, using polymerase to create double-stranded DNA around the probe, followed by direct reading of the

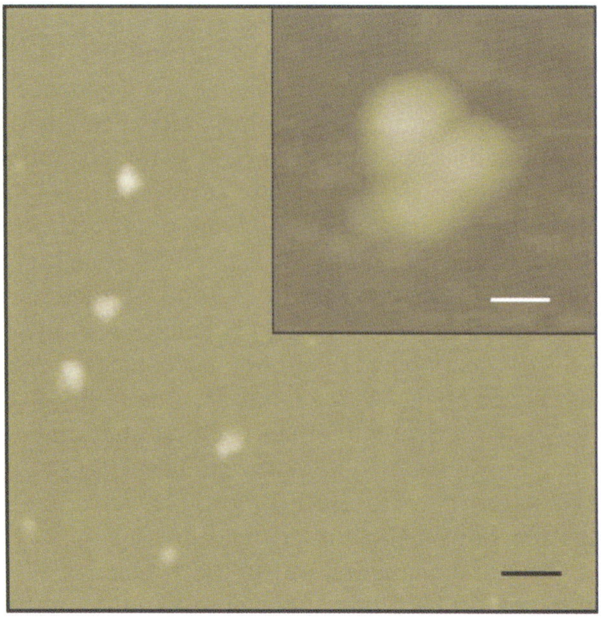

Fig. 15. Nanotube tip AFM images of the hSWI/SNF complex reveal a multi-lobed structure (scale bar is 50 nm). The inset shows a high-resolution image of a single hSWI/SNF complex (scale bar is 10 nm).

presence and spatial locations of the labels by AFM (Fig. 16). The oligonucleotide probes are designed such that, under appropriate hybridization conditions, binding does not occur in the presence of a single-base mismatch at polymorphic sites, i.e., labels are detected only at sequences fully complementary to the oligonucleotides. High-resolution SWNT probes are critical to our approach since they enable different size probes to be reproducibly distinguished as required for simultaneous detection of different polymorphic sites on single DNA fragments (Woolley et al., 2000). A similar approach may be possible using single molecule optical methods. However, an optical approach will require that fluorescent blinking and photobleaching of single labels be accounted for, while at the same time acquiring sufficient data to identify the centroid of a fluorescent peak to high accuracy.

Our new nanotube-AFM method has been tested by identifying the spatial location of specific sequences with excellent discrimination from corresponding single-base mismatches in the M13mp18 plasmid using seven base oligonucleotide probes. The essence of the experiments is shown in the image of a DNA molecule that was marked with a streptavidin-labeled GGGCGCG sequence (Fig. 17a). This image shows a DNA fragment with a 2200 nm contour length consistent with the 7249 base pairs of M13mp18, and a distinct streptavidin label 1080 nm from one end of the *Bgl* II digested DNA. Histogram summaries of results obtained from a number of streptavidin-labeled M13mp18 DNA molecules showed clear peaks at 0.48 (3512 bp) and 0.40 (2893 bp) from the fragment ends, for samples cut with *Bgl* II and *Bam* HI, respectively. In contrast, control experiments with unlabeled oligonucleotides did not exhibit clusters of labels, indicating that the histogram peaks are due to specific detection of streptavidin. These results show that the GGGCGCG site is at base 3390 (Fig. 17b), in good agreement with the known

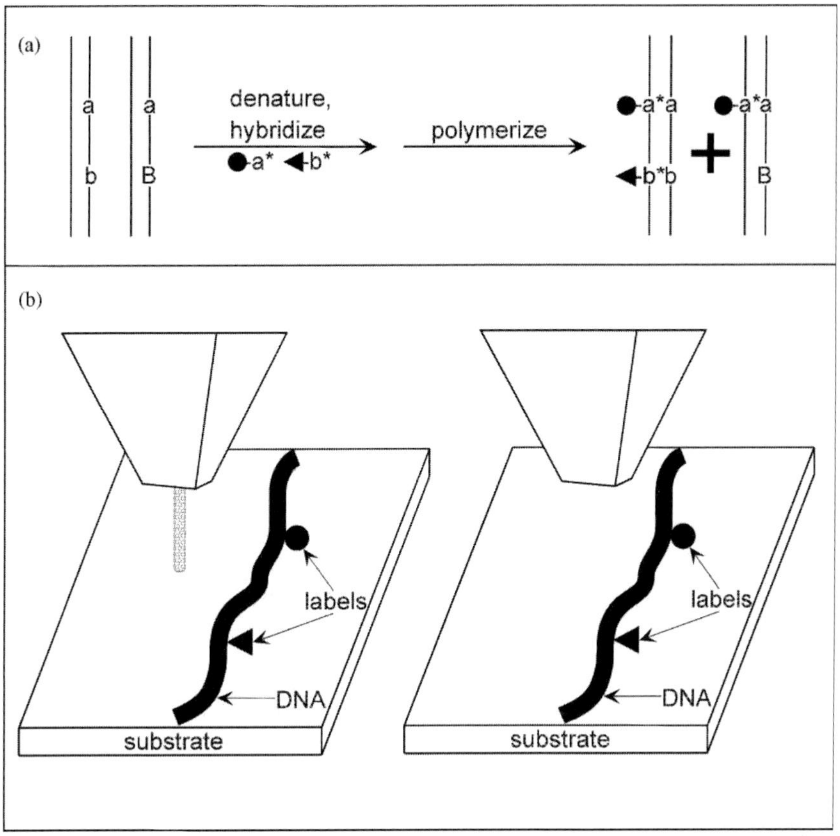

Fig. 16. (a) Schematic illustration of the method for labeling specific DNA sites for detection with SWNT AFM probes. Labeled oligonucleotide probes (●-a* and ▶-b*) are specifically annealed to their complementary target sequences (a and b) but not to sequences with a single-base mismatch (A and B) in the single-stranded DNA template. DNA polymerase and dNTPs are then used to synthesize double-stranded DNA fragments specifically labeled at a and b with ● and ▶, respectively. (b) Schematic illustrating the reading of the labeled DNA with an SWNT tip (left) and an Si tip (right). Only the nanotube has sufficient resolution to distinguish reproducibly the different labels.

location (base 3405). In addition, these data exhibited no evidence for labeling at the single-base mismatch sites located at 1115 and 3595, thus demonstrating the potential for SNP detection.

Because the streptavidin and IRD800 molecules can be readily distinguished on the basis of their heights and shapes using the SWNT tips, it is possible to detect two or more distinct sites using this method. Multiplexed detection has been tested by imaging M13mp18 labeled at GGGCGCG with IRD800 and at TCTCAGC with streptavidin using SWNT probes (Woolley et al., 2000). From histograms we have calculated that TCTCAGC occurs at bases 2024 and 4059, in good agreement with its known positions at 2013 and 4077, and that GGGCGCG is at base 3422, corresponding well with the expected value of 3405. These results demonstrate clearly the potential for multiplexed sequence detection in large DNA strands and open the possibility for profiling multiple polymorphic sites on DNA fragments in the 10 kilobase or larger size range.

The model experiments carried out on M13mp18 suggested that this approach could be used for identifying specific haplotypes associated with genetic disorders. We have demonstrated this

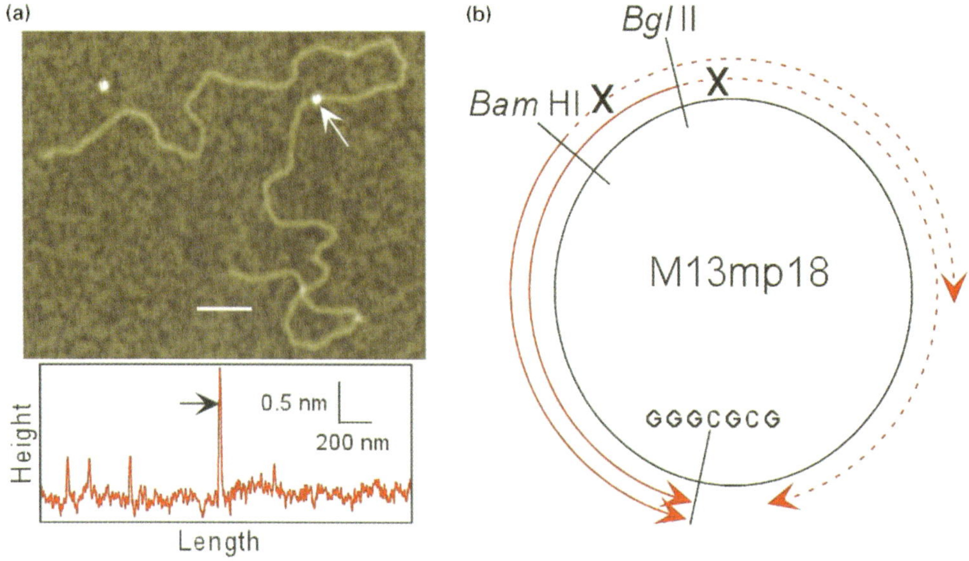

Fig. 17. Detection of labeled DNA sites with nanotube tips: (a) SWNT tip image and height profile along DNA, obtained with streptavidin-labeled GGGCGCG in M13mp18 digested with *Bgl* II; the arrow points to the streptavidin tag. Places where DNA strands cross each other (left side of height profile) are easily differentiated from labels. (b) Map of M13mp18 shows the location of GGGCGCG calculated from experimental data. Arrowheads indicate possible positions of the target sequence, based on the calculated distance from the restriction sites. Solid arcs indicate the correct paths, while incorrect paths are shown as dashed arcs (Woolley et al., 2000).

critical point by determining the haplotypes on a UDP-glucuronosyltransferase gene, UGT1A7 (Guillemette et al., 2000) whose enzyme product is involved in inactivation of carcinogens such as benzo[a]pyrene metabolites (Woolley et al., 2000). The UGT1A7 gene has two polymorphic sites (separated by 233 bp) that determine four alleles, each specifying different polypeptide chains (Fig. 18a). Importantly, individuals who are heterozygous at both sites have a single genotype, but one of two haplotypes, (*1/*3) or (*2/*4), which cannot be differentiated using conventional methods. This ambiguity is crucial in screening, since each allele exhibits substantially different enzymatic activity towards targeted carcinogens. Significantly, the haplotypes of several subject samples have been unambiguously shown to be (*1/*3) by direct inspection of AFM images of labeled DNA (Fig. 18b); that is, DNA molecules were only end-labeled with the streptavidin or the IRD800 probes.

We believe that direct haplotyping using SWNT AFM probes represents a significant advance over conventional approaches and could facilitate the use of SNPs for association and linkage studies of inherited diseases and genetic risk (Collins et al., 1997; Risch and Merikangas, 1996). To realize the greatest potential, will require that this methodology be extended to a high-throughput technique for genetic screening. The recent fabrication of multiple AFM tip arrays as large as 32×32 (Vettiger et al., 1999) could dramatically increase the number of samples typed in parallel, and the simplicity and distinctiveness of the AFM images of alternative haplotypes indicate that automated image analysis should be feasible. The implementation of these

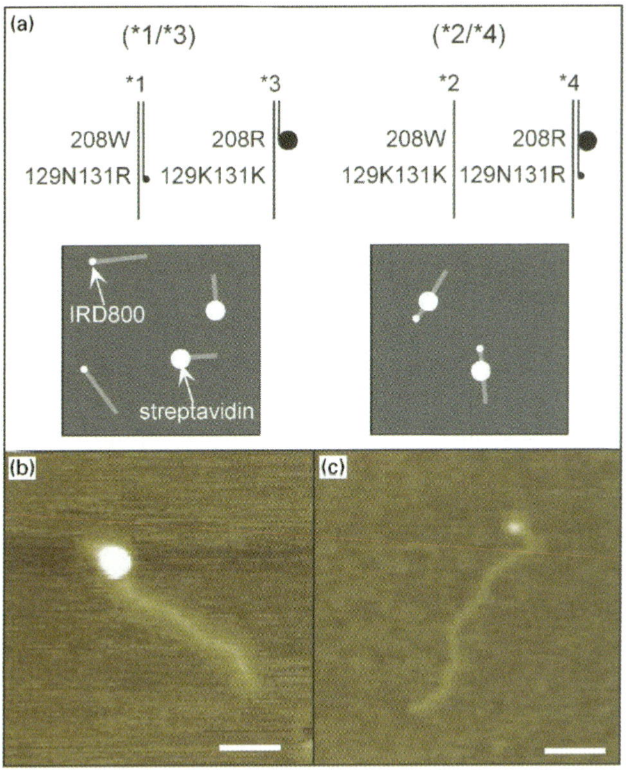

Fig. 18. Direct haplotyping of UGT1A7 using SWNT probes: (a) Schematic showing haplotypes, alleles, genotypes, and locations of probes in samples analyzed. The (*1/*3) and (*2/*4) haplotypes, which have the same genotype (heterozygous at both loci), are specifically labeled at the 129N131R and 208R sites with IRD800 (small filled circle) and streptavidin (large filled circle), respectively. The double lines in the stick representations indicate duplex DNA, while single lines corresponded to single-stranded DNA. AFM images of a (*1/*3) sample should have an approximately equal number of fragments that are ~210 nm long with IRD800 at one end, or ~140 nm long with streptavidin at one end. In contrast, a (*2/*4) sample should contain ~210 nm fragments with IRD800 at one end and streptavidin at ~70 nm. (b) Representative SWNT tip images of the *3 allele (streptavidin end-labeled, ~140 nm DNA) and the *1 allele (IRD800 end-labeled, ~210 nm DNA). The white bar corresponds to 50 nm in both images (Woolley et al., 2000).

improvements could facilitate high-throughput haplotyping using SWNT AFM probes. Lastly, the recent synthesis of carbon nanotubes with 0.25 nm radii (Sun et al., 2000), which are smaller than the spacing between DNA bases, indicates that further improvements in nanotube probes and labeling methods may even allow direct reading of the DNA sequence of fragments that are tens of kilobases in size.

6. Functional imaging with nanotube probes

The studies reviewed above have focused on the unique potential of carbon nanotube probes for reproducible, high-resolution structural imaging of proteins, protein assemblies, and protein–nucleic acid complexes. It is also possible to exploit the high normal force sensitivity of AFM for chemically and biologically sensitive measurements. In the past, the spatial resolutions of such measurements have been limited significantly by available probe tips. The well-defined chemistry possible at the ends of nanotube probes can overcome all previous limitations and thus opens up significant opportunities as described below.

6.1. Chemical force microscopy and force spectroscopy

In chemical force microscopy (CFM), the AFM tip is modified with specific chemical functional groups (Frisbie et al., 1994). This enables the tip to generate contrast dependent on the chemical properties of the sample from the friction signal in contact mode or the phase lag signal in tapping mode (Noy et al., 1998). Both of these signals are proportional to the adhesion between the tip and sample, and both can be measured simultaneously with topography. Functionalized tips have also been employed in force spectroscopy. In this mode of operation the tip is not scanned, but is brought into contact with a surface, then retracted. The forces applied to the tip during retraction are due to the interactions of tip and sample molecules. Force spectroscopy has been used to measure a variety of interactions including the intermolecular adhesion between fundamental chemical groups (Frisbie et al., 1994; Green et al., 1995; Noy et al., 1995, 1997b; Vezenov et al., 1997), the unfolding of protein molecules (Rief et al., 1997), antigen–antibody interactions (Hinterdorfer et al., 1996), and DNA stretching and unbinding (Noy et al., 1997a).

Despite the progress made in chemically sensitive imaging and force spectroscopy using silicon and silicon nitride tips, these probes have very important limitations. First, the tips have a large radius of curvature making it difficult to control the number of active tip molecules and limiting significantly the lateral resolution. Second, the orientation and often the spatial location of the attached molecules cannot be controlled. This leads to uncertainty in the reaction coordinate for force spectroscopy, and increased non-specific interactions. Carbon nanotube tips can overcome these limitations. They have small radii of curvature for much higher resolution and can be specifically modified only at their very ends. Hence, functionalization of nanotube probes creates tips that have as few as one (depending on size) active molecular sites localized in a relatively controlled orientation. Modified SWNT tips could lead to subnanometer resolution in chemical contrast and binding site recognition.

6.2. Nanotube tips in CFM and force spectroscopy

Nanotube tips etched in air are expected to have carboxyl groups at their ends based on bulk studies of oxidized nanotubes (Hiura et al., 1995), although conventional analytical techniques have insufficient sensitivity to prove this point for isolated tubes. Chemical species present at the ends of nanotube tips can be studied with great sensitivity by measuring the adhesion of a nanotube tip on a chemically well-defined surface (Fig. 19), for example, a hydroxyl (–OH) terminated self-assembled monolayer (SAM). We have demonstrated the presence of carboxyl groups at the open ends of manually assembled MWNT and SWNT tips by measuring force titrations as shown in Fig. 19b (Wong et al., 1998b, c). In the force titration, the adhesion force between a nanotube tip and an SAM surface terminating in hydroxyl groups is recorded as a function of solution pH, thus effectively titrating ionizable groups on the tip (Noy et al., 1997b;

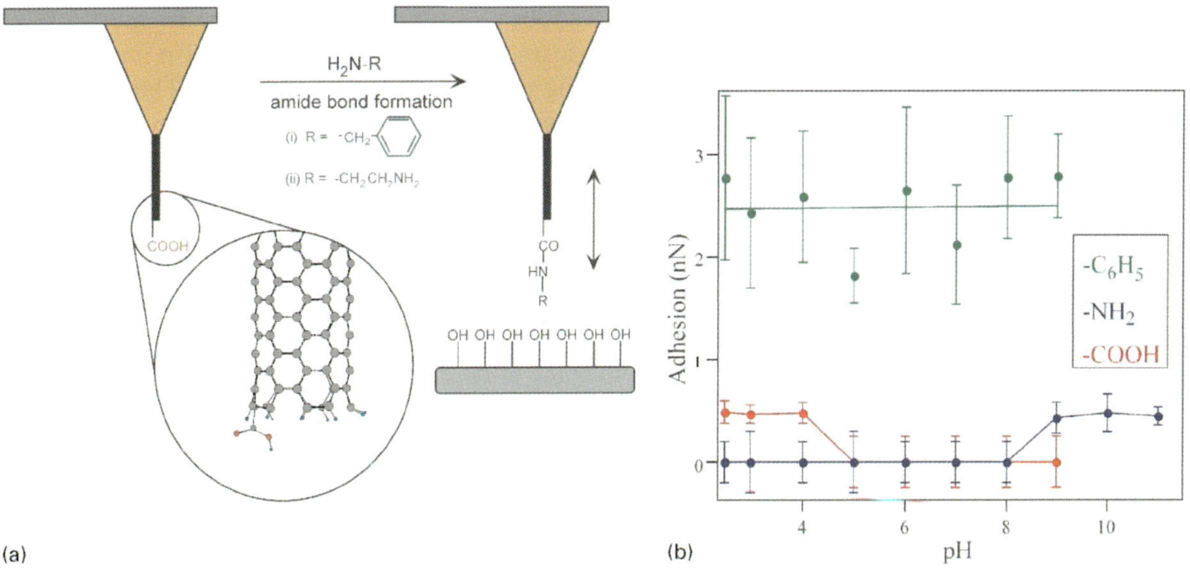

Fig. 19. Covalent functionalization of carbon nanotube AFM tips: (a) Schematic illustrating the modification of the end of an SWNT, which terminates in a carboxylic acid functional group. The terminal carboxyl group is coupled with an amine, which bears the desired functionality, to form an amide bond. The tip end functionality is then assessed by force spectroscopy measurements on a chemically well-defined surface (Wong et al., 1998b). (b) Adhesion force as a function of pH (force titration) between assembled SWNT bundle tips and a hydroxyl-terminated SAM (11-thioundecanol on gold-coated mica). Each data point corresponds to the mean of 50–100 adhesion measurements, and the error bars represent one standard deviation (Wong et al., 1998c).

Vezenov et al., 1997). Significantly, force titrations recorded between pH 2 and 9 with MWNT and SWNT tips exhibit well-defined drops in the adhesion force at ca. pH 4.5 that are characteristic of the deprotonation of a carboxylic acid. Moreover, the pH value where the adhesion drop occurs is similar to the pK_A of benzoic acid, thus suggesting that the carboxyl groups at the ends of nanotubes are solvent accessible.

The assembled SWNT and MWNT bundle tips also have been modified with organic and biological functionality by coupling organic amines to form amide bonds as outlined in Fig. 19 (Wong et al., 1998b, c). The success of this coupling chemistry was demonstrated by force titrations. Nanotube tips modified with benzylamine, which exposes nonionizable, hydrophobic functional groups at the tip end, yielded the expected pH-independent interaction force on hydroxyl-terminated monolayers (Fig. 19b). This covalent modification thus eliminates the prominent pH-dependent behavior observed with the unfunctionalized tips. Moreover, force titrations with ethylenediamine-modified tips exhibit no adhesion at low pH and finite adhesion above pH 7 (Fig. 19b). These pH-dependent interactions are consistent with our expectations for exposed basic amine functionality that is protonated and charged at low pH and neutral at high pH.

Covalent reactions localized at nanotube tip ends represent a powerful strategy for modifying the functionality of the probe. However, the linking atoms that connect the tip and active group introduce conformational flexibility that may reduce the ultimate resolution. In an effort to develop a chemically sensitive probe without conformation flexibility, we have also explored the modification of the tips during the electrical etching process (Wong et al., 1999). The hypothesis behind this approach was that radical species, which can react with diatomic molecules such as O_2, H_2, and N_2 are created during the electrical etching procedure used to optimize nanotube tip length. We tested this idea by carrying out the etching procedure in pure and mixed gas, and significantly, have demonstrated that these produce predictable changes in the chemical nature of the nanotube tip end. Specifically, force titrations carried out on tips modified in O_2, N_2, and H_2 exhibit behavior consistent with the incorporation of acidic, basic, and hydrophobic functionality, respectively, at the tip ends. We believe that these tips should enable chemical mapping at a resolution of a single SWNT tip, 0.5 nm or better.

In addition, these functionalized nanotube probes have been used to obtain chemically sensitive images of patterned monolayer and bilayer samples (Wong et al., 1998b, c). Tapping mode images recorded with –COOH- and benzyl-terminated tips exhibit greater phase lag on the –COOH and $-CH_3$ sample regions, respectively. In tapping mode AFM, we have shown that the differences in the tapping mode phase lag correlate directly with differences in intermolecular forces between the tip and sample: $\Delta\Phi_o(k/Q) \propto \Delta W_{st}$, where $\Delta\Phi_o$ is the change in phase lag, k is the spring constant, Q is the cantilever quality factor, and ΔW_{st} is the difference between the work of adhesion for the tip interacting with two chemically distinct sample regions (Noy et al., 1998). Because the adhesion force between the carboxyl-terminated nanotube tip and the COOH-terminated monolayer is greater than the CH_3-terminated region, and the adhesion between the benzyl-terminated tip and the CH_3-terminated region is greater than the COOH-terminated region, these results were consistent with chemically sensitive imaging and our model for the driven oscillator described above.

The "chemical resolution" of functionalized manually assembled MWNT and SWNT tips has been tested on partial lipid bilayers (Wong et al., 1998c). Significantly, these studies have shown that an assembled SWNT tip could detect variations in chemical functionality with resolution down to 3 nm, which is the same as the best structural resolution obtained with this type of tip. This resolution should improve with CVD SWNT tips, and recent studies bear this idea out (Cheung et al., 2000a). First, force titration data recorded on hydroxyl-terminated surfaces, which show finite adhesion below pH 5 and a repulsive interaction at higher pH, show (as expected) that these tips terminate in a carboxyl group when shortened in air. The magnitudes of the adhesion forces also indicate that only 1–4 carboxyl groups are active in the measurements. Second, these CVD SWNT tips can be used for chemically sensitive mapping as shown in Fig. 20. The high resolution of these tips is evident in the granular structure observed on samples prepared by standard lithography methods. This granularity corresponds to gold grain structure that typically cannot be resolved. We believe that these observations demonstrate that the CVD SWNT are sufficiently robust for solution work and that they have the potential for very high-resolution functional mapping.

In the future, we believe that these tips can be exploited in several ways for chemical or functional mapping in structural biology. For example, an important extension of our chemical mapping experiments is to protein surfaces, since the surface residues often define binding and orientation based on, for example, acidic or basic patches. Our structural studies of GroES (Cheung et al., 2000b), in which subunits were resolved with CVD SWNT tips (Fig. 10), represent

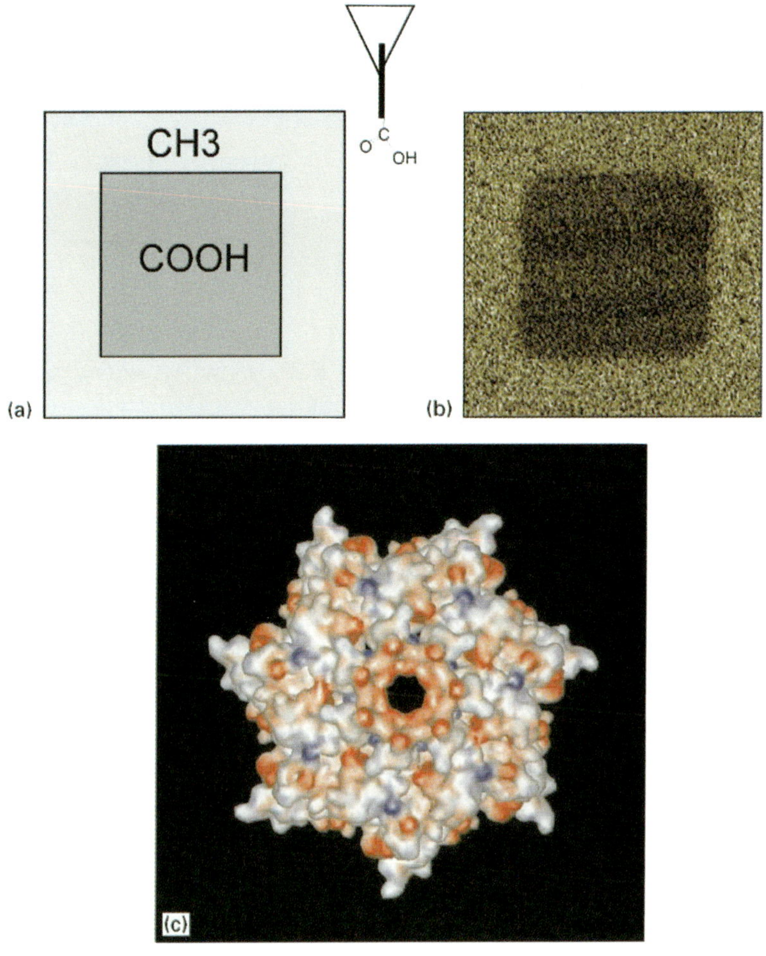

Fig. 20. Chemical mapping with CVD SWNT probes: (a) Schematic illustrating a carboxyl-terminated CVD SWNT tip imaging a patterned SAM that terminates in methyl and carboxyl domains. (b) Tapping mode phase image recorded on an SAM surface patterned as in (a). The darker contrast in the middle square is consistent with the stronger COOH–COOH vs. COOH–CH_3 interaction. (c) Plot of the electrostatic potential on the upper dome surface of GroES. Red corresponds to negative charge and blue to positive charge arising from acidic and basic residues, respectively.

a good case in which to test this point. Examination of the crystal structure and electrostatic potential for GroES (Hunt et al., 1996) shows that there is a high density of acidic residues (14 glutamates; 2 each from each subunit) and accumulation of negative charge density at the top of dome (Fig. 20c). Chemically sensitive mapping of the upper GroES surface using modified CVD SWNT tips, which are terminated with carboxyl or basic (obtained by coupling ethylenediammine or gas modification in N_2) groups, should then show repulsive or attractive interactions, respectively, at the top of the dome.

In addition, modified nanotube probes could be used to study ligand–receptor binding/unbinding with control of orientation, and to map the position of ligand–receptor binding sites in

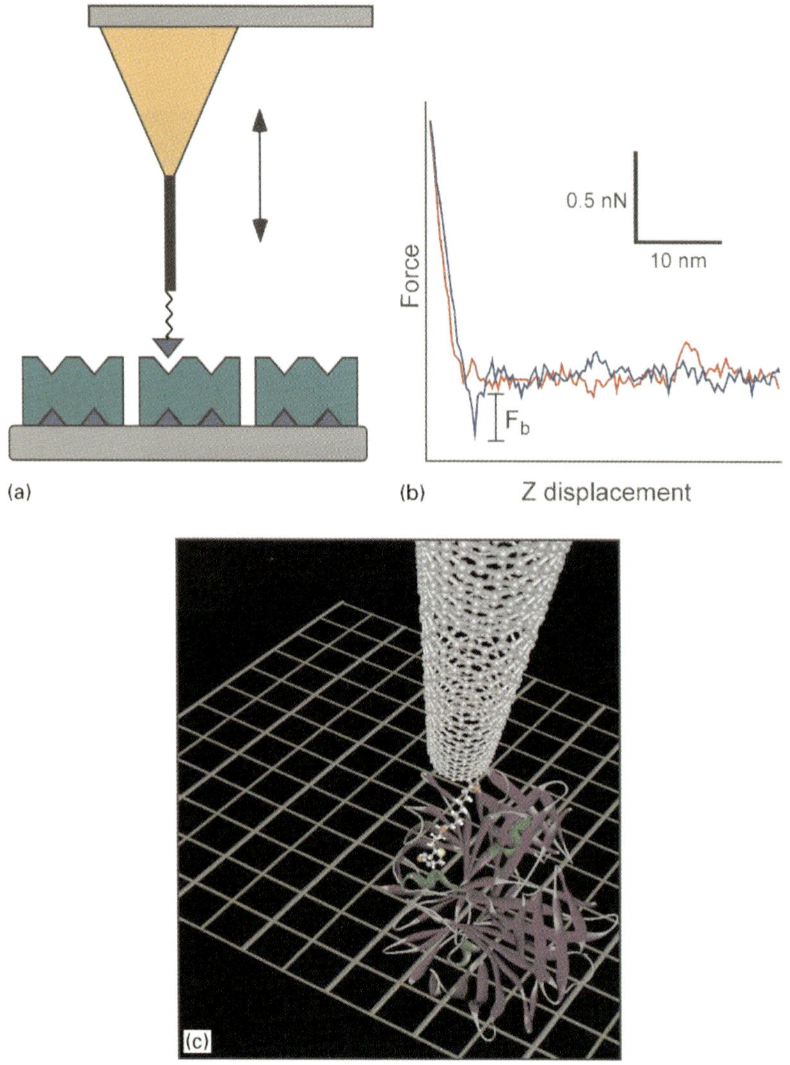

Fig. 21. Nanotube AFM tips for single-molecule force spectroscopy and mapping: (a) Schematic illustrating the experiment, where the lower gray rectangle corresponds to a mica substrate, the green blocks correspond to streptavidin molecules, the dark blue inverted triangle is biotin covalently linked to a carbon nanotube (heavy black vertical line). (b) Representative force–displacement curve recorded with a biotin-modified MWNT tip in pH 7.0 solution. The binding force, 200 pN, is consistent with unbinding of a single biotin ligand from streptavidin (Wong et al., 1998b). (c) Model of an SWNT tip (e.g., Fig. 7) that has been modified with biotin and is interacting with a streptavidin molecule.

proteins and on cell surfaces with nanometer or better resolution. To illustrate this point, we have examined biotin–streptavidin, which is a model ligand–receptor system that has been widely studied (Livnah et al., 1993). 5-(biotinamido)pentylamine was covalently linked to nanotube tips via the formation of an amide bond, and then force–displacement measurements were recorded on mica surfaces containing immobilized streptavidin as shown in Fig. 21 (Wong et al., 1998b). Force spectroscopy measurements show well-defined binding force quanta of ca. 200 pN per biotin–steptavidin pair. Control experiments carried out with an excess of free biotin in solution, which blocks all receptor sites of the protein, and with unmodified nanotube tips showed no adhesion, and thus confirm that the observed binding force results from the interaction of nanotube-linked biotin with surface streptavidin. A key feature of these results compared to previous work (Florin et al., 1994; Lee et al., 1994), which relied on nonspecific attachment of biotin to lower resolution tips, is the demonstration that a single active ligand can be specifically localized at the end of a nanotube tip using well-defined covalent chemistry. With the current availability of individual SWNT tip via surface CVD growth, it is now possible to consider without extrapolation the mapping of ligand binding sites for a wide range of proteins (Fig. 21c).

7. Summary and future prospects

AFM and related techniques show great promise as powerful tools for elucidating the structure and function of biological samples. The current limiting technology in AFM is the widely used, yet poorly defined Si or Si_3N_4 tip. The studies of carbon nanotube AFM probe tips described in this review demonstrate that nanotubes overcome the critical limitations of Si and Si_3N_4 tips, and may represent the ultimate AFM tip due to their cylindrical geometry, mechanical, and chemical properties. Our development of facile methods for the direct growth of nanotube tips by CVD allows the reproducible preparation of individual SWNT tips and represents a scalable approach that could make them widely available to the AFM and structural biology communities.

The availability of well-characterized individual SWNT tips will allow significant advances in AFM, as improvements in X-ray and electron sources have benefited X-ray diffraction and EM. Current progress on nanotube tips indicates that SWNT tips will allow sub-nanometer resolution on individual proteins in solution, and when the tip is functionalized provide similar resolution for mapping chemically distinct domains and ligand binding sites. We believe that the ability to obtain such functional information in vitro represents an important advantage compared to single particle cryoEM analysis, which has comparable structural resolution to SWNT probes. The range of new experiments that one can envision with SWNT probes, we believe, is truly unlimited, and likewise, the potential to contribute meaningful insight into the structure and function of complex protein and protein–nucleic acid systems is rich and exciting.

Acknowledgements

We thank Dr. J.D. Harper, Dr. P.T. Lansbury, Dr. G. Schnitzler and Dr. R.E. Kingston for valuable collaborations. C.M.L. acknowledges support of this work by the Air Force Office of Scientific Research and National Institutes of Health. J.H.H. acknowledges postdoctoral fellowship support from the National Institutes of Health.

References

Albrecht, T.R., Akamine, S., Carber, T.E., Quate, C.F., 1989. J. Vac. Sci. Technol. A 8, 3386–3396.
Arispe, N., Rojas, E., Pollard, H.B., 1993. Proc. Natl. Acad. Sci. USA 90, 567–571.

Beauchamp, J.C., Isaacs, N.W., 1999. Curr. Opin. Chem. Biol. 3, 525–529.
Bergkvist, M., Carlsson, J., Karlsson, T., Oscarsson, S., 1998. J. Colloid Interface Sci. 206, 475–481.
Brookes, A.J., 1999. Gene 234, 177–186.
Bustamante, C., Keller, D., Yang, G., 1993. Curr. Opin. Struct. Biol. 3, 363–372.
Bustamante, C., Rivetti, C., 1996. Annu. Rev. Biophys. Biomol. Struct. 25, 395–429.
Bustamante, C., Rivetti, C., Keller, D.J., 1997. Curr. Opin. Struct. Biol. 7, 709–716.
Cheung, C.-L., Chen, L., Lieber, C.M., 2000a. unpublished results.
Cheung, C.-L., Hafner, J.H., Lieber, C.M., 2000b. Proc. Natl. Acad. Sci. 97, 3809–3813.
Cheung, C.-L., Hafner, J.H., Odom, T.W., Kim, K., Lieber, C.M., 2000c. Appl. Phys. Lett. 76, 3136–3138.
Chiu, W., McGough, A., Sherman, M.B., Schmid, M.F., 1999. Trends Cell Biol. 9, 154–159.
Collins, F.S., Guyer, M.S., Chakravarti, A., 1997. Science 278, 1580–1581.
Cummings, B.J., Pike, C.J., Shankle, R., Cotman, C.W., 1996. Neurobiol. Aging 17, 921–933.
Dai, H., Hafner, J.H., Rinzler, A.G., Colbert, D.T., Smalley, R.E., 1996. Nature 384, 147–151.
Engel, A., Scheonenberger, C.-A., Muller, D.J., 1997. Curr. Opin. Struct. Biol. 7, 279–284.
Florin, E.-L., Moy, V.T., Gaub, H.E., 1994. Science 264, 415–417.
Frisbie, C.D., Rozsnyai, L.F., Noy, A., Wrighton, M.S., Lieber, C.M., 1994. Science 265, 2071–2074.
Fritz, J., Anselmetti, D., Jarchow, J., Fernandez-Busquets, X., 1997. J. Struct. Biol. 119, 165–171.
Fritz, M., Radmacher, M., Cleveland, J.P., Allersma, M.W., Stewart, R.J., Gieselmann, R., Janmey, P., Schmidt, C.F., Hansma, P.K., 1995. Langmuir 11, 3529–3535.
Fujihira, M., Ohzono, T., 1999. Jpn. J. Appl. Phys. 38 (Pt. 1), 3918–3931.
Glaeser, R.M., 1999. J. Struct. Biol. 128, 3–14.
Glusker, J.P., 1993. Bioanal. Instrum. 37, 1–72.
Green, J.B.D., McDermott, M.T., Porter, M.D., Siperko, L.M., 1995. J. Phys. Chem. 99, 10960–10965.
Grigorieff, N., 1998. J. Mol. Biol. 277, 1033–1046.
Guillemette, C., Ritter, J.K., Auyeung, D.J., Kessler, F.K., Housman, D.E., 2000. J. Biol. Chem., submitted for publication.
Guthold, M., Zhu, X., Rivetti, C., Yang, G., Thomson, N.H., Kasas, S., Hansma, H.G., Smith, H., Hansma, P.K., Bustamante, C., 1999. Biophys. J. 77, 2284–2294.
Hafner, J.H., Bronikowski, M.J., Azamian, B.R., Nikolaev, P., Rinzler, A.G., Colbert, D.T., Smith, K., Smalley, R.E., 1998. Chem. Phys. Lett. 296, 195–202.
Hafner, J.H., Cheung, C.-L., Lieber, C.M., 1999a. Nature 398, 761–762.
Hafner, J.H., Cheung, C.-L., Lieber, C.M., 1999b. J. Am. Chem. Soc. 121, 9750–9751.
Hafner, J.H., Cheung, C.-L., Lieber, C.M., 2000a. In preparation.
Hafner, J.H., Harper, J.D., Lansbury, P.T., Lieber, C.M., 2000b. J. Am. Chem. Soc., submitted for publication.
Hafner, J.H., Schnitzler, G., Cheung, C.L., Kingston, R.E., Lieber, C.M., 2000c. In preparation.
Han, W., Lindsay, S.M., Dlakic, M., Harrington, R.E., 1997. Nature 386, 563.
Hansma, H.G., Laney, D.E., Bezanilla, M., Sinsheimer, R.L., Hansma, P.K., 1995. Biophys. J. 68, 1672–1677.
Harper, J.D., Wong, S.S., Lieber, C.M., Lansbury, P.T., 1997. Chem. Biol. 4, 119–125.
Harper, J.D., Wong, S.S., Lieber, C.M., Lansbury, P.T., 1999. Biochemistry 38, 8972–8980.
Hassig, C.A., Schreiber, S.L., 1997. Curr. Opin. Chem. Biol. 1, 300–308.
Henderson, R., 1995. Q. Rev. Biophys. 28, 171–193.
Hinterdorfer, P., Baumgertner, W., Gruber, H.J., Schilcher, K., Schindler, H., 1996. Proc. Natl. Acad. Sci. USA 93, 3477–3481.
Hiura, H., Ebbesen, T.W., Tanigaki, K., 1995. Adv. Mater. 7, 275–276.
Hodge, S.E., Boehnke, M., Spence, M.A., 1999. Nat. Genet. 21, 360–361.
Hsia, A.Y., Masliah, E., McConLogue, L., Yu, G-Q., Tatsuno, G., Hu, K., Kholodenko, D., Malenka, R.C., Nicoll, R.A., Mucke, L., 1999. Proc. Natl. Acad. Sci. USA 96, 3228–3233.
Hunt, J.F., Weaver, A.J., Landry, S.J., Gierasch, L., Deisenhofer, J., 1996. Nature 379, 37–45.
Iijima, S., Brabec, C., Maiti, A., Bernholc, J., 1996. J. Chem. Phys. 104, 2089–2092.
Kadonaga, J.T., 1998. Cell 92, 307–313.
Kasas, S., Thomson, N.H., Smith, B.L., Hansma, H.G., Zhu, X.X., Guthold, M., Bustamante, C., Kool, E.T., Kashlev, M., Hansma, P.K., 1997. Biochemistry 36, 461–468.

Kim, S.H., 1998. Nat. Struct. Biol. 5, 643–645.

Kornberg, R.D., Lorch, Y., 1999. Cell 98, 285–294.

Krishnan, A., Dujardin, E., Ebbesen, T.W., Yianilos, P.N., Treacy, M.M.J., 1999. Phys. Rev. B 58, 14013–14019.

Kuhlbrandt, W., Williams, K.A., 1999. Curr. Opin. Chem. Biol. 3, 537–543.

Lambert, M.P., Barlow, A.K., Chromy, B.A., Edwards, C., Freed, R., Liosatos, M., Morgan, T.E., Rozovsky, I., Trommer, B., Viola, K.L., Wals, P., Zhang, C., Finch, C.E., Krafft, G.A., Klein, W.L., 1998. Proc. Natl. Acad. Sci. USA 95, 6448–6453.

Lansbury, P.T., 1996. Acc. Chem. Res. 29, 317–321.

Lee, G.U., Kidwell, D.A., Colton, R.J., 1994. Langmuir 10, 354–357.

Lemere, C.A., Bluzstzajn, J.K., Yamaguchi, H., Wisniewski, T., Saido, T.C., Selkoe, D.J., 1996. Neurobiol. Dis. 3, 16–32.

Li, J., Casell, A.M., Dai, H., 1999. Surf. Interface Anal. 28, 8–11.

Li, W.Z., Xie, S.S., Qian, L.X., Chang, B.H., Zou, B.S., Zhou, W.Y., Zhao, R.A., Wang, G., 1996. Science 274, 1701–1703.

Lin, H., Zhu, Y.J., 1999. Lal, R. Biochemistry 38, 11189–11196.

Livnah, O., Bayer, E.A., Wilchek, M., Sussman, J.L., 1993. Proc. Natl. Acad. Sci. USA 90, 5076–5080.

Lorch, Y., Cairns, B.R., Zhang, M., Kornberg, R.D., 1998. Cell 94, 29–34.

Lu, J.P., 1997. Phys. Rev. Lett. 79, 1297–1300.

Lyubchenko, Y.L., Schlyakhtenko, L.S., 1997. Proc. Natl. Acad. Sci. USA 94, 496–501.

Moechars, D., Dewachter, I., Lorent, K., Reverse, D., Baekelandt, V., Naidu, A., Tesseur, I., Spittaels, K., Van Den Haute, C., Checler, F., Godaux, E., Cordell, B., Van Leuven, F., 1999. J. Biol. Chem. 274, 6483–6492.

Mou, J., Czajkowsky, D.M., Sheng, S.J., Ho, R., Shao, Z., 1996. FEBS Lett. 381, 161–164.

Mou, J., Czajkowsky, D.M., Zhang, Y., Shao, Z., 1995. FEBS Lett. 371, 279–282.

Muller, D.J., Amrein, M., Engel, A., 1997. J. Struct. Biol. 119, 172–188.

Muller, D.J., Fotiadis, D., Engel, A., 1998. FEBS Lett. 430, 105–111.

Muller, D.J., Fotiadis, D., Scheuring, S., Muller, S., Engel, A., 1999a. Biophys. J. 76, 1101–1111.

Muller, D.J., Sass, H.-J., Muller, S.A., Buldt, G., Engel, A., 1999b. J. Mol. Biol. 285, 1903–1909.

Nikolaev, P., Bronikowski, M.J., Bradley, R.K., Rohmund, F., Colbert, D.T., Smith, K.A., Smalley, R.E., 1999. Chem. Phys. Lett. 313, 91–97.

Nishijima, H., Kamo, S., Akita, S., Nakayama, Y., 1999. Appl. Phys. Lett. 74, 4061–4063.

Noy, A., Frisbie, C.D., Rozsnyai, L.F., Wrighton, M.S., Lieber, C.M., 1995. J. Am. Chem. Soc. 117, 7943–7951.

Noy, A., Vezenov, D.V., Kayyem, J.F., Meade, T.J., Lieber, C.M., 1997a. Chem. Biol. 4, 519–527.

Noy, A., Vezenoz, D.V., Lieber, C.M., 1997b. Annu. Rev. Mater. Sci. 27, 381–421.

Noy, A., Sanders, C.H., Vezenov, D.V., Wong, S.S., Lieber, C.M., 1998. Langmuir 14, 1508–1511.

Orlova, E.V., Rahman, M.A., Gowen, B., Volynski, K.E., Ashton, A.C., Manser, C., van Heel, M., Ushkaryov, Y.A., 2000. Nat. Struct. Biol. 7, 48–53.

Pazin, M.J., Kadonaga, J.T., 1997. Cell 88, 737–740.

Perkins, S.J., Nealis, A.S., Sutton, B.J., Feinstein, A., 1991. J. Mol. Biol. 221, 1345–1366.

Rief, M., Gautel, M., Oesterhelt, F., Fernandez, J.M., Gaub, H.E., 1997. Science 276, 1109–1112.

Risch, N., Merikangas, K., 1996. Science 273, 1516–1517.

Scheuring, S., Ringler, P., Borgnia, M., Stahlberg, H., Muller, D.J., Agre, P., Engel, A., 1999. EMBO J. 18, 4981–4987.

Schnitzler, G., Sif, S., Kingston, R.E., 1998. Cell 94, 17–27.

Selkoe, D.J., 1999. Nature 399, A23–A31.

Shao, Z., Mou, J., Czajkowsky, D.M., Yang, J., Yuan, J.-Y., 1996. Adv. Phys. 45, 1–86.

Siegal, G., van Duynhoven, J., Baldus, M., 1999. Curr. Opin. Chem. Biol. 3, 530–536.

Sigler, P.B., Xu, Z., Rye, H.S., Burston, S.G., Fenton, W.A., Horwich, A.L., 1998. Annu. Rev. Biochem. 67, 581–608.

Silverton, E.W., Navia, M.A., Davies, D.R., 1977. Proc. Natl. Acad. Sci. USA 74, 5140–5144.

Singh, S., Paola, T., Bustamante, C.J., Keller, D.J., Capaldi, R., 1996. FEBS Lett. 397, 30–34.

Sobel, E., Lange, K., 1996. Am. J. Hum. Genet. 58, 1323–1337.

Stowell, M.H., Miyazawa, A., Unwin, N., 1998. Curr. Opin. Struct. Biol. 8, 595–600.

Sun, L.F., Xie, S.S., Liu, W., Zhou, W.Y., Liu, Z.Q., Tang, D.S., Wang, G., Qian, L.X., 2000. Nature 403, 384.

Thess, A., Lee, R., Nikolaev, P., Dai, H.J., Petit, P., Robert, J., Xu, C.H., Lee, Y.H., Kim, S.G., Rinzler, A.G., Colbert, D.T., Scuseria, G.E., Tomanek, D., Fischer, J.E., Smalley, R.E., 1996. Science 273, 483–487.

Treacy, M.M.J., Ebbesen, T.W., Gibson, J.M., 1996. Nature 381, 678–680.

Vesenka, J., Manne, S., Giberson, R., Marsh, T., Henderson, E., 1993. Biophys. J. 65, 992–997.

Vettiger, P., Brugger, J., Despont, M., Drechsler, U., Durig, U., Haberle, W., Lutwyche, M., Rothuizen, H., Stutz, R., Widmer, R., Binnig, G., 1999. Microelectron. Engng 46, 11–17.

Vezenov, D.V., Noy, A., Rozsnyai, L.F., Lieber, C.M., 1997. J. Am. Chem. Soc. 119, 2006–2015.

Viani, M.B., Schaffer, T.E., Chand, A., Rief, M., Gaub, H.E., Hansma, P.K., 1999a. J. Appl. Phys. 86, 2258–2262.

Viani, M.B., Schaffer, T.E., Paloczi, G.T., Pietrasanta, L.I., Smith, B.L., Thompson, J.B., Richter, M., Rief, M., Gaub, H.E., Cleland, A.N., Hansma, H.G., Hansma, P.K., 1999b. Rev. Sci. Instrum. 70, 4300–4303.

Walsh, D.M., Lomakin, A., Benedek, G.B., Condron, M.M., Teplow, D.B., 1997. J. Biol. Chem. 272, 22364–22372.

Wong, E.W., Sheehan, P.E., Lieber, C.M., 1997. Science 277, 1971–1975.

Wong, S.S., Harper, J.D., Lansbury, P.T., Lieber, C.M., 1998a. J. Am. Chem. Soc. 120, 603–604.

Wong, S.S., Joselevich, E., Woolley, A.T., Cheung, C.-L., Lieber, C.M., 1998b. Nature 394, 52–55.

Wong, S.S., Woolley, A.T., Joselevich, E., Cheung, C.-L., Lieber, C.M., 1998c. J. Am. Chem. Soc. 120, 8557–8558.

Wong, S.S., Wooley, A.T., Joselevich, E., Lieber, C.M., 1999. Chem. Phys. Lett. 306, 219–225.

Wong, S.S., Woolley, A.T., Odom, T.W., Huang, J.-L., Kim, P., Vezenov, D.V., Lieber, C.M., 1998d. Appl. Phys. Lett. 73, 3465–3467.

Woolley, A.T., Guillemette, C., Cheung, C.-L., Housman, D.E., Lieber, C.M., 2000. Nat. Biotech. 18, 760–763.

Wuthrich, K., 1995. Acta Crystallogr. D 51, 249–270.

Wuthrich, K., 2000. Nat. Struct. Biol. 7, 188–189.

Zhang, Y., Sheng, S.J., Shao, Z., 1996. Biophys. J. 71, 2168–2176.

Zlatanova, J., Leuba, S.H., van Holde, K., 1998. Biophys. J. 74, 2554–2566.